Natural Language Processing and Applications

Huaping Zhang • Jianyun Shang

Natural Language Processing and Applications

Springer

Huaping Zhang
Institute of Computer Science
Beijing Institute of Technology
Beijing, China

Jianyun Shang
Institute of Computer Science
Beijing Institute of Technology
Beijing, China

ISBN 978-981-97-9738-7 ISBN 978-981-97-9739-4 (eBook)
https://doi.org/10.1007/978-981-97-9739-4

The original submitted manuscript has been translated into English. The translation was done using artificial intelligence. A subsequent revision was performed by the author(s) to further refine the work and to ensure that the translation is appropriate concerning content and scientific correctness. It may, however, read stylistically different from a conventional translation.

Translation from the Chinese language edition: "自然语言处理与应用" by Huaping Zhang and Jianyun Shang, © Tsinghua University Press 2023. Published by Tsinghua University Press. All Rights Reserved.

Jointly published with Tsinghua University Press

The print edition is not for sale in China (Mainland). Customers from China (Mainland) please order the print book from: Tsinghua University Press.

This Springer imprint is published by the registered company Springer Nature Singapore Pte Ltd.
The registered company address is: 152 Beach Road, #21-01/04 Gateway East, Singapore 189721, Singapore

Preface

Natural language processing is a discipline that integrates linguistics, computer science, and mathematics, studying various theories and methods for effective communication between humans and computers using natural language. The goal of natural language processing is to enable machines to understand language as intelligently as humans do, with the ultimate goal of narrowing the gap between human communication (natural language) and computer understanding (machine language). Natural language processing is hailed as the "jewel in the crown of artificial intelligence." Microsoft's global vice president and renowned artificial intelligence expert Shen Xiangyang clearly stated at the China Computer Conference: "He who masters language masters the world." Natural language processing has become a difficult and hot spot in artificial intelligence research, harboring opportunities to change the future of the world. A report by Mordor Intelligence shows that the global natural language processing market was \$10.72 billion in 2020, and is expected to grow to \$48.46 billion by 2026, with a compound annual growth rate of 26.84%. With the global spread of the COVID-19 pandemic, the growth of natural language processing in the field of healthcare has been particularly rapid.

In recent years, there have been many excellent monographs or textbooks in the field of natural language processing, but most of them mainly introduce classic algorithms and related technologies of natural language processing, and there are not many that combine actual systems and application practices. I started offering the elective course "Big Data Analysis and Application" at Beijing Institute of Technology in 2016, and started offering the compulsory course "Big Data Processing Technology" for third-year students majoring in artificial intelligence at Beijing Institute of Technology in 2022, trying to combine natural language processing with big data and artificial intelligence. Using a research-based teaching method, I give classic task propositions of natural language processing, and students are required to give a review report in groups, detailing the classic algorithms of each technical point of natural language processing, reflecting the frontier progress of international academic research in the past 3 years, and finally giving an intuitive demonstration system and conducting experimental verification. The final exam of the course requires team cooperation to complete a natural language processing project with a certain degree of innovation, which is independently reviewed by experts from industry, academia, and research outside the school. After 6 years of continuous exploration, the course has received wide acclaim from more than 1000 students who have taken the course, with an average final score of 94.73. With the continuous improvement of the teaching practice system, I hope to share the results of more than 20 years of research in natural language processing and 6 years of teaching practice, and finally complete this textbook closely combined with natural language processing and application.

This book is divided into five parts: The Part I mainly includes an overview of natural language processing and its applications, classic deep learning platforms and algorithms, the latest developments in deep learning, and pre-trained language models; The Part II mainly includes web crawler technology, multi-format document parsing and management, speech-to-text recognition, image semantic representation and character recognition, Chinese word segmentation, and part-of-speech tagging; The Part III includes sentiment analysis, new word discovery, named entity recognition and keyword extraction, automatic construction, application of big data knowledge graphs, etc.; The Part IV includes information filtering, text classification, text clustering, text proofreading, automatic summarization; The Part V mainly introduces some distinctive natural language processing application projects and cases.

The feature of this book is that it integrates academic frontiers, practical applications, and teaching achievements, fully reflecting the international academic frontier progress in the direction of big data, artificial intelligence, and natural language processing, and incorporating the innovative achievements of the author team in the direction of natural language processing and applications for more than 20 years. The related achievements have successively won the first and second prizes of the Science and Technology Progress Award of the Xinjiang Uygur Autonomous Region, and the first prize of the Qian Weichang Chinese Information Processing Science and Technology Award. This book has absorbed the teaching achievements of "Big Data Analysis and Application" and "Big Data Processing Technology" for more than 6 years, and included several excellent projects of research groups as application cases. As a supporting website and download base for the achievements of this book, the NLPIR (Natural Language Processing and Information Retrieval) sharing platform provides demonstrations of actual results and downloads of various resources. This book can be used as a textbook for graduate students and senior undergraduates in the direction of natural language processing, and can also be used as a reference for researchers, engineering technicians, and enthusiasts in the direction of natural language processing.

The content of this book mainly involves the research results of the NLPIR laboratory of Beijing Institute of Technology where the author is located, and some chapters are from academic papers and graduate thesis published by the laboratory in the past 10 years. Zhang Huaping is responsible for the overall planning and task arrangement, Shang Jianyun is responsible for the coordination of this book, Wang Juan and Geleta Negasa Binegde completed the collation of all the initial drafts, and Nie Jingxin and Cao Zhejun translated all the figures and tables. This book uses the graduation thesis and published articles of graduate students Zhang Baohua, Jiang Qinghong, Cai Jiahao, Liu Ziyu, Liu Zihao, etc. under the guidance of the authors, and also uses the homework of some students of the graduate course "Big Data Analysis and Application" and the undergraduate course "Big Data Processing Technology" of Beijing Institute of Technology, which are marked in the corresponding parts. Each chapter has been carefully edited and arranged by Kang Kai, Wang Yanhao, Yang Manzhi, Zhang Xiaosong, Li Yulin, Zhang Junhui, Ma Yiyang, Zhang Hengyu, Gao Yuxiao, Zhao Qingqing, Yang Ziyuan, Liu Weikang, Zhang Hongbin, Yan Ruohao, Chen Lifeng, Li Jing, Cai Jiahao, Du Lun, Lei Peikun, Tang Zeyang, and other students. This book has received the support of the Basic Enhancement Plan Technology Field Fund (No.: 2021-JCJQ-JJ-0059), Beijing. This work was supported by the City Natural Science Foundation (Grant No.: 4212026) and the "Fourteenth Five-Year" Plan Textbook Project of Beijing Institute of Technology. During the planning and writing of this book, we received strong support and help from Bai Lijun and Professor Yang Fan of Tsinghua University Press, for which the authors express their heartfelt thanks. In the writing of this book and the research work of related scientific topics, we received support and help from many sides. The authors would like to express their sincere gratitude and high respect to the authors of the related literature and the teachers, colleagues, and project team members who have helped with this book. Due to the author's knowledge and level, there are inevitably inappropriate parts in the book, and we sincerely invite readers to criticize and correct them.

Beijing, China
July 2024

Huaping Zhang

Contents

Part II Information Processing

Part V Application

Part I
Basic Theory

Chapter 1
Overview of Natural Language Processing and Applications

Language is one of the essential characteristics that distinguish humans from other animals. Many aspects of human intelligence are closely related to language: human logical thinking is in the form of language, and the vast majority of human knowledge is also recorded and passed down in the form of language. This chapter gives the definition of natural language processing and introduces the complexity, importance, development history, and upstream and downstream tasks involved. Additionally, it presents an overview of the current state of Chinese NLP, covering task evaluation results, datasets, evaluation methods, and the progress of Chinese pre-trained language models.

1.1 Natural Language Processing

This section defines natural language processing technology, its complexity and importance, and the technical development path and upstream and downstream tasks of natural language processing.

1.1.1 Definition, Challenges, and Development History of Natural Language Processing

1.1.1.1 Definition

Natural Language Processing (NLP) is a science that integrates linguistics, computer science, and mathematics. It studies various theories and methods for effective communication between humans and computers using natural language. Natural language is the language people use daily and evolves with social development.

H. Zhang, J. Shang, *Natural Language Processing and Applications*,
https://doi.org/10.1007/978-981-97-9739-4_1

Processing refers to recognizing, analyzing, transforming, understanding, reasoning, and generating inputs of various granularities such as words, entities, sentences, chapters, topics, knowledge, etc. Natural language processing aims to make machines as intelligent as humans in understanding language and to bridge the gap between human communication (natural language) and computer understanding (machine language).

1.1.1.2 Challenges

The challenges of natural language processing are manifested in abstraction, combination, ambiguity, evolution, nonstandardization, subjectivity, knowledge, and difficulty in portability.

Natural language, a complex model of the underlying physical perception of the world, also represents human understanding of various things in the world and their relationships. It is a complete set of symbolic descriptions. This means natural language processing technology must deal with discrete abstract symbols, skipping perception.

The process focuses on various abstract concepts, semantics, and logical reasoning, which brings problems such as abstraction, combination, ambiguity, evolution, non-normativity, subjectivity, knowledge, and difficulty in transplantation, as shown in Table 1.1. Natural language processing has always been committed to solving these problems, aiming to make machines approach or reach human understanding, reasoning, and expression capabilities.

Nowadays, the Internet has brought a tremendous amount of text information. This information comes from different languages, contains rich knowledge, carries a lot of noise, contains rich rhetorical techniques, and constantly iterates expression methods. Natural language processing technology can fully liberate and develop productivity and improve the country's scientific and technological strength.

Table 1.1 Some difficulties in natural language processing

Challenges	Explanation	Examples
Abstractness	Many abstract nouns are difficult to understand and express	Value dignity rights
Combinability	Natural language can combine to produce a huge amount of semantics	Hard to say, not good to say, not saying good
Evolutiveness	Natural language constantly changes with social development	Ancient and modern different meanings
Nonstandardness	Compared to code, natural language is difficult for programs to understand	If, then → if … then …
Subjectivity	Different people have different understandings of the same expression	Add a suitable amount of MSG

1.1.1.3 Development History

The development history of natural language processing is shown in Fig. 1.1. This development process is accompanied by unifying empiricism and rationalism from division. Natural language processing development history can usually be divided into three stages: rule-based methods from the 1960s to the 1980s, statistical learning methods from the 1980s to 2010, and deep learning methods after 2010.

1.1.2 Upstream and Downstream Tasks of Natural Language Processing

1.1.2.1 Upstream Tasks

Linguists and symbolists perceive natural language understanding as a meticulously structured hierarchical process, as illustrated in Fig. 1.2. This process unfolds sequentially, with each step building upon the previous one. The next step can only be initiated after the language has been parsed through the upstream tasks.

Lexical analysis is the process of determining the morphemes of vocabulary and obtaining linguistic information from them. It includes Chinese word segmentation and part-of-speech tagging, among other things. The first task in processing Chinese is to segment the input string into separate words; this step is called word segmentation. Then, part-of-speech tagging assigns a part-of-speech category to each word.

Syntactic analysis analyzes the structure of sentences and phrases, as shown in Fig. 1.3. Everyday syntactic analysis tasks include the following two analyses:

1. Constituent syntactic analysis: Analyzes and identifies phrase structures and the hierarchical syntactic relationships between phrases
2. Dependency syntactic analysis: Identifies the interdependence between words

Semantic analysis aims to learn and understand the semantic content represented by a text, mapping natural language into machine-understandable language. It requires the model to understand natural language deeply, the scene, and specific reasoning ability. Semantic analysis can be further decomposed into lexical-level semantic analysis (word sense disambiguation and representation), sentence-level semantic analysis (semantic role labeling), and discourse-level semantic analysis (discourse structure analysis).

Discourse analysis refers to various analyses beyond the scope of a single sentence, including the relationships between sentences (paragraphs) and types of relationships.

Discourse analysis key components include division of semantic types, judgment of relationships between paragraphs, analysis of relationships between words across single sentences, and the inheritance and transition of topics. Standard discourse analysis techniques include anaphora resolution, discourse structure analysis and coherence theory.

Fig. 1.1 The development history of natural language processing

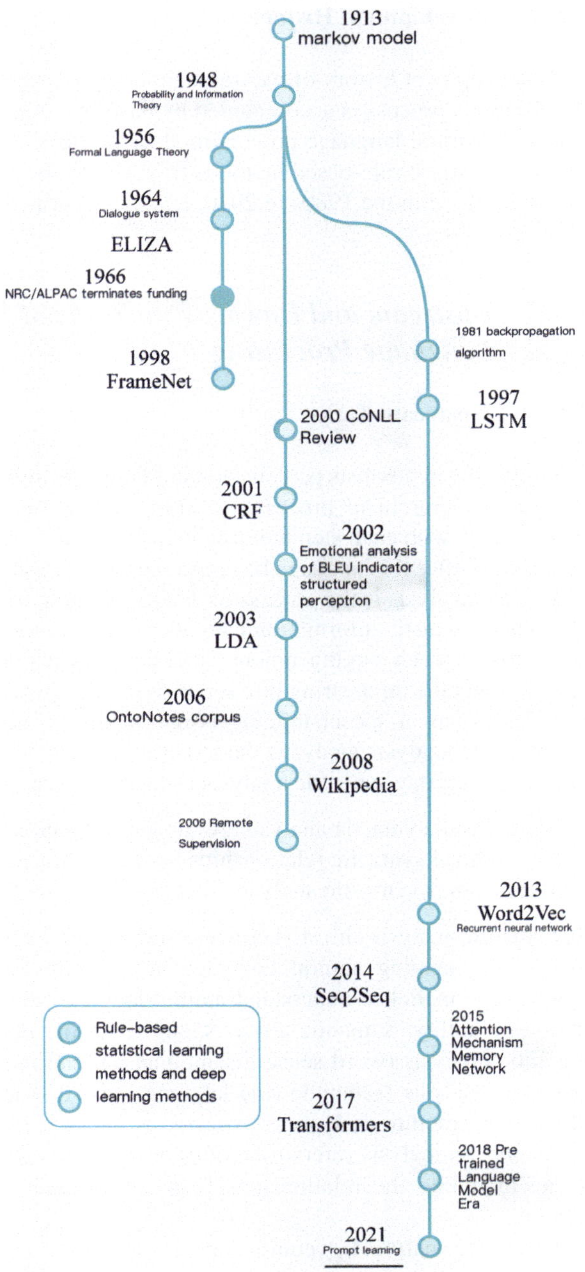

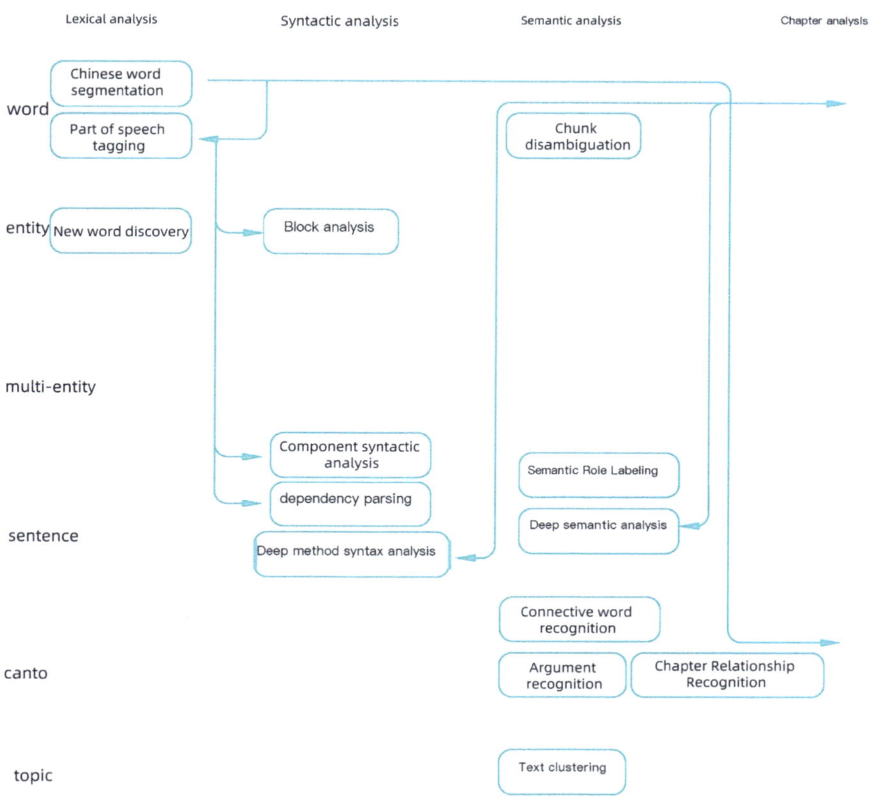

Fig. 1.2 Upstream tasks

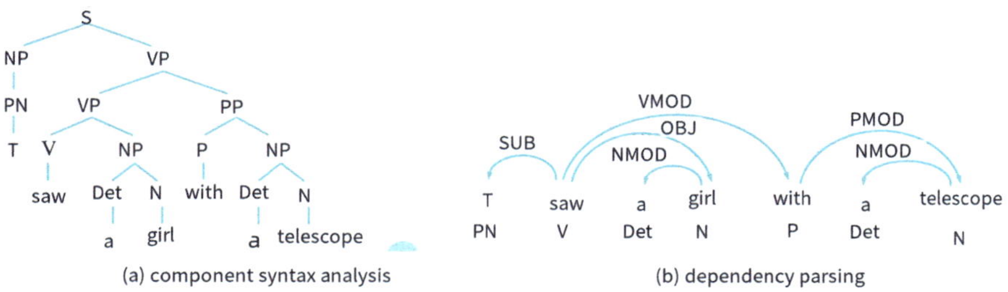

Fig. 1.3 Syntax analysis

1.1.2.2 Downstream Tasks

Natural language processing application technology usually depends on basic technology, as Fig. 1.4 shows. The downstream tasks of natural language processing include text classification, text clustering, text summarization, automatic question answering, machine translation, information extraction, information recommendation, information retrieval, etc.

Text classification: Assign documents to predefined categories based on the content or theme of the document.

Text clustering: Divide a collection of documents into several subsets based on the content or theme similarity between documents.

Text summarization: Obtain a concise description by compressing and refining the original text.

Automatic question answering: Use computers to automatically answer questions raised by users.

Machine translation: Use computers to automatically translate from one language to another.

Information extraction: Extract specified information from unstructured/semi-structured text, and convert it into structured information through information merging, redundancy elimination, and conflict resolution, mainly including entity extraction, attribute extraction, relationship extraction, event extraction, and aspect extraction.

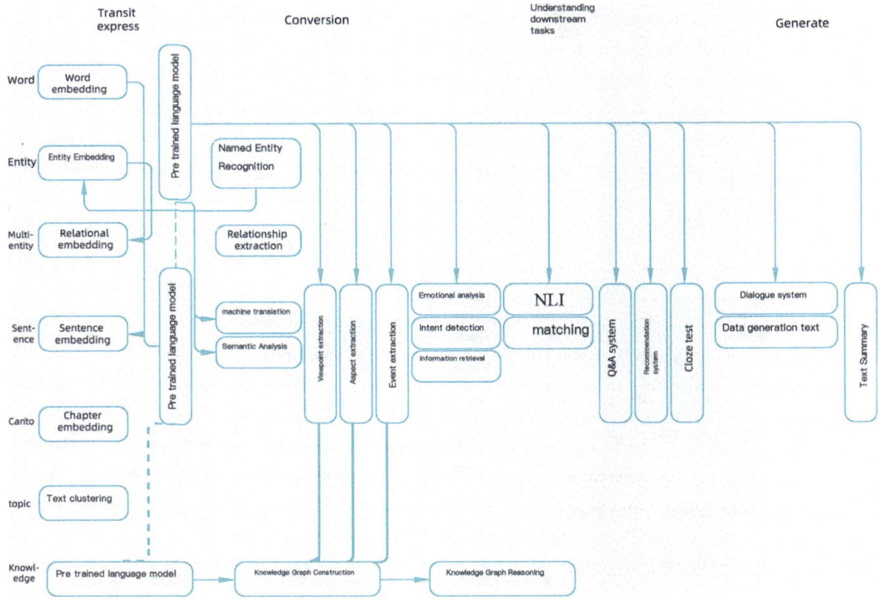

Fig. 1.4 Summary of natural language processing tasks

Information recommendation: Identify information that can satisfy the user's interests from large-scale information based on the user's habits, preferences, or interests.

Information retrieval: Organizing information in a certain way, satisfying user's information needs through search means.

1.2 Current Status of Chinese Natural Language Processing

1.2.1 Evaluation Results of Natural Language Processing Tasks

SOTA is an abbreviation for state-of-the-art, referring to the highest level of development in a specific moment for devices, technologies, or scientific fields. In natural language processing, if the proposed method's evaluation score surpasses the existing models, then this method can be called the SOTA method for a specific dataset. Table 1.2 lists the 12 mainstream natural language processing evaluation datasets.

It is well known that reproducing models and algorithms is also very important. Websites such as Paper with Code and Hugging Face promote models to increase their visibility and the development of language-related models.

Evaluations and competitions in natural language are also essential links in promoting model replacement and algorithm updates. Manual evaluation requires human annotators to perform large-scale quality checks on each model version. Although this method is accurate, labor-intensive quality check tasks consume much of the workforce. Automated evaluation methods such as GLUE and CLUE can quickly evaluate models, which is of great significance for model iteration.

1.2.2 Current Status of Chinese Task Datasets and Evaluations

Figure 1.5 shows the distribution of datasets in different languages in natural language processing. It can be seen that the number of Chinese datasets is minimal compared to English datasets, which indirectly indicates that the evaluation data for Chinese natural language processing is not comprehensive and insufficient.

There are four problems with the Chinese dataset in the current natural language processing evaluation: dataset bias, evaluation index distortion, unscientific evaluation tasks, and single evaluation technology [1]. Although the review of natural language processing is in a boom, the various problems exposed in various tasks show that the current evaluation of natural language processing lacks the constraints of principles and norms, which also leads to the review of natural language processing being unable to move toward a stable development stage.

Table 1.2 Twelve mainstream natural language processing evaluation datasets

Evaluating	Proposed time	Proposing institution	Task type	Scale	At the time of presentation best model	It is best to put to forward performance to the model	Best at the present performance of the model	Promote rate	Index
GLUE	April 2018	New York University	Comprehensive	Nine tasks	BiLSTM + Att + ELMO	70.0	90.7	0.6	F1
SuperGLUE	April 2019	New York University	Comprehensive	Eight tasks	Burt ft	71.5	89.3	0.9	F1
CLUE	November 2020	CLUE team	Comprehensive	Nine tasks	RoBERTa wwm-ext-large	74.9	80.1	5.2	EM,Acc
SQuAD1.1	October 2016	Stanford University	Reading comprehensive	100,000 questions	Logistic regression	51.0	89.9	1.1	F1
SQuAD1.1	June 2018	Stanford University	Reading comprehensive	150,000 questions	DocQA+ELMO	66.3	93.0	0.8	F1
WinGrande	November 2019	Allen studies sueodeng	Natural language reason	44,000 questions	RoBERTa	79.3	87.0	0.6	AUC
CoQA	March 2019	Stanford University	Reading comprehension	127,000 questions	Augmt.DrQA	65.4	90.7	1.1	F1
Commonsense QA	March 2019	Tel Aviv University	Reading comprehension	12,000 questions	BiDAF++	32.0	83.3	2.2	Acc
Race	December 2017	Carnegie Mellon University	Reading comprehension	100,000 questions	GA	44.1	91.4	1.2	Acc
Drop	April 2019	University of California	Reading comprehension	97,000 questions	NAQANet	47.0	90.1	2.1	F1

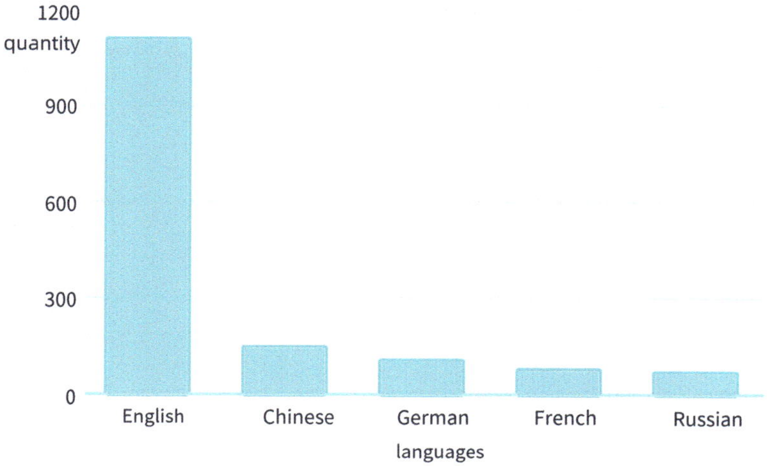

Fig. 1.5 Current distribution of datasets in different languages in the field of natural language processing

Table 1.3 Highly recognized evaluations in recent years

Ranking	Model	Research institution	Evaluation time	Score 1.1
1	HUMAN	CLUE	19-12-01	86.678
2	Shennong	Yunxiaowei AI	21-10-15	84.236
3	ShenZhou	QQ Browser Lab	21-09-19	83.872
4	Mengzi	Lanzhou Technology	21-09-14	81.092
5	BERT	BERT	21-10-18	78.618

The evaluation acceptance in Table 1.3 is relatively high. Although some indicators have reached a very high level, most of the algorithms have not been publicly verified, which will also affect the choice of algorithms and the progress of scientific research by researchers.

1.2.3 Current Status of Chinese Pre-trained Language Models

Since the second half of 2018, developing pre-trained language models has shown an explosive trend, with an increasing number of researchers and a more comprehensive range of research directions. As can be seen from Fig. 1.6, the research on pre-trained language models is mainly in English, and there are very few Chinese pre-trained language models. Many models that perform well on English datasets still need Chinese versions, which impedes Chinese natural language processing researchers.

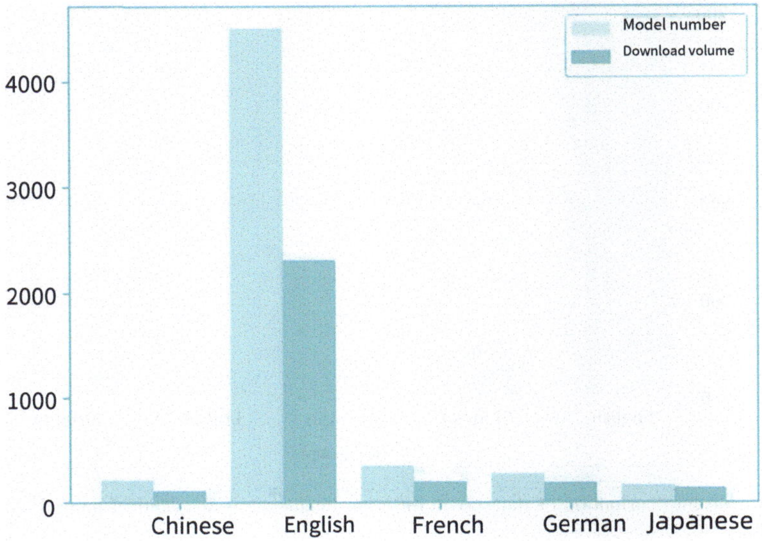

Fig. 1.6 Number and download volume of pre-trained language models

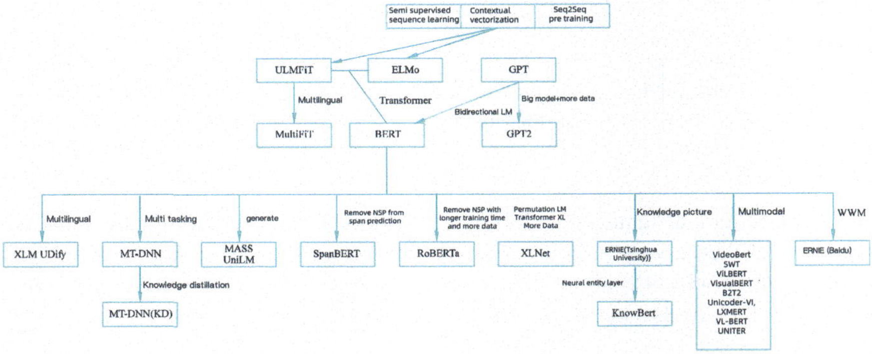

Fig. 1.7 Pre-trained language models. (Source: https://huggingface.co/models)

The development of English pre-trained language models is rapid, and Fig. 1.7 shows the mainstream pre-trained language models. Many Chinese natural language processing researchers have trained Chinese pre-trained language models under the framework of English pre-trained language models using large-scale Chinese datasets, such as bert-base-Chinese, Chinese-roberta-wwm-ext, Albert-tiny-Chinese, etc.

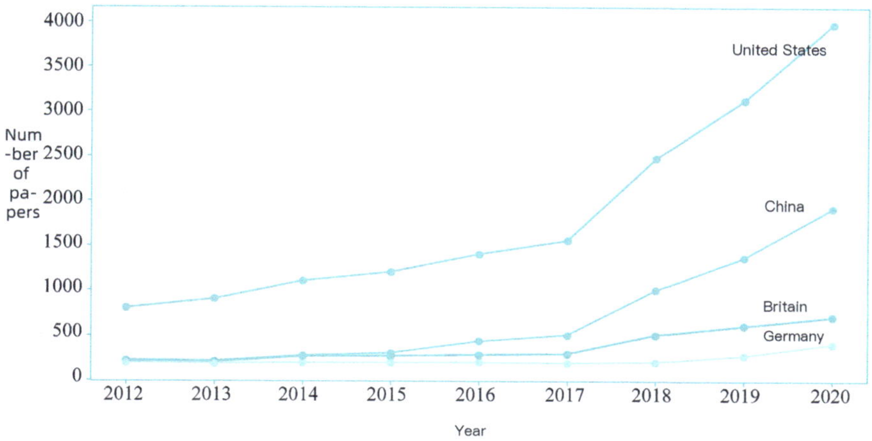

Fig. 1.8 Number of papers published in 13 top natural language conferences by the United States, China, the United Kingdom, and Germany. (Source: https://github.com/thunlp/PLMpapers)

1.2.4 Current Status of China's Influence

From 2012 to 2020, the number of papers published by the United States, China, the United Kingdom, and Germany at 13 top natural language conferences is shown in Fig. 1.8. It can be seen that the number of papers published by American researchers still maintains a leading position, while the number of papers published by Chinese researchers at top conferences has exploded from 2017 to 2020.

In natural language processing research, various technology began to train Chinese pre-trained language models, such as BERT, T5, etc., when conducting model training. A large number of technical websites have already set up unique Chinese versions for Chinese researchers to learn and improve too.

1.3 Development Trends of Natural Language Processing

The development trend of natural language processing can be summarized as processing from manual to automation, application from general to scenario-based, and algorithms from single to platform-based.

Table 1.4 Development history of machine learning

Normal form	Intertask relationship	Time
Characteristic engineering	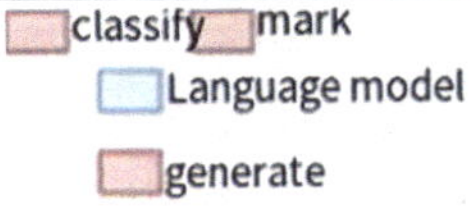	1913–2001 2002–2011
Architecture engineering	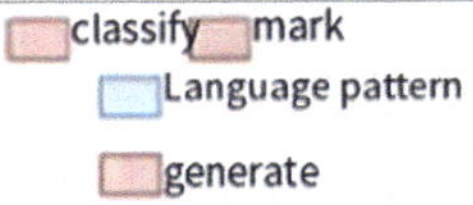	2011–2018
Target engineering	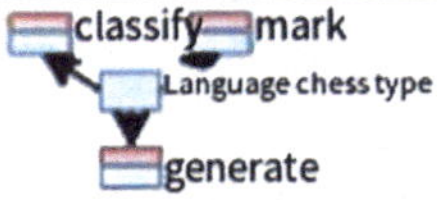	2018–2021
Paradigm fusion	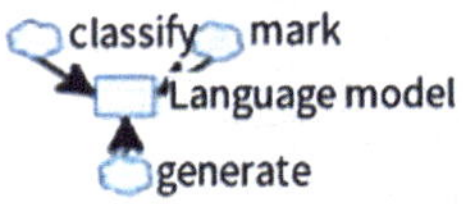	2021–2024

1.3.1 Processing from Manual to Automation

With the advancement of deep learning, the efficiency and performance problems of rule-based and statistical learning methods that rely on human-designed rules and pattern finding have gradually emerged. Deep learning eventually eliminates them. Later, in the field of deep understanding, architecture engineering and objective engineering, which require a lot of manual design and resource training, were also criticized. The academic and industrial circles are calling for the next revolution from manual to automation. Table 1.4 shows the development history of machine learning from feature engineering to paradigm integration.

1.3.1.1 Feature Engineering

Early natural language processing models heavily relied on feature engineering. Natural language processing researchers or engineers use their domain knowledge to define and extract significant features from raw data and provide models with appropriate inductive bias to learn from these limited data. Standard features include sentence length, word frequency, etc. Feature engineering modeling can be done by establishing rules or using typical machine learning methods.

1.3.1.2 Architectural Engineering

The advent of neural network models has significantly influenced the field of natural language processing. It has led to a shift in focus from the design of features and the learning of weight parameters to the design of models and the training of model

Table 1.5 Comparison of feature engineering and architecture engineering

	Rule method	Statistical learning method	Deep learning method
Example	Check Sentiment Dictionary	Naive Bayes	Word2Vec + DNN
Decision	mean ($score_{senti}$)	argmax $P(W\vert C_i)P(C_i)$	argmax(softmax(O))

parameters. This shift, known as architecture engineering, involves designing a suitable network with specific prior knowledge. By learning abstract features end-to-end, the generalization ability of the model has been significantly improved.

The comparison between feature engineering and architecture engineering is shown in Table 1.5.

1.3.1.3 Objective Engineering

With the emergence of pre-trained language models, a new norm has formed in natural language processing. The previous fully supervised mode (feature engineering and architecture engineering) has gradually been replaced, transforming into a pre-trained and fine-tuning paradigm. In this new paradigm, a model with a fixed architecture is pre-trained as a language model to predict the likelihood of text data appearing.

Since the pre-trained language model uses self-supervised learning, it can learn the knowledge of unlabeled corpora. The model uses abundant raw text data on the Internet to acquire expertise and complete downstream tasks. Next, additional parameters are introduced and fine-tuned using the objective function of the specific task, making the pre-trained language model adaptable to different downstream tasks.

In this paradigm, researchers have shifted their focus to objective engineering, and they only need to design a small network according to the task.

Excellent performance can be achieved by selecting the training objectives for the pre-trained and fine-tuning stages.

1.3.1.4 The Next Paradigm

The third revolution in natural language processing is on the horizon, and the most promising is prompt learning. Prompt learning does not adjust the pre-trained language model to downstream tasks through objective engineering but redefines the downstream functions to make them look more like tasks solved by prompts in the original training.

In this way, only the appropriate prompts must be selected; without additional specific task training, the pre-trained language model can generate the expected output. The advantage of this method is that the pre-trained language model trained in a completely unsupervised manner can solve many tasks with appropriate prompts. The pre-trained language model can solve multiple zero-sample and few-sample problems simultaneously using the unlabeled corpus during pre-training.

In addition to prompt learning, machine reading comprehension and natural language inference models are adaptable and may continue to simplify human work while achieving better performance.

1.3.2 Applications from General to Scenario-Based

When natural language processing technology moves from the academic greenhouse to the industrial field, it encounters various problems and difficulties. These problems are usually classified as sample, domain, language, modality, and device problems.

1.3.2.1 Sample Problem

The ability to learn and generalize based on small samples is an important dividing line between human and artificial intelligence. Humans can often recognize new things based on one or a few samples, while machine learning algorithms usually need hundreds or thousands of supervised samples to achieve generalization.

If there is sufficient data annotation, the neural network based on deep learning can effectively extract target features. However, more training data is needed in actual production activities. The parameter learning in the neural network is complex to generalize on limited samples, and the neural network is prone to overfitting. Therefore, the problem of few samples has become a vital and challenging point for deep learning to move toward practical applications [2].

1.3.2.2 Domain Problem

Domain adaptation and domain problems [3] are opposite solutions. The former is currently widely studied in academia, and the latter is the mainstream method in the industry. The assumption that the training set and the test set are independently and identically distributed is a significant challenge to meet in actual applications, due to the inherent differences in the source and target domains. When these differences in distribution exist, the model obtained from the source domain cannot achieve good prediction results in the target domain, leading to the problem of the gap between domains.

Domain adaptation is a machine learning technique when the training set and the test set do not meet the condition of independent and identical distribution. The source domain represents a different field from the test samples, with rich supervised annotation information; the target domain represents the field where the test samples are located, with no labels or only a few labels. The source and target domains often belong to the same type of task, but their distributions are different. The goal is to let the source domain and the target domain share the same features and categories and use the information-rich source domain samples to improve the performance of the target domain model under different feature distributions.

Domain adaptation involves mapping the data of the source domain and the target domain to a feature space and finding a measurement criterion in the feature space that minimizes the difference in feature distribution. This allows the discriminator trained on the data features of the source domain to be used on the target domain data. The main types of domain adaptation methods include those based on domain distribution differences, adversarial methods, reconstruction-based methods, and sample generation-based methods.

In an ideal situation, a model performing well in any domain is the best solution. However, even under academic datasets, such domain adaptation models cannot achieve satisfactory results (under natural conditions, the performance will be worse). Therefore, in actual engineering, we have to settle for the next best thing, introduce relevant knowledge according to the needs of the actual business domain, and build models suitable for specific domains.

In education, medical care, and law, which require high accuracy and reasoning ability, it is usually necessary to establish a domain knowledge graph and then perform rule-based or deep learning-based reasoning on the knowledge graph. In automotive systems emphasizing safety, deep learning accuracy and real-time requirements are usually high, and rules must be combined to ensure safety. Each domain must establish a unique dictionary and preprocess according to actual needs.

Although this dramatically reduces the degree of code and model reuse, making the model meet the actual requirements is a necessary and effective measure under current conditions. Researchers also look forward to models with generalization ability and transfer learning and are actively exploring vertical domain models.

1.3.2.3 Language Issues

There are many languages but few available corpora, making model training difficult. There are over 6000 languages worldwide, but only 15 are commonly used. In terms of unsupervised data, most languages have less than 10 GB of corpus resources, as shown in Fig. 1.9; in terms of supervised data, most companies have

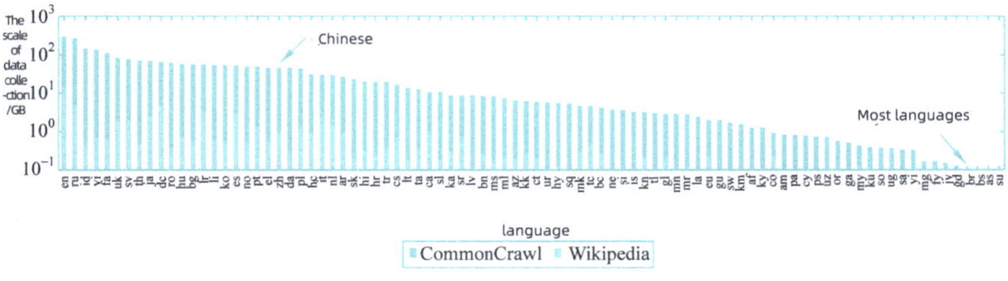

Fig. 1.9 Corpus resources in different languages

some business data accumulation in Chinese/English, but the accumulation in new languages is zero.

The diversity of languages increases the maintenance difficulty and slows down the expansion speed. Recruiting data annotators for minor languages also presents particular challenges, indirectly limiting supervised data acquisition. The lack of data makes training an effective monolingual model on minor languages easier. Algorithm engineers need help understanding the target language, and the quality of translation tools is low, slowing down the development and debugging speed. In engineering, the benefits of model optimization mostly come from data production and cleaning. Still, this process becomes difficult in a multilingual setting, and engineers need to understand the target language, making this process almost impossible to carry out. On the one hand, after the model is online, the models corresponding to different languages need to be optimized and bad cases handled separately, and on the other hand, if there is no suitable reuse mechanism, the cycle to support a new language will be relatively long, making it difficult to adapt to rapidly expanding business needs. External translation software can only alleviate this problem to a certain extent; its effect on minor languages is not reliable, and the act of translation itself dramatically affects the efficiency.

1.3.2.4 Modality Problem

As shown in Fig. 1.10, the concept of multimodality originated from the research on information representation methods in the field of computer–human–computer interaction, where the term “modality” represents and exchanges information on a specific physical medium. Learning from multiple modalities of information sources makes it possible to capture the correspondence between modalities and gain a deep understanding of natural phenomena. Current multimodal machine learning faces many challenges, including representation, translation, alignment, fusion, and joint learning [4].

Fig. 1.10 Multimodality problem

1.3.2.5 Device Problem

In deep learning research, researchers try to use robust, complex model networks to pursue higher performance. Engineerers are trying to apply algorithms on more stable, efficient hardware platforms to improve efficiency. Although complex models perform well, they depend on storage space and computing resources, making their application effectively on hardware platforms challenging, so model compression is significant. The current three directions of deep learning model compression and acceleration algorithms are accelerating network structure design, model pruning and sparsification, and quantization acceleration [5].

1.3.3 From Single Algorithm to Platform

Deep learning needs writing code on the algorithm platform to avoid affecting the progress of the underlying algorithm implementation. Currently, the emergence of a large number of underlying platforms and algorithm platforms has provided help for researchers and production enterprises. In the continuous competition, several product ecosystems have been formed.

1.3.3.1 Underlying Platforms

Thanks to the open-source ecosystem model created by various parties in the early development of the deep learning framework to better promote technological development, various deep learning frameworks are flourishing, rapidly promoting the application of deep learning technology in the industrial field. Common deep learning platforms include Pytorch, TensorFlow, and PaddlePaddle, as shown in Fig. 1.11.

PyTorch's open-source toolkit uses the Python scripting language, which is generally used for natural language processing and computer vision. It has powerful GPU capabilities, high memory usage efficiency, and dynamic computation graphs, making it popular for assisting in developing dynamic neural networks. It can build graphics and visualize data according to user requirements.

The TensorFlow toolkit was developed in 2015 and is rated the easiest to use and deploy in machine learning and deep learning.

Fig. 1.11 Mainstream frameworks

One of the tools is TensorFlow, which was initially created by the Google Brain team to serve its research and production goals. Because it provides many free tools, libraries, and community resources, it is now widely accepted by companies such as Uber, Twitter, and eBay.

PaddlePaddle (PP) is a platform developed by Baidu that supports parallel distributed computing. It was open-sourced to the professional community in 2016. With advanced deep learning features and end-to-end development toolkits, it is favored by deep learning developers in various industries. It supports the deep learning algorithms of many Baidu products.

1.3.3.2 Algorithm Platforms

The NLPIR Big Data Semantic Intelligence Analysis Platform was developed by the NLPIR Laboratory of Beijing Institute of Technology over 20 years. This platform, designed for big data content processing, integrates 13 functions of precise network collection, natural language understanding, text mining, and network search technology, providing client tools, cloud services, and secondary development interfaces. With 20 years of technical precipitation and application verification, the NLPIR platform has won unanimous certification and praise from 300,000 institutions and 400,000 users worldwide. It is a significant weapon for semantic intelligence analysis in the era of big data. The customer fields served by the NLPIR platform include government agencies, research institutes, universities, financial risk control, media publishing, and other enterprises.

The LTP platform is a complete open Chinese natural language processing system developed by the Social Computing and Information Retrieval Research Center of Harbin Institute of Technology over 10 years. LTP has formulated a language processing result representation rule based on XML, and on this basis, it provides a complete set of rich, efficient, and high-precision Chinese natural language processing modules, including lexical analysis, syntactic analysis, semantic analysis, and other core Chinese processing technologies, which have achieved excellent results in domestic and international technical evaluations.

1.4 Challenges Faced by Chinese Internet Natural Language Processing

Compared to general news long text natural language, the Chinese Internet natural language faces practical challenges such as adversarial information, multilingual interaction, and social evolution. The specific analysis is as follows.

1.4.1 Information Adversarial

The era of information explosion is also an era of information overload. Researchers are increasingly feeling overwhelmed by the plethora of information. Fraud, pyramid schemes, online gambling, and spam advertisements are ubiquitous, so focusing on the supervision of information content is necessary. Current technical means can only perform keyword matching, and the existing review mechanism also relies on a large workforce, making it impossible to conduct real-time verification. Problems include insufficient technical accumulation, high workforce costs, low efficiency, and weak supervision. The importance and urgency of information security prevention strongly call for a new revolution in information filtering technology.

1.4.2 Multilanguage Interaction

The direction of natural language processing needs national strategic macroguidance. Especially in China, the study of minority languages has the following significance:

1. Ethnic cultural heritage. These current and historical texts carry the civilization of the Chinese nation for thousands of years, and the wealth and resources of the country and the nation are important areas and content that Chinese information processing must pay attention to. Our government has 55 ethnic minorities, accounting for 8.4% of the total population. There are 72 languages currently used by ethnic minorities and countless ancient languages that have been extinct.
2. International market competition. Ethnic language information processing is integral to China's "Belt and Road" initiative. It significantly impacts the language information processing technology of countries within and around China. The languages of the countries along the "Belt and Road," including the languages of the ethnic minorities in the region, may reach about 200. However, only 20 related languages are currently offered as majors in our universities, which obviously cannot keep up with the country's current needs.
3. National information security. Ethnic language information processing (especially in Tibetan, Mongolian, Uighur, Kazakh, and Korean) is related to our country's discourse and dominance in this field and concerns people's livelihood and national interests.

The processing of minority language information also includes basic theoretical research, coding formulation, the establishment of large-scale corpora, system development, etc. Artificial intelligence has brought solutions to the study and application in this field, and natural language processing is the jewel in the crown of artificial intelligence and the golden key to solving multilingual problems. For machine translation and dialogue systems of resource-scarce minor languages, more human-annotated data can be added based on active learning methods, and unannotated data can be used through unsupervised and semi-supervised methods or other natural language processing tasks, or even other language information can be used through multitask learning methods.

1.4.3 Social Evolution

1.4.3.1 Topic Evolution

Unlike traditional news media, modern social networks focus more on user participation. The sources of information on the Internet are diverse, including hot topics of public concern, and of course, there may also be sensitive topics related to public safety and social stability. The development of events will change accordingly under the influence of many factors such as time and culture, which is topic evolution. Topic evolution reflects the process of a topic from its inception, rising in popularity, declining in popularity, to its end. Over time, the intensity and content of the topic will change, that is, there is a migration of issues.

1.4.3.2 Language Evolution

The discovery of new words through unsupervised methods can effectively obtain the domain features of the text. New words are inevitable in the development and evolution of any language. Language evolution has two forms: temporal longitudinal evolution and domain horizontal evolution. A specific example of temporal longitudinal evolution is Internet slang, such as the recently famous "lying flat," "yards," "Versailles," etc. These words fully reflect the social and cultural phenomena within a period, directly reflecting social hotspots and trends. Domain horizontal evolution refers to the unique professional terms in different fields, such as "lifelong learning," "CNN," etc., in artificial intelligence. In addition, it is often difficult to find common vocabulary for minor languages. If the new word discovery algorithm is applied to this field, relevant vocabulary information can be mined to fill in the data gap. At present, word embedding is usually obtained by pre-trained on unlabeled corpora. Most current language representation methods will ignore new words (out-of-vocabulary words). However, new words in the language also contain much information, and discarding them will reduce model performance.

References

1. Dong Qingxiu, Sui Zhifang, Zhan Weidong, et al. Problems and counter measures in natural language processing evaluation[J]. Chinese Journal of Information, 2021, 35(6): 1–15.
2. Li Xinye, Long Shenpeng, Zhu Jing. A survey of few-shot learning based on deep neural networks[J]. Computer Application Research, 2020,37(8): 2241–2247.

3. Ramponia, Plank B. Neural Unsupervised Domain Adaptation in NLP— A Survey[C]. Proceedings of the 28th International Conference on Computational Linguistics, 2020: 6838–6855.
4. Wang Y. Survey on Deep Multi-modal Data Analytics: Collaboration, Rivalry, and Fusion[J]. ACM Transactions on Multimedia Computing, Communications, and Applications(TOMM), 2021, 17(1) : 1–25.
5. Choudhary T, Mishra V, Goswami A, et al. A Comprehensive Survey on Model Compression and Acceleration[J]. Artificial Intelligence Review, 2020, 53(7): 5113–5155.

Chapter 2
Deep Learning Classic Platform and Algorithm for Natural Language Processing

This chapter introduces widely used deep learning platforms and algorithms within academia and industry contexts. It covers two key areas: deep learning platforms and classic algorithms in the field of artificial intelligence. This chapter also introduces prominent classic deep learning frameworks like TensorFlow, PyTorch, and PaddlePaddle, highlighting their unique features and roles in facilitating deep learning development. Additionally, it discusses classic algorithms like CNN, LSTM, RNN, GAN, etc., and their applications for the two major tasks of computer vision and natural language processing.

2.1 Classic Deep Learning Platform

The characteristic of deep learning requiring large computations determines that only hardware such as GPU/TPU can be used. Ordinary users find it challenging to use CUDA and other controls for GPU directly; therefore, deep learning frameworks are needed to help people quickly use resources such as GPU for deep learning development. Using deep learning platforms, users can utilize hardware resources, simplify the construction of computation graphs and partial derivative operations, efficiently run deep learning algorithms, and lower the entry barrier for deep learning.

The development status of deep learning platforms is shown in Fig. 2.1. After 2015, major mainstream frameworks began to appear, including the well-known TensorFlow, PyTorch, and Baidu's PaddlePaddle.

The 2021 market share of open-source deep learning frameworks in China is shown in Fig. 2.2, with TensorFlow, PyTorch, Caffe, and PaddlePaddle accounting for over 80% of the user share. In addition, MindSpore, released by Huawei in 2020, is also starting to emerge.

H. Zhang, J. Shang, *Natural Language Processing and Applications*,
https://doi.org/10.1007/978-981-97-9739-4_2

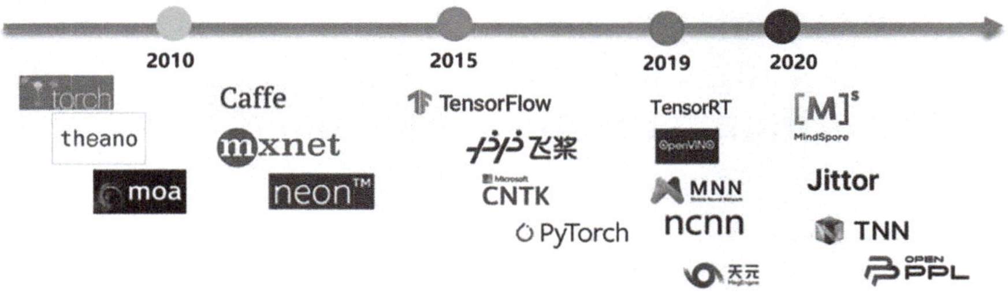

Fig. 2.1 Development status of deep learning platforms

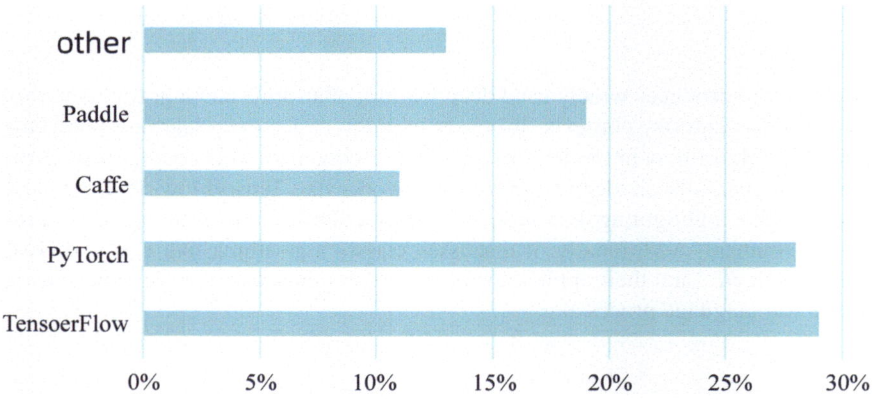

Fig. 2.2 Market share of open-source deep learning frameworks in China in 2021

2.1.1 TensorFlow Platform

TensorFlow is a deep learning framework developed by Google; it is an end-to-end open-source machine learning platform.

TensorFlow is named after its principle: Tensor represents tensor, and Flow represents flow. Thus, TensorFlow is a deep learning framework that uses data flow graphs for numerical computation and is widely used in various deep learning fields.

In 2015, Google released TensorFlow and open-sourced its code. In 2017, the official version of TensorFlow1.0 was released. In 2019, TensorFlow2.0 was released, which is more streamlined, straightforward, and extensible than TensorFlow1.0. TensorFlow is constantly improving, with new versions adding support for the Windows environment and expanding API, etc.

As one of the most widely used deep learning frameworks, TensorFlow has extensive applications in industry and academia. For example, LAIX uses TensorFlow to train AI teachers; Airbnb uses TensorFlow for image classification; the system automatically classifies the pictures uploaded by the landlord into the living room, kitchen, bedroom, etc.

2.1.1.1 Features of TensorFlow

TensorFlow is a deep learning framework that supports Linux, Windows, and macOS platforms and mobile devices and other platforms. TensorFlow provides a vibrant set of deep learning-related APIs. In addition, TensorFlow allows the construction of computational networks in the form of computational graphs and can conveniently operate on these networks. Users can write their upper-level structures and libraries in Python based on TensorFlow. Suppose TensorFlow does not provide the APIs needed by users. In that case, they can write their low-level functionality in C++ code and add the newly written functionality to TensorFlow through custom operations.

Compared to TensorFlow1.0, the most apparent difference in TensorFlow2.0 is the default use of a dynamic graph mechanism. The dynamic graph mechanism means that when you press the enter key after writing a line of code, you can immediately know the result, facilitating users' understanding and debugging of the program. Because TensorFlow 2.0 uses a dynamic graph mechanism, there is no need to execute data flow graphs in a session, so TensorFlow 2.0 uses functions instead of sessions. The static graph mechanism of TensorFlow1.0 also has certain advantages. For example, under the same training conditions, the CPU usage of TensorFlow2.0 during training is 32.3%, while the CPU usage of TensorFlow1.0 during training is 63%, which can better utilize the advantages of hardware.

2.1.1.2 Main Concepts of TensorFlow

TensorFlow has many features and is one of the most widely used deep learning frameworks. It is portable and can run on CPUs, GPUs, and mobile and embedded devices. TensorFlow supports multiple popular languages, such as Python and Java. TensorFlow uses TensorBoard to visualize the training process, model, and data. Visualization can be performed according to Fig. 2.3. TensorFlow has a high degree

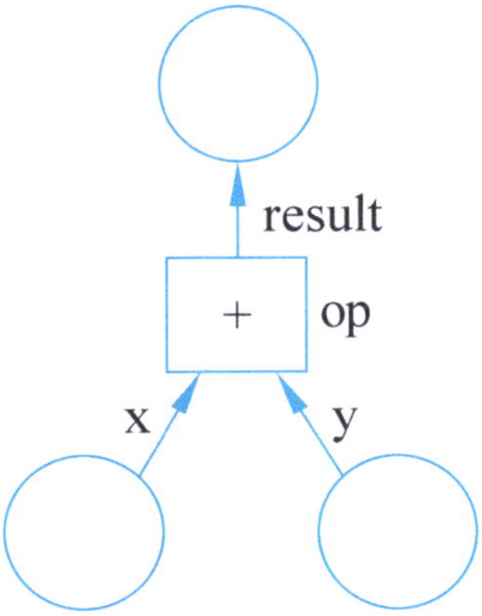

Fig. 2.3 Data flow graph $x + y$ = result

of flexibility, and if calculations can be represented as a data flow graph, then TensorFlow can be used.

The most crucial concept in TensorFlow is the data flow graph. Every calculation in TensorFlow represents a data flow graph. It includes two elements: x and y,

Figure 2.3 shows that in TensorFlow, data is Tensor, a multidimensional array represented by arrow lines in the data flow graph.

2.1.1.3 Practical Tools of TensorFlow

TensorBoard is a visualization tool of TensorFlow, which can visualize the running state of the TensorFlow program through the log files' output during the running process of the TensorFlow data flow graph state. TensorBoard and TensorFlow programs run in different processes; TensorBoard will automatically read the latest TensorFlow log files and present the latest state of the current TensorFlow program.

TensorFlow Lite can help users use TensorFlow on mobile and embedded ends. Run TensorFlow on a computer and use TensorFlow Lite to use TensorFlow on a mobile. The specific principle is to first train the network on the computer, then use the converter to convert the model into the TensorFlow Lite format, and then read and run the model on the phone.

2.1.2 Pytorch

In January 2017, the Facebook Artificial Intelligence Research (FAIR) team open-sourced PyTorch on GitHub and quickly topped the GitHub popularity list. As a framework that was only released in 2017 and had advanced design concepts, PyTorch's history can be traced back to Torch, born at New York University in 2002. Torch used a not-very-popular language, Lua, as its interface. Lua is simple and efficient, but because not many people use it, many people are deterred by the need to learn a new language to master Torch, even though Lua is a language simpler than Python.

Considering Python's leading position in computational science and its ecological integrity and easy-to-use interface, almost any framework must provide a Python interface. In 2017, the team behind Torch launched PyTorch. PyTorch is not simply a wrapper for Torch with a Python interface but a reconstruction of all modules above Tensor, and the addition of the most advanced automatic differentiation system not only can achieve powerful GPU acceleration but also supports dynamic neural networks, which many mainstream deep learning frameworks (such as TensorFlow, etc.) do not support. PyTorch can be seen as a GPU-supported NumPy or a robust deep neural network with automatic differentiation capabilities.

Apart from Facebook, PyTorch has already been adopted by organizations such as Twitter, Carnegie Mellon University, and Salesforce. PyTorch has become the most popular dynamic graph framework at present.

PyTorch is a rare, concise, elegant, efficient, and fast framework. Among the current open-source frameworks, no other framework can surpass PyTorch in terms of flexibility, ease of use, and speed in these three aspects.

PyTorch's design pursues minimal encapsulation, trying to avoid reinventing the wheel. Unlike TensorFlow, which is filled with new concepts such as session, graph, operation, name_scope, variable, Tensor, layer, etc., PyTorch's design follows the following three levels of abstraction from low to high: tensor → variable (autograd) → nn. Module, representing high-dimensional arrays (tensors), automatic differentiation (variables), and neural networks (layers/modules), and these three levels of abstraction are closely linked.

They can be modified and operated simultaneously. Another benefit of the concise design is that the code is easy to understand. PyTorch's source code is only about 1/10 of TensorFlow's, and the lower level of abstraction and more intuitive design make PyTorch's source code very easy to read. PyTorch's source code is easier to understand than many framework documentations.

PyTorch's flexibility does not come at the expense of speed. In many evaluations, its speed performance surpasses frameworks like TensorFlow and Keras. Developers describe PyTorch as "a tensor and dynamic neural network in Python with powerful GPU acceleration capabilities." The same algorithm implemented in PyTorch is often faster than in other frameworks.

PyTorch is the most elegantly object-oriented design among all frameworks. PyTorch's object-oriented interface design comes from Torch, and Torch's interface design is known for its flexibility and ease of use. Torch initially inspired the author of Keras to develop Keras. PyTorch inherits the mantle of Torch, and the API's design and the modules' interface are incredibly consistent with Torch. PyTorch's design aligns most closely with people's thinking, allowing users to focus as much as possible on implementing their ideas. What you think is what you get, without considering too many constraints of the framework itself.

2.1.3 Paddle Paddle

PaddlePaddle is a deep learning framework developed by Baidu in August 2016. Its design purpose is "committed to simplifying the innovation and application of deep learning technology." It is an industrial-grade deep learning framework with modules from development to deployment to specific inference products. Users only need to design the model, and after further optimization by PaddlePaddle's own PaddleSim, it can be deployed in locations such as servers, mobile ends, and web front ends through the corresponding modules of PaddlePaddle. Moreover, the model can be developed and trained by PaddlePaddle itself or converted from models produced by third-party frameworks (such as TensorFlow) through the X2Paddle tool. The following are the features of the PaddlePaddle platform.

2.1.3.1 High Encapsulation

To make the model more straightforward, PaddlePaddle has highly encapsulated its API while retaining the use of basic APIs. For example, PaddlePaddle adopts two methods during model training and testing: using the API under the paddle models, such as Model. Fit is used for training and testing; the other uses the underlying API to decompose the high-level API and execute a custom training and testing process. This way, it is easier for beginners, and those with a sure foundation can use this framework more flexibly.

2.1.3.2 Simple Dynamic–Static Conversion

The API of PaddlePaddle2.0 supports the conversion from a dynamic graph to a static graph. You only need to add a function decorator @paddle.jit.to_static in front of the function to be converted to a static graph, and this function can perform dynamic–static conversion. This function will also convert the dynamic graph to a static one during training, reducing training resource consumption.

2.1.3.3 Official Model Library Support

Since its launch, PaddlePaddle has continuously launched a large number of official model libraries, including classic networks in vision, such as AlexNet and ResNet, as well as commonly used BERT and Baidu's improved version of BERT, ERNIE, in NLP. PaddlePaddle provides official models in the fields of intelligent vision, intelligent text processing, and intelligent recommendation, which can be used with slight modifications.

2.1.3.4 Visualization of the Training Process

PaddlePaddle's visualization analysis tool, VisualDL, can help users understand the structure and training process of deep learning models more clearly and intuitively.

2.2 Deep Learning Classic Algorithm

Deep learning aims to learn sample data's inherent rules and representation levels. The information obtained in these learning processes considerably helps interpret text, images, and sound. Like humans, it aims to enable machines with analytical learning abilities to recognize text, photos, and sound data. Deep learning is a complex machine learning algorithm, and the results achieved in speech and image recognition far exceed previous related technologies.

2.2.1 *Convolutional Neural Networks*

2.2.1.1 The Origin of Convolutional Neural Networks

In the 1960s, Hubel and Wiesel discovered during their research on neurons in the cat's cerebral cortex used for local sensitivity and direction selection that their unique network structure could effectively reduce the complexity of feedback neural networks. They then proposed convolutional neural network (CNN) [1], a type of feedforward neural network that includes convolutional computations with a deep structure and is one of the representative algorithms of deep learning. CNN has become a research hotspot in many scientific fields, especially in pattern classification. It has been widely applied because it avoids complex pre-processing of images and can directly input original photos. Essentially, CNN is a multilayer structure built by mimicking the information processing process of visual cells, known as the Hubel–Wiesel structure.

2.2.1.2 Overview of Convolutional Neural Networks

Convolutional neural networks (Fig. 2.4) and traditional neural networks (Fig. 2.5) have many similarities. They both mimic the structure of the human nervous system, composed of neurons with learnable weights and bias constants. Each neuron can receive input signals and, after computation, output scores for each category. However, the input of convolutional neural networks is generally images. Convolutional networks successfully reduce the dimensionality of large-scale image recognition problems through a series of methods, ultimately making it trainable.

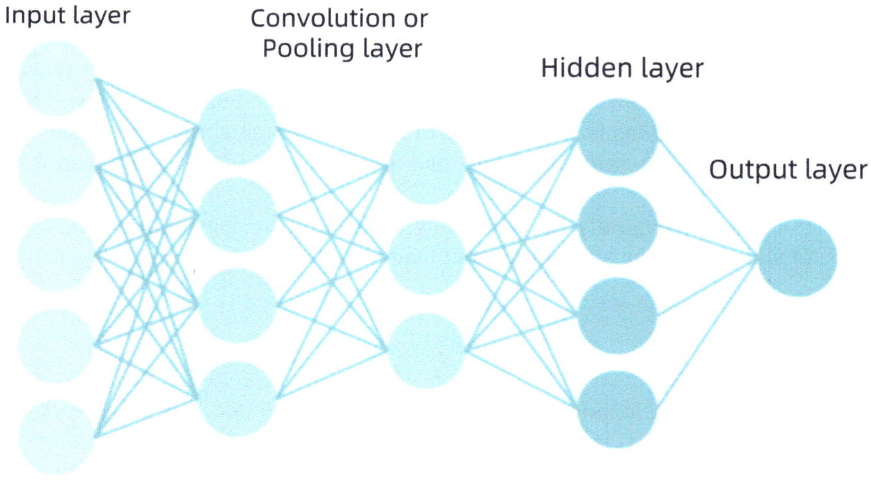

Fig. 2.4 Schematic diagram of a convolutional neural network

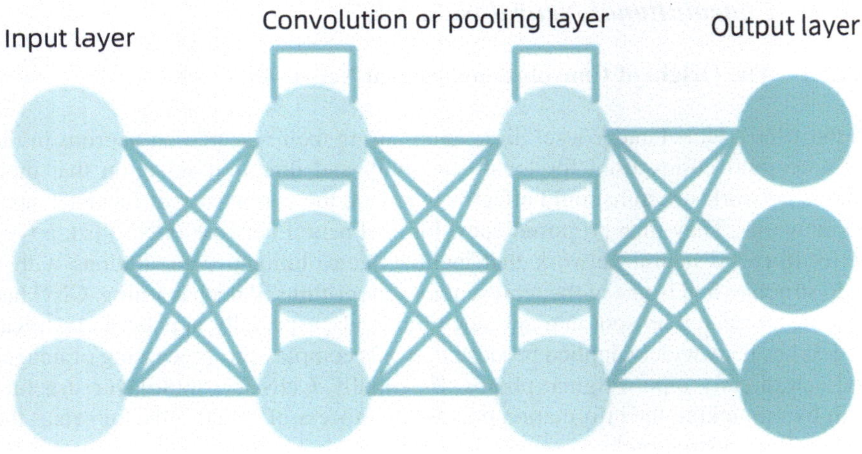

Fig. 2.5 Schematic diagram of the traditional neural network

Convolutional neural networks utilize this feature, designing neurons with three dimensions: width, height, and depth.

Convolution

The initialization of convolutional neural networks mainly involves initializing the weights and bias amounts of each convolution kernel in the convolution layer. The input sources of the convolution layer include the input images and the feature maps generated after convolution or pooling operations. The convolution operation mainly uses a fixed-size convolution kernel, slides on the input image or feature map at a confident stride, and maps the feature map to the next layer's feature map through inner product operations and nonlinear functions. The number of feature maps output by the convolution layer is determined by the number of convolution kernels defined by humans, and the size of the convolution kernel determines the size of production, the stride of sliding, and the size of the input feature map from the previous layer.

The output size is jointly determined by the size of the convolution kernel, the stride of sliding, and the size of the input feature map from the previous layer.

Pooling

Pooling, also known as downsampling, allows convolutional neural networks to train the corresponding classifiers using the features obtained through convolution. However, the size of the input image is large. In that case, the features obtained solely through convolution operations often have high dimensions, so overfitting can quickly occur during the classifier's training. The pooling operation is a downsampling process of the feature map of the previous layer. A low-dimensional feature representation can be obtained by aggregating the statistics of adjacent small areas in the feature map of the previous layer. Through pooling, the purpose of describing large-size feature maps with small-size feature maps can be achieved, effectively improving the performance of the classifier and preventing overfitting. Everyday pooling operations include average pooling and max pooling, which refer to mapping each aggregated area's average value and maximum value to the next layer's feature map, respectively.

Fully Connected

Similar to the layers in traditional neural networks, the output of the neurons in the next layer of the fully connected layer is related to the input of all neurons in the previous layer. The fully connected layer allows the network's parameters to converge quickly on the training samples.

Nonlinear Functions

The human brain's understanding of the objective world is not linear but a complex nonlinear mapping. Therefore, neural networks often simulate the nonlinear cognitive behavior of the human brain through nonlinear activation functions in their design. The commonly used nonlinear activation functions mainly include Sigmoid and Rectified Linear Unit (ReLU).

2.2.2 Recurrent Neural Networks

2.2.2.1 Development History of Recurrent Neural Networks

In 1982, John Hopfield invented a single-layer feedback neural network, known as the Hopfield neural network, to solve combinatorial optimization problems. This is the earliest form of Recurrent Neural Networks (RNNs). In 1986, Michael I. Jordan defined the concept of recurrence and proposed the Jordan neural network. In 1990, American cognitive scientist Jeffrey L. Elman simplified the Jordan neural network and used the BP (Back Propagation) algorithm for training, resulting in the simplest RNN [2] model containing a single self-connected node. However, due to the problems of gradient vanishing and gradient exploding, the training of RNN was tough, and its application needed to be improved. It was not until 1997 that Jurgen, the director of the Artificial Intelligence Research Institute, Schmidhuber, proposed Long Short-Term Memory (LSTM) [3]. LSTM, using gated units and memory mechanisms, significantly alleviated the early training problems of RNN. Also, in

1997, Mike Schuster proposed the bidirectional RNN model. These two models significantly improved the early structure of RNN, expanded the application range of RNN, and laid the foundation for developing subsequent sequence modeling. Although RNN achieved good results in some sequence modeling tasks at this time, there was little progress in the following years due to the large consumption of computing resources.

In 2010, Tomas Mikolov improved the feedforward neural network language model (NNLM) [4] proposed by Bengio, who proposed the RNN-based Language Model (RNN LM), and used it in speech recognition tasks, significantly improving the recognition accuracy. On this basis, Tomas Mikolov proposed Word2Vec in 2013. Unlike NNLM and RNNLM, the goal of Word2Vec is not to focus on building language models but on how to use language models to learn the semantic vectors (distributed representation) of each word. The concept of distributed representation originated from Hinton's work in 1986. Word2Vec triggered a wave of deep learning in natural language processing and inspired new fields such as knowledge representation and neural symbolic representation.

On the other hand, in 2014, Bengio's team and Google almost simultaneously proposed the Seq2Seq architecture. RNN was used for machine translation. Bengio's team suggested the attention mechanism shortly after, which improved the Seq2Seq architecture. Since then, machine translation has wholly entered the era of neural machine translation (NMT), which is not only simple in process but also far superior to the effects of statistical machine translation. Almost all mainstream machine translation systems use neural machine translation technology. In addition, the attention mechanism is also widely used in various tasks based on deep learning. In 2017, Facebook's Artificial Intelligence Laboratory proposed a Seq2Seq architecture based on convolutional neural networks, replacing RNN with CNN with gated units, significantly accelerating the model's training speed. Shortly after, Google proposed the Transformer architecture, replacing the original RNN and CNN with the self-attention mechanism, further reducing the complexity of the model.

Regarding word representation learning, the Allen Institute for Artificial Intelligence proposed a context-related representation learning method, ELMo 2018, using a bidirectional LSTM language model to learn different vector representations for words in other contexts, improving six natural language processing tasks. On this basis, the OpenAI team proposed the pre-trained language model GPT [5], replacing LSTM with a Transformer to train language models. Unlike the previous method of learning word vectors as features when applied to specific tasks, GPT directly connects the Softmax as the task output layer to the last layer of the pre-trained language model and then fine-tunes the model, achieving better results in multiple tasks. Shortly after, Google proposed the BERT [6] model, extending the unidirectional language model in GPT to a bidirectional language model—the Masked Language Model (MLM), and introduced the sentence prediction task in the pre-trained model. The BERT model achieved excellent results in 11 natural language processing tasks, marking another milestone in deep learning in natural language processing.

2.2.2.2 Overview of Recurrent Neural Networks

Recurrent neural networks were proposed based on the view that "human cognition is based on experience and memory." It not only considers the input at the previous moment but also gives the neural network the "memory" function of the previous content, that is, the current output of a sequence is also related to the earlier production. The specific manifestation is that the network will remember the last information and apply it to the calculation of the current output, that is, the nodes between the hidden layers are no longer unconnected but connected, and the input of the hidden layer includes not only the production of the input layer but also the output of the hidden layer at the previous moment.

First, look at a simple recurrent neural network, as Fig. 2.6 shows. This neural network consists of an input layer, a hidden layer, and an output layer. The hidden layer uses two arrow lines to represent the cyclic update of data, realizing the time memory function. The input layer receives the input, the hidden layer is activated, and the output is from the output layer. Next, we will build a deeper network with multiple hidden layers. Here, the input layer receives the input, the first hidden layer activates this input, sends these activated inputs to the next hidden layer, and continuously activates to get the output.

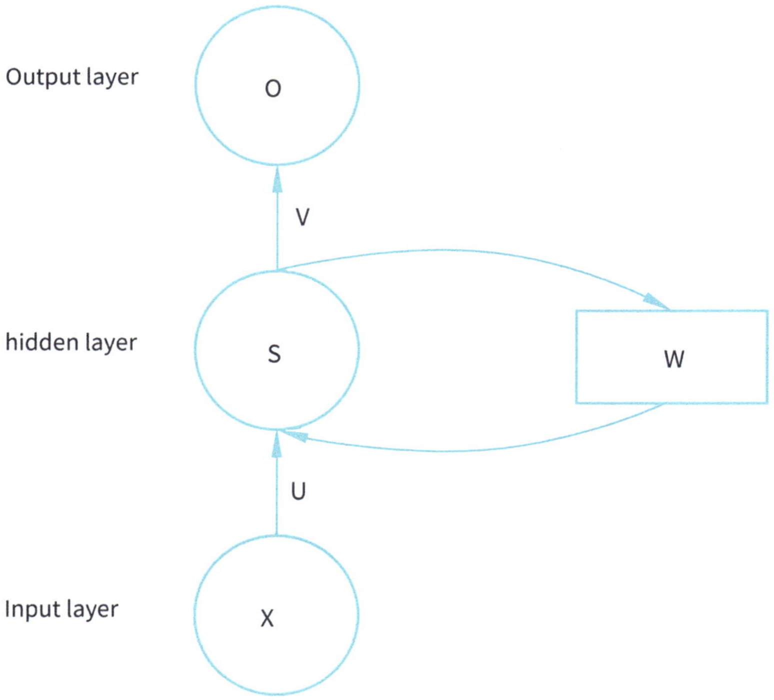

Fig. 2.6 Simple recurrent neural network

2.2.2.3 Seq2Seq

Seq2Seq is a sequence-to-sequence model. Its task mainly has two characteristics: One is that the input and output are of variable length, and the other is that there is a sequential relationship between the input and output elements. This situation generally appears in machine translation tasks. If a Chinese sentence is translated into English, the length of this English sentence may be shorter than the Chinese sentence or longer than the Chinese sentence.

The Seq2Seq model was proposed by Cho et al. and Sutskever et al. The basic idea is to infer the corresponding output sequence from the global information of the input sequence. This model consists of an encoder and a decoder, as shown in Fig. 2.7. In the work of Cho et al., both the encoder and the decoder use RNN. The encoder encodes a variable-length sequence X into a fixed semantic vector. The decoder extracts semantic information from this vector and outputs another variable-length sequence Y; a word vector represents each word in the sequence.

Among them, the encoder encodes the sequence into a fixed-length vector that can map the overall characteristics of the sequence, which is called a semantic vector here; the decoder restores the fixed-length vector obtained by the encoder to the corresponding sequence data, generally using the same structure as the encoder.

The simplest way to obtain the semantic vector is to directly use the hidden state of the last input as the semantic vector, change the last hidden state to get the semantic vector, or change all the hidden states of the input sequence to get the semantic vector.

The main shortcomings of the Seq2Seq model are twofold: First, its accuracy from encoding to decoding largely depends on a fixed-length semantic vector. There is a problem of information loss in the compression process from the input sequence to the semantic vector, and in a slightly longer sequence, the input information at the beginning is easily covered by the input information at the back. Second, during decoding, the context vector used in each moment's output during the decoding process is the same, without distinction, that is to say, the prediction vector used when predicting each word in the prediction result is the same, which will also cause problems for decoding.

Fig. 2.7 Structure of the Seq2Seq model

2.2.2.4 Attention Mechanism

In the traditional Seq2Seq [7] model, the encoder must compress all input sequence information into one. It is sometimes difficult to compress all the input sequence information into such a fixed-length vector in the context vector. Bengio and others proposed that the attention model allows the Seq2Seq model to obtain more input sequence information during encoding.

The attention mechanism [8] is a mechanism that focuses the model's attention on the current translation word. For example, when translating "I have a cat" to "I," attention should be placed on the "I" of the original sentence. When translating to "cat," attention should be placed on the "cat" of the original sentence. After using the attention mechanism, the input of the decoder is no longer a fixed context vector but will calculate the current context vector based on the current translation information.

During the decoding phase, it is necessary to calculate the relevance of the current decoder output hidden state and each hidden state of the encoder. According to this relevance, the secret state of each input sequence is weighted and summed, and the weighted sum vector is fused with the hidden state of the decoder at the current moment. This vector is used to predict the output at the final position.

2.2.3 Generative Adversarial Networks

2.2.3.1 Overview of Generative Adversarial Networks

GAN [9] is the abbreviation of generative adversarial network. Its basic idea is an adversarial game. A simple understanding of the generative adversarial network is that it can be seen as a game process, with the generative and discriminative models as the two sides of the game. For example, in the process of criminals making counterfeit money and banks identifying counterfeit money, the generative model is equivalent to the party making counterfeit money, whose purpose is to generate more realistic and unrecognizable counterfeit money based on the observed money situation and bank's identification technology; the discriminative model is equivalent to the party identifying counterfeit money, whose purpose is to identify the counterfeit money made by criminals as much as possible. Through the competition and continuous improvement of both sides, the ability of criminals to make counterfeit money and the ability of banks to identify counterfeit money have been improved, and finally, the counterfeit money created by criminals cannot be determined by the bank, that is, the counterfeit money maker has successfully deceived the counterfeit money identifier.

The generative adversarial network is a generative model consisting of a generator and a discriminator. The generator generates images, and the discriminator judges whether this image comes from a real dataset or is generated by the generator. Through the discriminator's scoring of the generated images, the generator can better optimize the quality of the generated images, making the images generated by the generator increasingly close to the pictures of the actual dataset. The above game scenario will continue until both the generator and the discriminator can no longer improve themselves, and at this point, the generative model will become a reasonably perfect model.

For the generator, the input is an n-dimensional vector, which can be random. It can also be generated according to a particular distribution, and the output is an image. The generator model can be the simplest fully connected neural network, or it can be a deconvolution neural network, etc. For the discriminator, the input is an image, and the discriminator judges whether it is an actual image or a generator-generated image by scoring this image. Therefore, the discriminator is a binary classification model, a fully connected or convolutional neural network.

The training of the generative adversarial network is an iterative process. First, initialize the generator and discriminator. In each training iteration, first input m random vectors to the generator to generate m images (generated samples) and then sample m authentic images (actual samples) from the real dataset. Fix the discriminator, input the actual images with positive labels and the generated images with negative labels into the discriminator, and train the discriminator to judge the exact images and generated images more accurately. Then, fix the discriminator and train the generator so that the data distribution of the generated samples is as close as possible to the data distribution of the actual samples, equivalent to making the discriminator make as many mistakes as possible. In this way, iterative training is carried out continuously so that the generation ability of the generator becomes stronger and stronger, and the data distribution of the images it produces is closer and closer to the data distribution of authentic images; at the same time, the discriminator's ability to distinguish whether an image is an actual image or a generated image also gets stronger and stronger.

As shown in Fig. 2.8, since the generative adversarial network was proposed in 2014, people have conducted extensive research on it and proposed many different types of generative adversarial networks, such as BGAN, CGAN, DCGAN, WGAN, StyleGAN, etc., which are widely used in various scenarios.

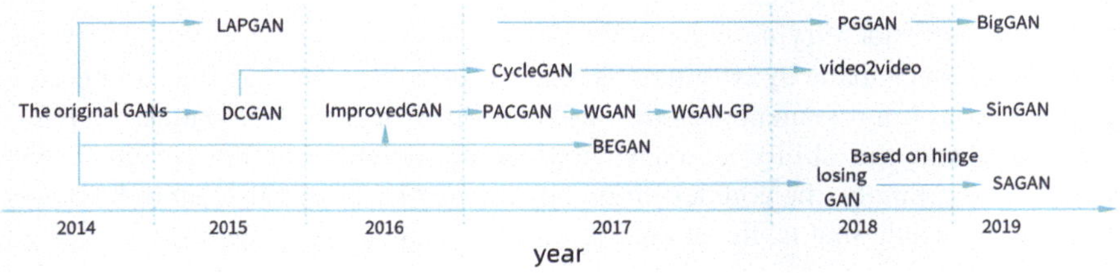

Fig. 2.8 Milestone generative adversarial networks and their variants

2.2.3.2 Theoretical Basis of Generative Adversarial Networks

The purpose of the generative adversarial network is to let the generator learn the data distribution P_g of the actual sample. Assume the input noise follows a distribution $P_z(z)$, and the generator G maps the input random noise to the sample data space, denoted as $G(z, \theta_g)$. G is a differentiable function, a multilayer perceptron with parameters θ_g. Also, define a multilayer perceptron $D(x, \theta_d)$ with parameters θ_d, it outputs a single value, representing the probability that x comes from the actual data distribution P_g rather than the generated data distribution. The objective function of the generative adversarial network is shown in Eq. (2.1):

$$\min_G \max_D V(D,G) = E_{x\sim p_{\text{data}}(x)}\left[\log D(x)\right] + E_{z\sim p_z(z)}\left[\log\left(1-D\left(G(z)\right)\right)\right] \quad (2.1)$$

As shown in Fig. 2.9a, when in the initial state, the data distribution generated by the generator and the actual data distribution are quite different, and the probability of the discriminator distinguishing the sample could be more stable. Hence, the discriminator must first be trained to determine the sample better. After training the discriminator multiple times, the sample state shown in Fig. 2.9b is reached; at this point, the sample distinction is very significant and promising. Then, the generator is trained. After training the generator, the sample state shown in Fig. 2.9c is reached; at this point, the data distribution generated by the generator is closer to the actual data distribution than before. After multiple rounds of iterative training, we hope to eventually reach the state shown in Fig. 2.9d, where the data distribution of the generated samples approximates the data distribution of the actual samples, and the discriminator cannot distinguish whether the samples are generated or objective (the discrimination probability is 0.5). In other words, the generator can generate very realistic samples.

In Fig. 2.9, the thick dashed line represents the data distribution of actual samples, the thin dashed line represents the distribution of the discriminator's discrimination probability, and the solid line represents the data distribution of generated samples. z represents noise, and z to x represents the distribution mapping after passing through the generator.

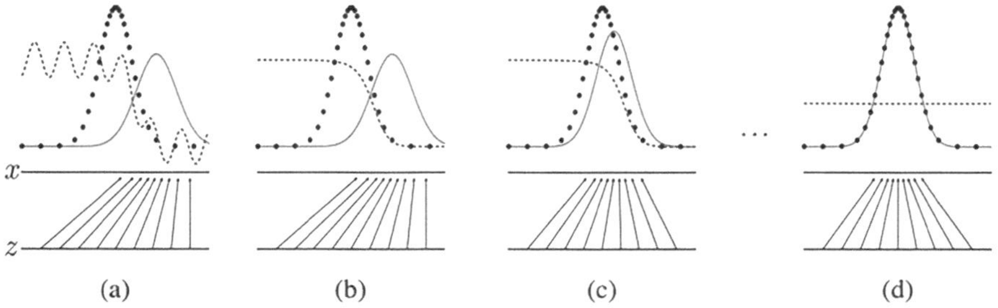

Fig. 2.9 Training situation of generative adversarial networks

2.2.3.3 Application Examples of Generative Adversarial Networks

Generative adversarial networks have applications in many areas, such as image generation, image style transfer, image translation, image video repair, and text-to-image generation. Generative adversarial networks, such as face generation, were first used for image generation. Figure 2.10 is the process of generating faces using generative adversarial networks. In recent years, the number of researchers in generative adversarial networks has increased. With the deepening of research on generative adversarial networks, the quality of image generation is getting higher and higher, and even the faces generated by computers are difficult to be distinguished by manual methods. Figure 2.11 is a face generated using generative adversarial networks.

From the original generative adversarial networks, which had difficulty generating 32 × 32 resolution (unit: pixel) images, to the program that produces 2 K resolution images that are difficult to distinguish between true and false, generative adversarial networks have shown their unique advantages and gradually penetrated multiple fields. In addition, generative adversarial networks are also very suitable for image style conversion, such as converting sketch images to natural objects, converting grayscale images to color images, etc., as shown in Fig. 2.12.

Generative adversarial networks can not only be applied to image processing but also have some applications in the field of natural language processing. For example, BR-CSGAN can be used for neural machine translation, and some generative adversarial network models can be used for unsupervised speech recognition and text summary generation.

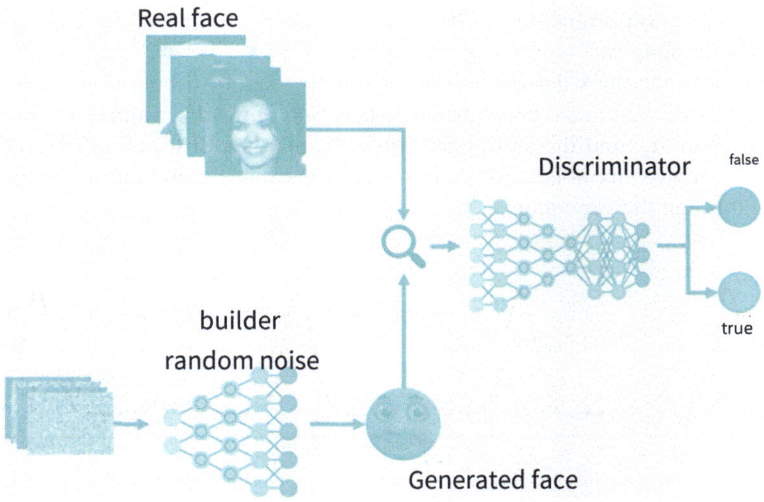

Fig. 2.10 Process of generating faces using generative adversarial networks

Fig. 2.11 Faces generated using generative adversarial networks

Fig. 2.12 Image style transfer using generative adversarial networks

References

1. Krizhevsky A, Sutskever I, Hinton G E. Imagenet Classification with Deep ConvolutionalNeural Networks[J] . Advances in NeuralInformation Processing Systems, 2012, 25: 1097–1105.
2. Sadeghis, Ramanathan K. A Hubel Wiesel Model of Early Concept Generalization Based on Local Correlation of Input Features[C]. The 2011 International Joint Conference on Neural Networks. IEEE, 2011: 709–716.
3. Gers F A, Schraudolph N N, Schmidhuber J. Learning Precise Timing with LSTM Recurrent Networks[J] . JournalofMachine Learning Research, 2002, 3(Aug) : 115–143.
4. Mikolov T, Karafiát M , Burget L, et al. Recurrent NeuralNetwork Based Language Model[C] . Interspeech. 2010, 2(3) : 1045–1048.
5. Raffel C, Shazeer N, Roberts A, et al. Exploring the Limits of Transfer Learning with a Unified Text-to-TextTransformer[J] . JournalofMachine Learning Research, 2020, 21: 1–67.
6. Devlin J, Chang M W , Leek, etal. BERT: Pre-trained ofDeepBidirectionalTransformersfor-Language Understanding. In: Proceedings of the 2019 Conference of the North American Chapter of the Association for ComputationalLinguistics: Human Language Technologies, Volume1(Long andShortPapers) , 2019, 4171–4186.
7. Liz, Caij, He S, et al. Seq2Seq dependency parsing[C]. Proceedings of the 27th International Conference on Computational Linguistics. 2018: 3203–3214.

8. Vaswania, Shazeer N, Parmar N, et al. Attention is All You Need[J]. Advances in Neural Information Processing Systems, 2017,30: 5998–6008.
9. Goodfellow I, Pouget-Abadie J, Mirza M, et al. Generative Adversarial Nets[J]. Advances in Neural Information Processing Systems, 2014, 27: 2672–2680.

Chapter 3
Advances in Deep Learning for Natural Language Processing

This chapter explores the challenges faced by traditional deep learning and the cutting-edge advancements in deep learning for natural language processing (NLP), focusing on three key areas: data, training, and applications. The data-oriented aspect will focus on self-supervised learning, hint learning, and semi-supervised learning, aiming to solve the problem of complex labeling and achieve data self-supervision; the training-oriented element will focus on meta-learning, multitask learning, lifelong learning, and MRC framework, aiming to achieve model multi-problem transfer and even multidomain transfer, solving the problem of high cost of developing neural networks; and the application-oriented aspect will start with model compression, model security, and model interpretability, mainly discussing the existing issues in the application field of the model and the latest solutions.

3.1 Bottlenecks Encountered in Traditional Deep Learning

3.1.1 Overview of Deep Learning

Two significant changes have occurred in the development of natural language processing since the twenty-first century, with the development of high-performance computing systems. Table 3.1 presents the four paradigms in this process.

With the rise of machine learning, natural language processing has undergone its first significant change. This change can be further divided into three stages: the feature engineering stage, the architecture engineering stage, and the objective engineering stage. The traditional machine learning method is the feature engineering stage, where fully supervised learning is prevalent, and annotated corpora, as a necessary condition for supervision, are challenging to obtain. The research work at this stage focuses on manually extracting features to induce machine learning models better to learn from limited data. The architecture engineering stage also

H. Zhang, J. Shang, *Natural Language Processing and Applications*,
https://doi.org/10.1007/978-981-97-9739-4_3

Table 3.1 The four paradigms in the development process of natural language processing

Paradigm	Engineering	Task relation
a. Fully supervised learning (non-neural network)	Features (e.g., word identity, part-of-speech, sentence length)	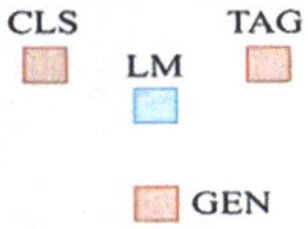
b. Fully supervised learning (non-neural network)	Architecture (e.g., convolutional, recurrent， self-attentional)	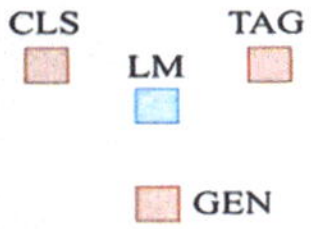
c. Pre-train, fine-tune	Objective (e.g., masked language modeling; next sentence prediction)	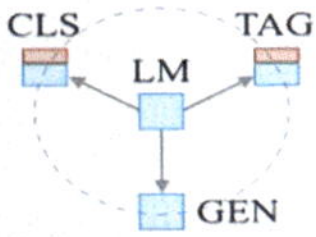
d. Pre-train, prompt, predit	Prompt (e.g., cloze, prefix)	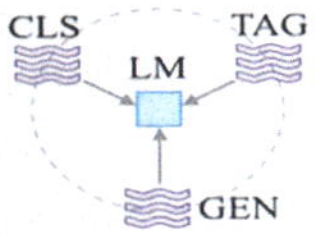

corresponds to the early stage of neural networks. One of the outstanding advantages of neural networks is that they do not require manual feature design, and neural networks will automatically extract features. The research work at this stage focuses on designing neural network structures that are more suitable for tasks, so the model's feature extraction ability is more vital. In the objective engineering stage, pre-trained language models rise. Fully supervised learning is always limited by the limited learning corpus, and the unsupervised learning mode of pre-trained language models can learn unrestrictedly on rich, unannotated corpora. Therefore, "pre-trained + fine-tuning" has become mainstream, and research has shifted to designing training objectives for pre-trained and fine-tuning.

In 2021, the development of natural language processing entered a new stage amid the second significant change, namely the introduction of prompt learning. The "pre-trained + fine-tuning" process replaces the "pre-trained + prompt learning + prediction" process. In this period, instead of adapting the pre-trained language model to downstream tasks through target engineering, downstream tasks are redesigned to make them look more like the tasks solved in the original language model training with the help of text prompts. For example, when classifying the sentiment of social media posts, for "I missed the bus today," you can continue to prompt "I missed the bus today. I felt so," and let the language model fill in the blank with an emotional word. You can also prompt, "English: I missed the bus today. French:"The language model can fill in the blank with a French translation. In this way, by choosing the appropriate prompt, you can manipulate the model's behavior so that the pre-trained language model can predict the expected output, sometimes without additional task-specific training. The advantage of this method is that, given a set of appropriate prompts, a language model trained entirely in an unsupervised manner can be used to solve many tasks. However, like most conceptual enticing prospects, there is a pitfall here. This method introduces the necessity of prompt engineering: finding the most suitable prompt to enable the language model to solve the current task.

3.1.2 Problems Encountered by Traditional Deep Learning

Although deep learning applied to natural language processing has achieved the best results in current evaluation tasks, it is not universal. After years of development, its bottleneck has become apparent, and the related problems can be summarized as data, training, and application.

3.1.2.1 Data Aspect

The problems of traditional deep learning in the data aspect are as follows:

1. Small sample problem. If there is sufficient data annotation, the neural network based on deep learning can achieve unparalleled advantages in extracting target features. When the training data is minimal, the parameters in the neural network are difficult to generalize, and the entire neural network will show overfitting. Therefore, the small sample problem has always been a focus and difficulty in deep learning research. However, this problem can be solved to some extent by active learning, semi-supervised learning, unsupervised learning, and self-supervised learning.
2. Insufficient annotated data. The premise of deep learning is a large amount of annotated data. Although there are some methods to reduce the dependence on data, such as transfer learning, small sample learning, unsupervised learning, and weakly supervised learning, their performance cannot be compared with supervised learning.
3. Noise. In existing data, text is the most unstructured form, containing various kinds of noise.
4. Diverse data structures, especially the large amount of unstructured data such as tweets, posts, chat conversations, news, blog articles, etc. This problem can be solved by graph neural network (GNN), multimodal, etc.

3.1.2.2 Training Aspect

The problems of traditional deep learning in training are as follows:

1. Multitasking. Natural language processing includes various tasks such as lexical analysis, syntactic analysis, semantic analysis, etc., which have difficulties, high

annotation costs, and poor annotation consistency, making large-scale annotated data often difficult to obtain. These tasks seem different on the surface but have a close internal connection. Good performance in one task significantly promotes its generalization ability in related tasks, so multitask learning can naturally be applied to these tasks to improve the analysis accuracy of each task.
2. Multimodel. The existing natural language processing tasks usually have one task corresponding to one model, but a lot of information can be generalized. It has been a long-standing dream of many researchers to have a general model to complete all tasks. This dream is gradually becoming a reality with the development of pre-trained language models.
3. Multilingual. Many languages exist, but few available corpora make model training difficult. Over 6000 languages are worldwide, but only 15 are commonly used.
4. Lack of common sense. The current machine text generation models can write an article convincing many people. Still, they are imitating the situations seen during the training phase, and there is a big gap in human performance.
5. Catastrophic forgetting, that is, training the model with new information, will interfere with the knowledge learned before. This phenomenon often leads to a sudden drop in performance, or in the worst case, the old knowledge is entirely covered by the new knowledge. For deep neural network models that learn from fixed training data, the primary defect of deep models is that the information that increases over time cannot be fully utilized.

3.1.2.3 Application Aspect

The problems of traditional deep learning in application are as follows:

1. The model size is enormous. The pre-trained language model based on large-scale unlabeled corpora has pushed natural language processing research to a new height and formed a new pattern of "pre-trained + fine-tuning." However, from BERT with more than 100 million parameters, to the GPT-3 model with parameters gradually breaking through the 100 billion level, to the currently largest language model MT-NLG with over 530 billion parameters jointly launched by Microsoft and Nvidia, as the scale of language models becomes larger and larger, not to mention pre-trained a brand new language model, even fine-tuning existing large models requires computational resources that individuals and general institutions cannot afford. This problem can be solved by prompt learning.
2. Domain problem. Most deep learning techniques require the assumption that the training set and test set are independently and identically distributed, but this assumption is problematic to satisfy in actual applications. When there is a difference in distribution between the training set and the test set (e.g., transferring the model from the financial field to the legal field), the model will not achieve good prediction results.

3. Interpretability problem. Many of today's deep neural networks cannot make decisions in an entirely understandable way from a human perspective. Even though current models bring the results of graphic recognition and voice recognition close to perfection, people always hold a bit of caution toward these predictions because they need to understand what the projections are based on and do not know when errors might occur. This is also why almost all current models cannot be deployed in some critical areas with high-performance requirements (such as transportation, healthcare, law, finance, etc.).

3.2 Frontier Progress in Data-Oriented Deep Learning

Deep neural networks based on supervised learning have succeeded dramatically in the past decade. However, supervised learning heavily relies on manual labeling and is susceptible to human attacks; these flaws prompt people to look for new solutions.

3.2.1 Active Learning

It is well known that deep learning methods, mainly based on supervised learning, often hope to optimize a large number of parameters through a large amount of annotated data, so that the model learns how to extract high-quality features.

However, in recent years, due to the rapid development of Internet technology, we are in an era of information flood, with a large amount of unlabeled data. Acquiring a large amount of high-quality annotated data sets requires a large workforce, which is not allowed in some areas that require high professional knowledge, especially in voice recognition, information extraction, medical imaging, etc. For example, the annotation and description of lung lesion images of COVID-19 patients require experienced clinicians, and it is impossible to ask them to complete a large amount of medical image annotation work. Therefore, a method is needed to minimize the annotation cost as much as possible while the model meets the performance requirements. Active Learning (AL) is a method that tries to maximize the performance gain of the model by marking the minimum amount of samples. Active learning can actively select the most valuable unlabeled samples for annotation to achieve the model's target performance with as few annotated samples as possible.

Active learning can be divided into membership query synthesis, stream-based selective sampling, and pool-based active learning from the application scenario. Membership query synthesis refers to the learner being able to request to query the label of any unlabeled sample in the input space, including samples generated by the learner. The difference between stream-based selective sampling and pool-based active learning mainly lies in the former making independent judgments on whether to query the label of unlabeled samples in the data stream for each sample. At the same time, the latter can choose the best query sample based on the evaluation and ranking of the entire data set.

In comparison, the pool-based active learning application scenario is more common in paper applications. Still, the stream-based selective sampling application scenario is obviously more suitable for small mobile terminal devices requiring timeliness. Figure 3.1 shows the cycle framework of pool-based active learning. In the initial state, one or more samples are randomly selected from the unlabeled pool U and provided to an annotator.

The annotator queries the labels, obtains the annotated dataset L, and then trains the model on L in a supervised learning manner. Then, using the new knowledge, the annotator selects the following sample to query, adds the newly queried sample to L, and trains the model. This process is repeated until the annotation budget is exhausted or a preset termination condition is reached.

Natural language processing has always been a very challenging task. It aims to enable computers to understand complex human language, helping humans handle various tasks related to natural language. Insufficient data labeling is also a key challenge faced by natural language processing tasks. Below are some active learning methods applied in the field of natural language processing.

3.2.1.1 Sentiment Analysis

Sentiment analysis is a typical task in natural language processing, aiming to enable computers to understand a description in natural language and extract and analyze its emotional tendency information. Shalini and others [1] used relevant tweets from

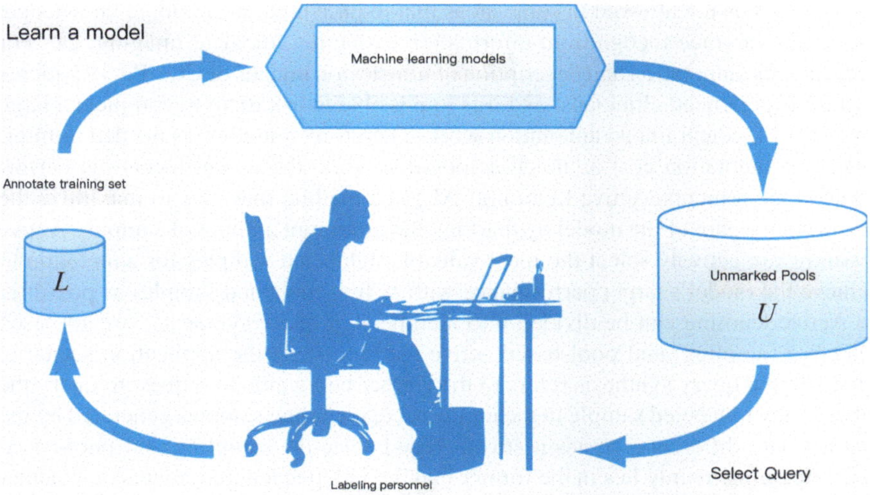

Fig. 3.1 Framework of pool-based active learning cycle

disaster-stricken areas to extract information for identifying infrastructure damage during earthquakes. They combined models based on RNN and GRU with active learning, using active learning-based methods to pre-train language models and retrieve tweets about infrastructure damage from different regions, thereby significantly reducing the workload of manual annotation.

3.2.1.2 Question Answering and Summarization

Question-answering systems and automatic summarization are everyday tasks in natural language processing. Active learning has achieved impressive results in these fields. However, the performance of these applications still depends on massive labeled datasets, and active learning is expected to bring new hope to this challenge. Asghar and others [2] used an online active learning strategy combined with deep learning models, interacting with real users and learning incrementally from user feedback in each round of dialogue to achieve open dialogue. Automatic summarization aims to extract the most essential information from significant texts. Hanbay and others [3] proposed a novel active learning strategy neural network (ALPNN) for identifying concepts and relationships in extensive EEG reports. This can help humans extract practical clinical knowledge from many EEG reports.

3.2.1.3 Other Fields

In the field of knowledge graphs, an active learning framework for knowledge graph pattern expansion can generate new semantic types for knowledge graph patterns without relying on a set of target patterns and human user observations. The cost of label annotation in speech and audio is also relatively high. Maldonado and others [4] found that models trained on a corpus of thousands of recordings collected from a few speakers could not be generalized to new fields. Therefore, they studied the practical scheme of training deep neural networks for speech emotion recognition tasks using active learning in situations where annotation resources are limited.

3.2.2 *Self-Supervised Learning*

Supervised learning is facing a bottleneck. It heavily relies on expensive manual annotations and suffers from generalization errors, spurious correlations, and adversarial attacks. Therefore, people expect neural networks to achieve more with fewer labels, fewer samples, and fewer experiments.

As an alternative method, self-supervised learning has extremely high data efficiency and good generalization ability in representation learning, thus attracting wide attention from researchers. Many advanced models (especially in the field of natural language processing) are following this paradigm. Self-supervised representation learning uses the input data as supervision, benefiting almost all downstream tasks. In the invited speech of AAAI2020, Turing Award winner Yann LeCun described self-supervised learning as "predicting any unobserved or hidden input part (or feature) from any observed or unhidden input part."

It can be concluded that self-supervised learning should have the following two characteristics: one, using a "semi-automatic" process to obtain labels from the data itself, the other, predicting data from other parts through local data.

Since self-supervised learning does not have manually annotated labels, it can be considered unsupervised learning. However, unsupervised learning is mainly used to detect specific data patterns (such as clustering, community discovery, or anomaly detection). The purpose of self-supervised learning is to restore a certain sense of supervised learning.

The key to the success of self-supervised learning is to find a way to utilize a large amount of unlabeled data available in the era of big data so that learning algorithms can get rid of human supervision and return to self-supervision of data, which has been applied in many fields. In natural language processing, there is a lot of information in unlabeled data, especially the inherent symbiotic relationship in the data. For example, in the incomplete sentence "I like apples," a well-trained language model considering the collinearity of "apples" and "eating" will predict the blank as "eating" through the Cloze Test [5].

3.2.2.1 Self-Supervised Learning in Word Embedding

CBOW and Skip-Gram [6] are pioneering works in natural language processing in terms of self-supervised learning in word embedding. The purpose of CBOW is to predict the input token based on the context tokens, while Skip-Gram aims to predict the context tokens based on the input token. FastText [7] uses the CBOW structure, while Word2Vec and GloVe use both. Since 2013, research on word embedding representations has made it unnecessary to initialize word embeddings randomly. Still, instead, knowledge has already been learned in the corpus, ushering in a new era of natural language processing.

3.2.2.2 Self-Supervised Learning in Pre-trained Language Models

Generative self-supervised learning is widely used in pre-trained language models, and some pre-trained tasks are shown in Table 3.2. The representatives of autoregressive language models are GPT and GPT-2 [8], which use the decoder architecture of Transformer as the language model.

The representation of denoising autoencoders, such as the Mask Language Model (MLM), makes the model robust enough to introduce noise. It randomly masks some tokens from the input and then predicts them based on their context information. BERT is the most representative product in this field, and [MASK] was introduced to shield some tokens during training. SpanBERT [9] chooses to mask

Table 3.2 Some pre-trained tasks

Operational	Element	Original text	Corrupted text
Mask	One token Two tokens One entity	Jane will move to New York. Jane will move to New York. Jane will move to New York.	Jane will [Z] to New York. Jane will [Z] to New York. Jane will [Z] to New York.
Replace	One token Two tokens One entity	Jane will move to New York. Jane will move to New York. Jane will move to New York.	Jane will [X] to New York. Jane will [X] [Y] to New York. Jane will [X] to New York.
Delete	One token Two tokens	Jane will move to New York. Jane will move to New York.	
Permute Rotate Concatenation	One token None Two languages	Jane will move to New York. Jane will move to New York. Jane will move to New York.	New York. Jane will move to to New York. Jane will move Jane will move to New York. [/s]

random continuous spans, ERNIE learns entity-level and phrase-level knowledge by masking entities or phrases, and ERNIE [10] further integrates knowledge from the knowledge graph into the language model.

Contrastive self-supervised learning is also used in pre-trained language models, such as Next Sentence Prediction (NSP). It requires the model to distinguish between the following and randomly sampled sentences. However, some later experiments showed that NSP needs more help for performance. Therefore, NSP loss was removed in RoBERTa. ALBERT proposed the sentence order prediction (SOP) task.

The pioneering work of generative–contrastive self-supervised learning is ELECTRA [11]. It proposed replacing token detection (RTD) with GAN. The structure is pre-trained on the language model, requiring the prediction of which words have been replaced.

3.2.2.3 Other Self-Supervised Learning in the Field of Natural Language Processing

Wang et al. [12] use contrastive learning technology to pre-train with an unsupervised corpus, allowing the model to understand event knowledge and event structure better. The authors use AMR parsing signals as unsupervised signals for pre-trained (AMR annotation is done automatically by the machine), and the results show that

AMR is not limited to event types and event patterns (schema), thus solving the problem of weak generalization. Although AMR annotation itself is noisy, according to the author's subsequent experiments, the impact of AMR noise on the experimental results is minimal.

3.2.3 *Prompt Learning*

Traditional supervised learning trains models to receive inputs and predict outputs, while prompt learning predicts by directly simulating the probability of text occurrence.

To use these models to perform prediction tasks, the original text x is modified into a text prompt with some unfilled slots utilizing a template, and then the language model is used to fill in the unfilled information probabilistically to obtain the answer. From this, the final output y can be derived, as explained in Table 3.3.

This framework allows (pre-trained) models to learn in advance on a large amount of unlabeled corpus (or supervised corpus of other tasks). Then, by defining a new prompt, the model can complete few-shot and zero-shot tasks and quickly adapt to new domains.

Prompt learning can be divided into several steps. The necessary steps are as follows:

1. Choose a pre-trained language model.
2. Design or train a template through prompt engineering.

Table 3.3 Terms in prompt learning

Name	Symbol	An example	Description
Input	x	I Love this movie.	One or multiple texts
Exportation	y	+ + (very positive)	Output label or text
Prompt function	$f$prompt(x)	[X] Overall, it was a [Z] movie.	A function that converts the input into a specific form by inserting the input x and adding a slot [Z] where answer z maybe filled later
Tips	x'	I love this movie. overall, it was a [Z] movie.	A text where [X] is instantiated by input x but answer slot [Z] is not.
Prompt candidate	$f$fill (x', z)	I love this movie. overall, it was a bad movie.	A prompt where slot [Z] is filled with any answer
Answerprompt	$f$fill (x)	I Love this movie. Overall, it was a goodmovie.	A prompt where slot [Z] is filled with a true answer
Answer	Z	.. good. . , fantastic. boring	A token, phrase, or sentence that fills [Z]

3. The final labels are mapped to a series of tokens that are easy for the model to understand through answer engineering.

The optional steps are as follows:

1. Ensemble of prompt learning models.
2. Update the parameters of the prompt/pre-trained language model.

Unlike the "pre-trained + fine-tuning + prediction" model, prompt learning can transfer the downstream task to the language model task through paradigm (task) transfer. The "pre-trained + fine-tuning + prediction" paradigm transfers the pre-trained language model task to the target downstream task by designing a small network (target engineering).

Section 3.2.2 introduces self-supervised learning and the self-supervised learning methods used by various pre-trained language models. A large part of the ability of pre-trained language model prompt learning depends on self-supervised learning during the pre-trained process. At the same time, when choosing a pre-trained language model for prompt learning, the model type also needs to be considered. Among the four pre-trained language models, it is generally believed that the left-to-right language model is suitable for prompt learning generation tasks, MLM is ideal for understanding tasks, and pre-trained language models with both encoders and decoders are suitable for translation and summarization. If event extraction tasks are also considered as a type of summary, they may be ideal for the latter two pre-trained language models.

Prompt engineering is creating a prompt that can enable downstream tasks to achieve a high performance. In a previous work, this involves the creation of prompt templates, usually manually searching for the best template for each model and task. The templates for everyday tasks are shown in Table 3.4. In addition to first considering the shape of the prompt (cloze or prefix), it is also necessary to decide whether to create the required prompts manually or automatically.

Prompt engineering designs appropriate inputs for the pre-trained language model in prompt learning. In contrast, answer engineering aims to find a mapping between an answer space and the original output to produce an effective prediction model. Like prompt learning, answer engineering also needs to determine the shape of the answer (token, fragment, or sentence) and choose the answer design method (manual or automatic).

3.2.4 Graph Neural Networks

The representation of natural language reflects a unique view of natural language and fundamentally affects how natural language is processed and understood. Figure 3.2 shows that natural language was usually represented as a bag of words or word vectors in early natural language processing research. Text sequences such as BoW (Bag of Words) and TF-IDF (Term Frequency–Inverse Document Frequency)

Table 3.4 Templates for common tasks

Type	Task	Input ([X])	Template	Answer ([Z])
Mask	Sentiment	I love this topic	[X] the movie is [Z].	Great, fantastic.
	Topics	He prompted the LM.	[X] the text is about [Z].	Sports Science ....
	Intention	What is text fare to Denver?	[X] the question is about [Z].	Quantity City .....
Text-span CLS	Aspect sentiment	Poor service but good food.	[X] what about service [Z].	Bad Terrible ...
Text-pair CLS	NLI	[X1]: an old man with.... [X1]: an old man with....	[X1] ? [Z],[X2]	Yes No ...
Tagging	NER	[X1]: mike went to Paris. [X2]: Paris	[X1] [Z] is a [X2] entity	Organization location
Text generation	Summarization	Las Vegas police....	[X] TL: DR: [Z]	the victim... A woman...
	Translation	Je vous aime.	French: [X] English: [Z]	I love you. I fancy you. .....

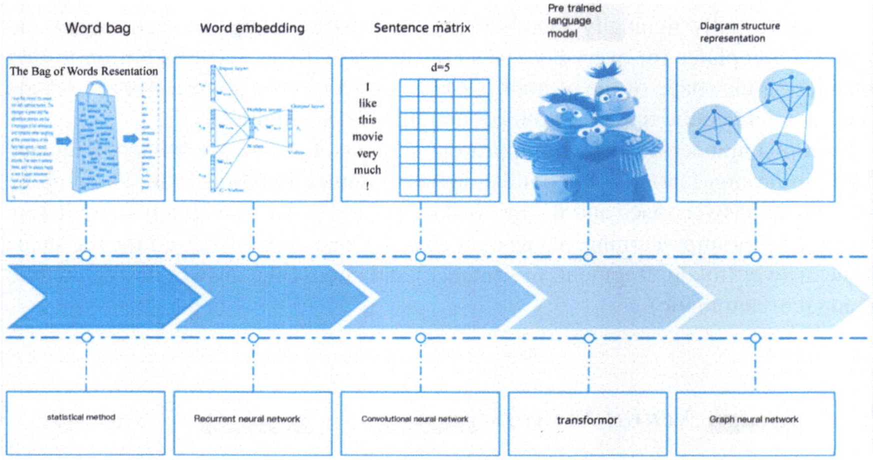

Fig. 3.2 Evolution of natural language representation methods

were considered tokens. With the success of word embedding technology, sentences are usually represented as a sequence composed of tokens, and some famous deep learning technologies, such as RNN and CNN, are widely used for text sequence modeling. Later research works analogized text to images encoded into matrix representation; in recent years, pre-trained language models have also emerged. However, these existing studies have yet to thoroughly utilize the semantic structure information in the text. Graphs are a universal and powerful form of representation, such as common syntax trees, semantic parsing graphs, etc. Graph-based natural language representation can capture richer semantic structure information between text elements, and graph-structured data can encode complex pairwise relationships between entity tags to learn more information-rich representations. Graph neural network (GNN) is a universal graph-based learning framework that can learn embeddings for each node in the graph and aggregate node embeddings to produce graph embeddings.

GNN research used for natural language processing tasks is mainly divided into three aspects: graph construction, graph representation learning, and graph-based encoder–decoder models, as shown in Fig. 3.3.

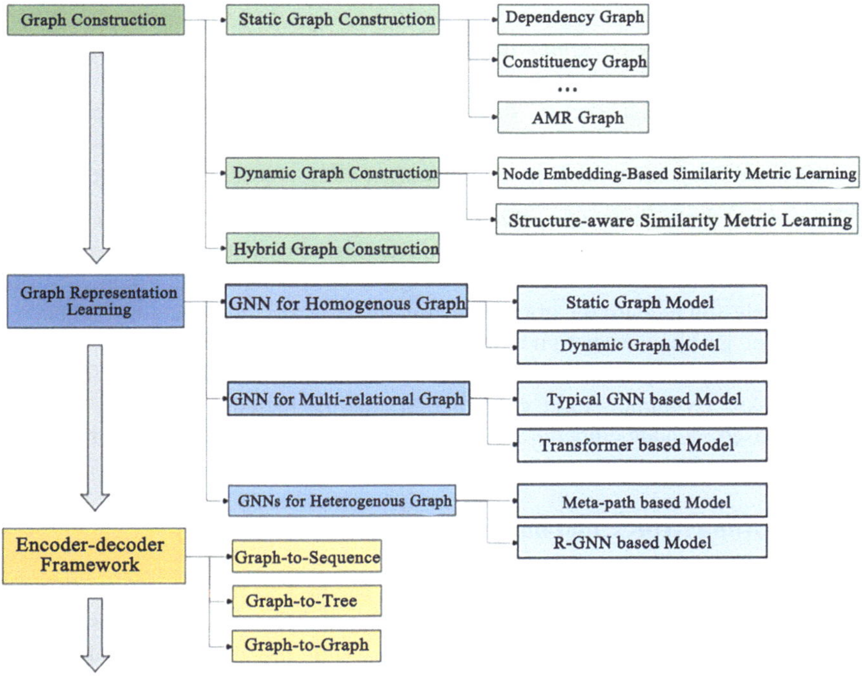

Fig. 3.3 Organization structure of GNN

3.2.4.1 Graph Construction

Graphs are ubiquitous in natural language processing. Although viewing text as sequence data may be the most obvious, the practice of representing text as various graphs in natural language processing has a long history. Standard text or knowledge graph representations include dependency graphs, constituent graphs, AMR graphs, IE graphs, lexical networks, and knowledge graphs. In addition, a text graph containing multiple hierarchical elements can be constructed, such as documents, paragraphs, sentences, and words. There are two main methods for graph construction: static graph construction and dynamic graph construction.

The purpose of the static graph construction method is to build a graph structure during preprocessing, usually using existing relationship parsing tools (such as dependency analysis) or manually defined rules. Conceptually, the static graph contains different domain/external knowledge hidden in the original text sequence, enriching the content of the original text with rich, structured information. Static graph construction mainly includes dependency graph construction, constituent graph construction, knowledge graph construction, similarity graph construction, etc.

Although the construction of static graphs has the advantage of encoding the prior knowledge of the data into the graph structure, it has several limitations:

1. A large workforce and domain knowledge are required to construct a reasonably performing graph topology.
2. Manually constructed graph structures are prone to errors (such as noise or incompleteness).
3. Since the graph construction stage and the graph representation learning stage are irrelevant, errors introduced in the graph construction stage cannot be corrected and may accumulate in later stages, leading to performance degradation.

To address these challenges, recent research on natural language processing has GNNs that have begun to explore the construction of dynamic graphs without the need for manual intervention or domain expertise. Most dynamic graph construction methods aim to learn the graph structure dynamically (i.e., weighted adjacency matrix). The graph construction module can be optimized together with the subsequent graph representation learning module in an end-to-end manner for downstream tasks. Dynamic graph construction mainly includes similarity measure learning based on node embedding, structure-aware similarity learning, etc.

3.2.4.2 Graph Representation Learning

Graph representation learning aims to find a method to incorporate the graph's structure and attribute information into low-dimensional embeddings through machine learning models. The graph constructed from the original text data may be homogeneous or heterogeneous.

Most graph neural networks, such as GCN, GAT, and GraphSage, are designed for homogeneous graphs. However, they need to adapt better to many natural language processing tasks. For example, given a natural language text, the dependency graph constructed contains multiple types of relationships, and traditional GNN methods cannot directly utilize it. Therefore, the arbitrary graph should first be converted into a homogeneous graph, including static and dynamic graphs, and then the bidirectional encoding of the graph neural network should be considered.

Most real-world graphs, such as knowledge graphs and AMR graphs, have multiple types of nodes and edges, which are called heterogeneous graphs. In addition to transforming heterogeneous graphs into relational graphs, it is sometimes necessary to fully utilize the category information of nodes and edges.

3.2.4.3 Encoder–Decoder Models Based on GNN

The encoder–decoder architecture is one of the most widely used machine learning frameworks in natural language processing, such as the Seq2Seq model. Given GNN's considerable advantages in modeling graph structure data, many researchers have recently developed encoder–decoder frameworks based on GNN, including graph-to-sequence (Graph2Seq), graph-to-tree, and graph-to-graph models.

Graph-to-sequence models usually adopt a GNN-based encoder and an RNN/Transformer-based decoder. Compared with the Seq2Seq paradigm, the graph-to-sequence paradigm is better at capturing the rich structural information of the input text and can be applied to any graph structure data.

Compared with graph-to-sequence models that consider the structural information at the input end, many natural language processing tasks also include outputs represented by complex structures (such as trees) with rich structural information at the output end. For syntactic parsing, semantic parsing, and math word problems, considering the structural information of the output is a natural choice. For this reason, some graph-to-tree models have been proposed, which add structural information at both the input and output ends, making the information flow in the encoding–decoding process more complete.

Graph-to-graph models are encoder–decoder models usually used to solve graph transformation problems. The graph encoder generates the latent representation of each node in the graph or a graph-level representation for the entire graph through GNN. Then, the graph decoder generates and outputs the target graph based on the node-level or graph-level representation from the encoder.

Currently, in academia or industry, many typical natural language processing applications use GNN, including natural language generation, machine reading comprehension, and question-answering.

Dialogue systems include text classification, text matching, topic modeling, sentiment classification, knowledge graphs, information extraction, semantic and syntactic parsing, inference, and semantic role labeling.

Fu et al. [13] proposed an end-to-end relationship extraction model, GraphRel, that uses a Graph Convolutional Network (GCN) to learn named entities and relationships jointly. Compared to previous baselines, this method considers the interaction between named entities and relationships through relation-weighted GCN, thus extracting relationships better. It is the most advanced method for relationship extraction.

Kim et al. [14] introduced a new algorithm for solving the Textbook Question Answer (TQA) task, building context graphs from text and images. They also proposed an f-GCN module based on the graph convolutional network, which is very effective for TQA problems.

3.2.5 *Multimodal Learning*

Multimodal learning is the technology that allows machines to obtain information from multiple domains, such as text, images, speech, videos, etc., to achieve information conversion and fusion, thereby improving model performance. It is a typical interdisciplinary field. People live in an environment where multiple domains intermingle. The sounds heard, the physical objects seen, the smells smelled, etc., are all modal forms of various domains. To allow deep learning algorithms to understand the surrounding world more comprehensively and efficiently, machines need to be able to learn and integrate these multidomain signals.

For this reason, researchers have begun to focus on integrating data from multiple domains to achieve the complementarity of various heterogeneous information. For example, research on speech recognition shows that the visual modality provides information about the movement of the mouth and pronunciation, including opening and closing, which helps improve speech recognition performance. Therefore, the comprehensive semantics provided by multiple modes are very valuable for deep learning.

Multimodal learning mainly includes five research directions, which are also the challenges currently faced by researchers: representation, translation, alignment, fusion, and co-learning.

3.2.5.1 Representation

The first and most crucial difficulty is representing and combining data from multiple modalities to maximize their complementarity and redundancy. Representation refers to the unified representation of information from each modality, usually as real-valued vectors. There are two methods of representation: joint representation and cooperative representation, as shown in Fig. 3.4.

To obtain recognition results, joint representation maps data sources from various modalities into the same space, such as inputting voice and images. In contrast, collaborative representation maps each modality into independent spaces, but there are constraints between these spaces.

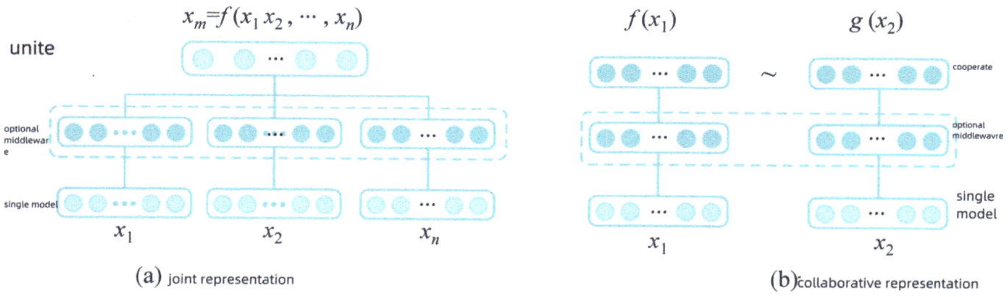

Fig. 3.4 Structure of joint representation and cooperative representation

3.2.5.2 Transformation

Transformation refers to mapping information from one modality to another. For example, given an image, we hope to get a sentence describing the image, or given a text description, generate an image that matches it. The models mainly used for transformation are example-based models and generative models.

Example-based models establish dictionaries between multiple modalities to form corresponding relationships, and generative models train the model to generate mapping capabilities.

3.2.5.3 Alignment

Modality alignment is one of the critical technologies of multimodal fusion, which refers to finding the corresponding relationship between subcomponents from two or more modalities. For example, given an image and a title, we hope to see the corresponding relationship between the area in the image and the words or phrases in the title.

Multimodal alignment methods are divided into explicit alignment and implicit alignment. Explicit alignment focuses on the alignment of subcomponents between modalities, while implicit alignment is the latent alignment of data during deep learning model training.

3.2.5.4 Fusion

Multimodal fusion integrates information from multiple modalities to complete classification or regression tasks. Its value is as follows:

1. Introducing multiple modalities when observing the same phenomenon may lead to more robust predictions.
2. Exposure to information from multiple modalities can capture complementary details, especially when this information is not "visible" under a single modality.
3. A multimodal system can still work when a particular modality is missing.

Fusion is mainly divided into two categories: model-free methods and model-based methods. Model-free methods do not depend on a specific machine learning algorithm and are divided into feature fusion, decision fusion, and hybrid fusion; model-based methods explicitly complete fusion in construction.

3.2.5.5 Collaborative Learning

Collaborative learning assists the learning of modalities with scarce data sources through modalities with rich data sources.

The data forms mainly include the following:

1. Parallel data (same data set, one-to-one correspondence between instances)
2. Nonparallel data (different data sets and instances do not overlap, but general concepts or categories overlap)
3. Mixed data (instances or concepts are connected by a third data set)

Multimodal learning has been applied in various fields, such as medical question answering, fake news detection, etc. These methods combine multimodal information instead of single information to train models, which is superior to existing methods.

Vu et al. [15] proposed a novel visual question-answering method that allows querying images through written questions. Experiments on various medical and natural image datasets show that this method achieves the same or higher accuracy compared to existing methods by fusing image and question features in a novel way.

Qi et al. [16] proposed a novel Multi-View Neural Network (MVNN) framework to fuse frequency domain and pixel domain visual information, which can be used for fake news detection. They designed a CNN-based network to automatically capture the complex patterns of phony news images in the frequency domain. They used a multibranch CNN–RNN model to extract visual features from different semantic levels in the pixel domain, dynamically fusing the frequency and pixel domain's feature representations using an attention mechanism. Extensive experiments on real-world datasets show that the performance of MVNN is superior to that of existing methods.

Deng et al. [17] proposed a deep, dense fusion network with multimodal residuals to integrate multimodal information, including language, acoustic speech, and visual images for sentiment analysis. Its performance is superior to the current 11 most advanced baselines.

3.3 Advances in Deep Learning for Training

3.3.1 Multitask Learning

Traditional machine learning models often only focus on one task, while Multitask Learning (MTL) trains multiple tasks simultaneously. MTL can utilize the information in various functions simultaneously, so compared to single-task models, it often achieves better results. Multitask learning has also been widely applied in natural language processing. The following will briefly sort out and introduce the application of deep learning-based multitask learning algorithms in natural language processing.

Multitask learning methods are usually divided into hard and soft parameter sharing.

Hard parameter sharing is often used as the baseline system for multitask learning. No matter how many tasks, its basic structure is the same: the underlying parameters are shared uniformly, and the top-level parameters are independent, as shown in Fig. 3.5. Since most parameters are shared, the probability of model overfitting will decrease. The more parameters are shared, the smaller is the likelihood of overfitting; the fewer parameters are shared, the closer it is to multiple single-task learning being carried out separately.

The current research focus is on soft parameter sharing. Its basic structure is as follows: the lower level shares some parameters, and some unique parameters are not shared; the top level has its own parameters, as shown in Fig. 3.6. The shared

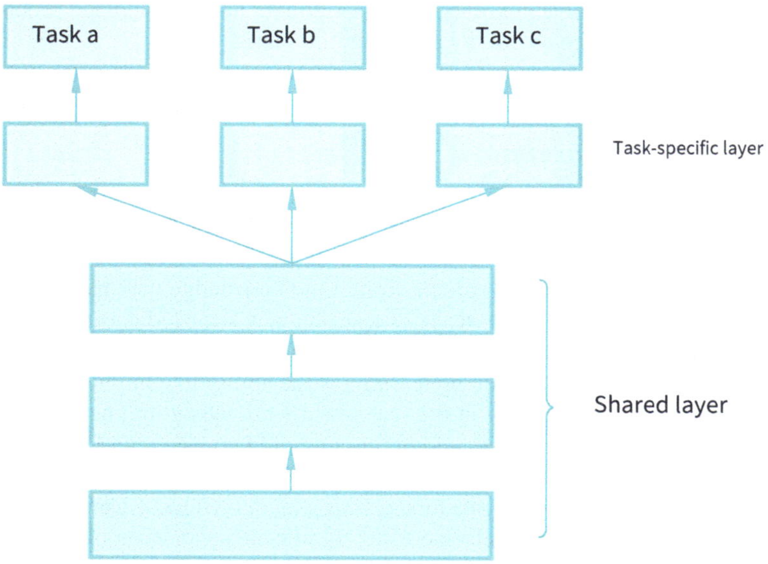

Fig. 3.5 Hard parameter sharing structure

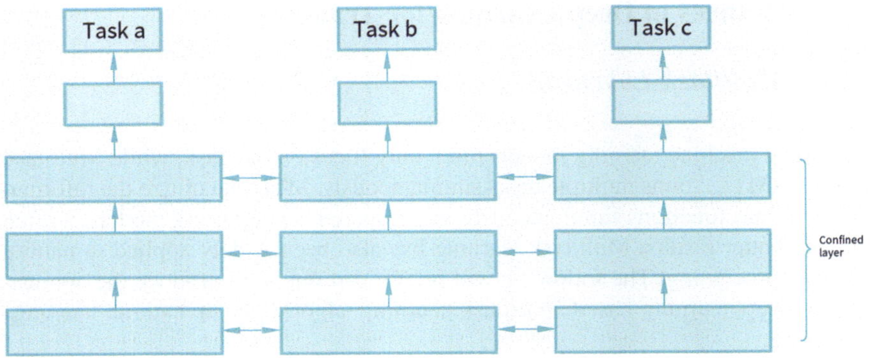

Fig. 3.6 Soft parameter sharing structure

and nonshared parameters at the bottom level are fused together and sent to the top level to predict multiple different targets and finally optimized together.

The directions for improving multitask learning are as follows:

1. Model structure design. That is, deciding which parameters to share and which parameters not to share. There are two ways: to distinguish the shared layer, that is, to find a way to give each task a unique shared layer fusion method, and the other is to design the fusion method for different tasks and shared layer levels.
2. Objective loss design and optimization improvement. Different tasks' data distribution and importance differ, and the losses of multiple tasks are not necessarily suitable for the target task when directly added together.
3. Design more reasonable auxiliary tasks. A standard natural language processing method uses the language model as an auxiliary task.

3.3.2 *Lifelong Learning*

Lifelong learning is also known as continuous learning or incremental learning. Humans can continuously acquire, adjust, and transfer knowledge throughout their lives. Although people tend to gradually forget the knowledge they have learned in their lifetime, only in very few cases does learning new knowledge catastrophically affect the knowledge already learned. This learning ability is referred to as lifelong learning ability. Specifically, lifelong learning is the ability to continuously process the continuous information flow in the real world while absorbing new knowledge and retaining and even integrating and optimizing old knowledge [18].

In machine learning, lifelong learning solves a common defect in model training: catastrophic forgetting. General machine learning models (intense learning methods based on backpropagation) usually perform significantly worse on old tasks when trained on new tasks.

One of the main reasons for catastrophic forgetting is that traditional models assume that data distribution is fixed or stationary. Training samples are independently and identically distributed, so the model can repeatedly see the exact data for all tasks. However, when the data become a continuous data stream, training data distribution is nonstationary. When the model continuously acquires knowledge from the nonstationary data distribution, the new knowledge will interfere with the old knowledge, leading to a rapid decline in model performance or even completely covering or forgetting the old knowledge learned before.

To overcome catastrophic forgetting, the model must demonstrate the ability to integrate new knowledge from new data, refine existing knowledge (plasticity), and prevent significant interference of new inputs with existing knowledge (stability). These two conflicting demands constitute the stability–plasticity dilemma.

The simplest solution to catastrophic forgetting is to retrain the network parameters using all known data to adapt to the changes in data distribution over time. Although retraining the model completely solves the problem of catastrophic forgetting, this method could be more efficient and dramatically hinders the model's real-time learning of new data. The main goal of lifelong learning is to find the most helpful balance point in the stability–plasticity dilemma under limited computational and storage resources.

Lifelong learning currently has a definition, so it is easy to confuse concepts such as online learning, transfer learning, and multitask learning, especially the difference between lifelong learning and online learning. Online learning usually requires each sample to be used only once, and all data come from the same task. In contrast, lifelong learning is multitasking; it allows multiple processing of current task data before entering the next task. Generally speaking, lifelong learning has the following characteristics:

1. While learning new knowledge, it can retain most of the knowledge learned before, that is, the model performs well on both old and new tasks.
2. Computing power and memory should remain fixed or grow slowly with the increase in the number of categories. The ideal situation is that once a task is completed, all observation samples of the task are discarded.
3. The model can continuously learn new knowledge from new tasks and new data. When new tasks appear at different times, the models are all trainable.

3.3.3 Paradigm Shift

In science and philosophy, a paradigm usually refers to a set of concepts or ways of thinking used to solve problems in a field. In natural language processing, a paradigm is the machine learning framework used to solve a class of natural language processing tasks determined by the model's input, output, and structure. For example, the SeqLab paradigm is usually used for named entity recognition tasks:

1. The input is a piece of text.
2. The output is the label of each word in the text.
3. The model uses a sequence labeling architecture.

With the development of pre-trained language models, the dream of using a single model to complete all-natural language processing tasks is approaching realization. Deploying a universal model based on a large-scale pre-trained language model has the following advantages:

1. No longer need a large amount of annotated data. Since the universal model generally adopts pre-trained and multitask training, the demand for annotated data has dramatically decreased.
2. Strong generalization ability. Compared to training a task-specific model, which directly transforms the target task into a general paradigm, the model can be directly applied to unseen tasks.
3. Convenient deployment. The inference of the general model as a commercial black-box API only needs to change the input and output to meet user needs.

Natural language processing tasks usually have one or more commonly used paradigms. A paradigm can solve one or more natural language processing tasks and can be instantiated into multiple deep learning models. Paradigm transfer is using the paradigm of a type of task to solve a different kind of task, such as using the Machine Reading Comprehension (MRC) paradigm to solve the named entity recognition task. The following is a detailed introduction to the MRC paradigm.

The so-called machine reading comprehension is to give an article and a question based on the article and let the machine answer the question after reading the article. There are four everyday tasks of MRC: cloze, multiple choice, fragment extraction, and free answering, as shown in Fig. 3.7. The difficulty of constructing these four tasks gradually deepens; the requirements for natural language understanding are getting higher and higher; the flexibility of the answer is getting higher and higher; and the actual application scenarios are becoming more and more extensive.

Before 2016, machine reading comprehension was mainly a statistical learning method. After the release of the SQuAD dataset, some attention-based matching models appeared, such as match-LSTM and BiDAF. After 2018, various pre-trained language models appeared, such as BERT and ALBERT. The development of the machine reading comprehension dataset has extensively promoted the development of machine reading comprehension capabilities. Among them, SQuAD is a dataset about machine reading comprehension launched by Stanford University. In 2018, the natural language understanding team of Stanford University launched the 2.0 version of this dataset. In the new dataset version, some questions without answers were added.

There are four possible future development trends of MRC:

1. MRC based on external knowledge. In human reading comprehension, people will use common sense or accumulated background knowledge to respond when some questions cannot be answered based on the given text. Still, external knowledge is not well utilized in the machine reading comprehension task. The

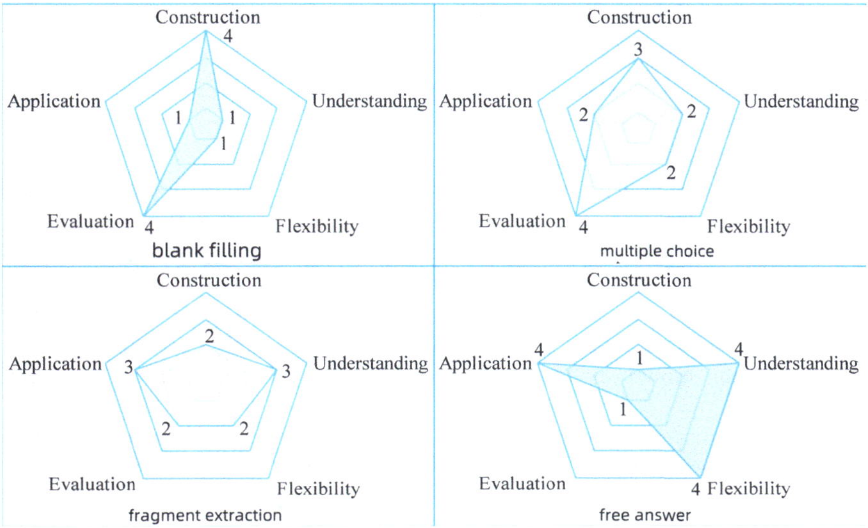

Fig. 3.7 The four common tasks of MRC

challenge is retrieving relevant external knowledge and the fusion of external expertise.

2. MRC with unanswerable questions. The machine reading comprehension task has a potential assumption that there must be a correct answer in the given article, but this does not match the actual application. Some questions need to be accurately answered by the machine. This requires the machine to judge whether it can answer the question based on the given article. If not, mark it as unanswerable and stop.

   Stop answering, otherwise, answer. Its challenge is the discernment of unanswerable questions and the recognition of interfering answers.
3. Multidocument MRC. The questions are designed based on the corresponding articles in machine reading comprehension tasks. However, when people ask questions, they usually pose a question and then use relevant available resources to obtain the clues needed to answer it. It is no longer just given one article but requires the machine to answer the question based on multiple articles. Its challenges are as follows: ① retrieval of relevant documents, ②interference of noisy documents, ③ no answer in the retrieved documents, ④ there may be multiple answers, and ⑤ need to aggregate various clues.
4. Dialogue reading comprehension. When given an article, the questioner first poses a question; the respondent provides an answer; and the questioner then poses another related question based on the answer. Multiround question-and-answer dialogue can be seen as the above process iterated multiple times. Its challenge is the use of dialogue history information and anaphora resolution.

3.4 Advanced Developments in Deep Learning for Applications

3.4.1 Model Compression

Model compression is also known as knowledge distillation. With the continuous development of deep learning, the scale of deep learning training is constantly expanding, and the magnitude of the model is gradually increasing, which makes it difficult for larger deep learning models to be deployed on low-resource devices. Therefore, a method is needed to reduce the magnitude of the model while allowing the reduced model to complete its intended tasks and maintain the recognition effect of the original model as much as possible; this is the concept of model compression.

Model compression methods can be divided into five types: low-rank decomposition, pruning, quantization, weight sharing, and knowledge distillation.

3.4.1.1 Low-Rank Decomposition

The weight matrix of the deep learning neural network is regarded as a full-rank matrix. Composing the original weight matrix using multiple low-rank matrices can simplify the neural network, that is, low-rank decomposition. A combination of various low-rank matrices can represent the original dense weight matrix, and the low-rank matrix can be decomposed into the product of numerous small-scale matrices, reducing the neural network's scale. For a two-dimensional matrix's low-rank decomposition, singular value decomposition is a feasible and convenient decomposition method.

3.4.1.2 Pruning

A neural network model comprises many interconnected floating-point neurons, and each layer of neurons passes information down through weights. However, in a layer of neurons, some nodes with tiny weights impact the model's information load, so these neurons with small weights can be pruned. This pruning work reduces the model scale while maintaining the original model accuracy; this method is called pruning.

3.4.1.3 Quantization

Generally, neural networks' parameters are represented by 32-bit floating-point numbers. However, such high precision is optional. Through quantization operations, the original 32-bit floating-point numbers can be represented by 0255, sacrificing a certain degree of accuracy to quantize the model and thereby reducing its scale. Quantization methods can be roughly divided into binary quantization and ternary quantization.

3.4.1.4 Weight Sharing

When building a neural network, some information appears and is reused multiple times globally. If these pieces of information are clustered to mine shared weight coefficients and allowed to share weights in a categorical manner, the model can ultimately be compressed. This compression method is called weight sharing.

3.4.1.5 Knowledge Distillation

The Softmax function is usually used for classification in the model's output process. Therefore, researchers have proposed a method to simplify the model using soft targets, namely knowledge distillation. The soft target refers to the classification result generated by introducing the temperature parameter T into the Softmax function. The result generated by the Softmax without introducing T is called the challenging target. Compared to the stiff target, the resulting curve of the soft target is smoother.

The steps of knowledge distillation are as follows:

1. Use regular labels (i.e., challenging targets) to train a large NET-Teacher model.
2. Use the previously trained NET-Teacher on the transfer set to train and generate corresponding soft targets.
3. Train a small-scale model NET-Student, fitting both soft and hard targets, adding a loss function corresponding to the soft target, and adjusting the weight of the two functions through the Lambda function.
4. Remove the soft target; only keep the challenging target; and form the final version of the small model, thus completing the knowledge distillation process.

3.4.2 Explainable Learning

Deep learning has achieved the best results in many judgment and prediction tasks. Still, there is always a problem: many current deep neural networks need help understanding the model's decisions from a human perspective, that is, the ability of machine learning to explain. Explainable learning models aim to provide human-understandable explanations for the conclusions generated by machine learning, providing users with the evidence chain for the machine to reach this conclusion. Machine learning models undergo extensive training, and the model itself implies knowledge hidden in the model in massive parameters. In contrast, explainable learning models can explicitly represent knowledge in a human-understandable way, solving the problem of model explainability, which is beneficial for users to apply and deploy the model in real scenarios with more confidence.

Based on the pattern of the generated explanation, explainable learning models can be divided into rule-based, latent semantic-based, attribute-based, and instance-based models.

1. Rule-based explainable learning models are models based on rules, recording the deduction process, which can serve as the explanation output of the model.
2. The interpretable learning model based on latent semantics analyzes the explanations that can support the model output from the weights of the hidden layer neurons of the deep neural network.
3. The interpretable learning model based on attributes needs to calculate significance scores to find the characteristics with the strongest correlation with the model output in the input to explain the model output.
4. The interpretable learning model based on instances finds multiple instances similar to the model output in the existing database, provides instance support for the model output, and uses this as an explanation.

Here are some general methods. Before modeling, choose interpretable models, such as decision tree models, linear regression models, logistic regression models, generalized linear regression models, generalized additive models, Bayesian instance models, etc.; after modeling, use interpretability methods, mainly for deep learning models with black-box properties, mainly including hidden layer analysis methods, simulation/proxy model methods, and sensitivity analysis methods.

Taking BERT as an example, some explanations can be generated through visualization methods, as shown in Fig. 3.8. The BERT model is complex, and it is challenging to understand the meaning of the weights it learns directly. The interpretability of deep learning models could be more robust, but they can be understood through some visualization tools. Tensor2Tensor provides excellent tools for visualizing attention.

3.4.3 Adversarial and Algorithm Security

With the continuous development of deep learning, neural networks have been widely used in audio and video recognition, natural language processing, and game theory. Therefore, ensuring the safety and robustness of deep learning algorithms is crucial. However, deep learning models are vulnerable to adversarial sample attacks. Attackers can add specific disturbances to benign data to generate adversarial samples. These adversarial samples with slight disturbances will not affect human judgment but cause deep learning models to produce incorrect results. At the same time, the successful implementation of adversarial attacks in scenarios such as autonomous driving further demonstrates the feasibility of adversarial attacks in the real

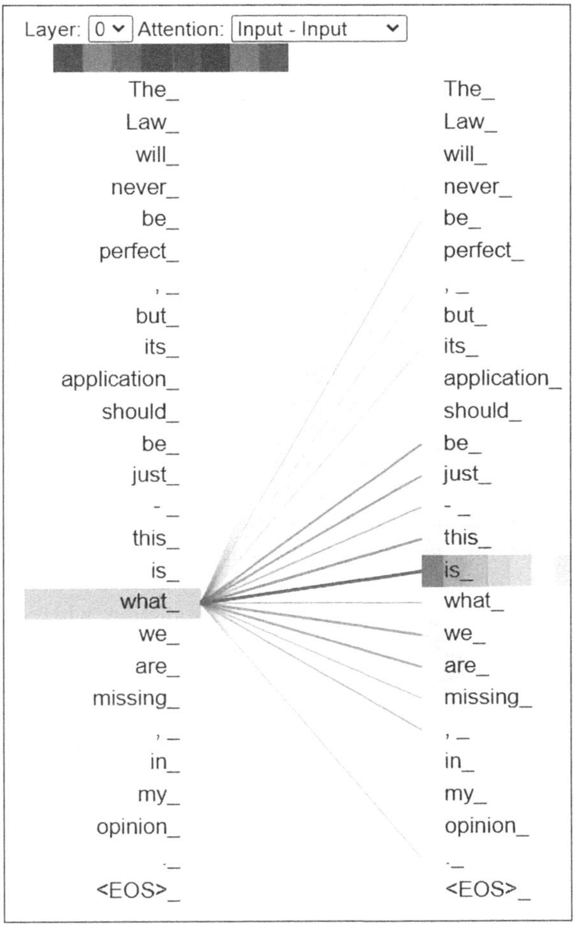

Fig. 3.8 Visualization of the model

world. Therefore, research on adversarial attacks and adversarial defense techniques has attracted more and more attention.

There are two reasons why neural networks can be disturbed: one is that the part of the neural network that contains semantic information is not a single node but is contained in the entire network, and the second is that the mapping from input to output in the neural network is mostly discontinuous. Experiments can prove that images can deceive neural networks through certain modifications, resulting in incorrect output.

Attacks on neural networks can be divided into white-box attacks and black-box attacks. A white-box attack refers to an attack launched when the attacker knows the underlying structure of the neural network, so the attacker of a white-box attack must understand the structure of the neural network, find the part that is easiest to attack based on the structure of the neural network, and then launch an attack.

White-box attacks are mostly exploratory attacks carried out by companies to ensure the security of neural networks. When the attacker does not understand the underlying network structure, the attack launched is a black-box attack. Black-box attacks pass the training samples to the network, obtain outputs through the network, build the corresponding inference network, and then develop adversarial samples against the inference network, ultimately using these samples to launch adversarial attacks on the original model. This type of attack does not require an understanding of the neural network structure but only the development of the corresponding inference network for adversarial attacks. Hence, most attacks on neural networks are black-box attacks. Currently, there are three methods for neural networks to defend against adversarial attacks:

1. Modify training samples. Adding more adversarial samples to the training set can effectively avoid some attacks. However, the classification boundary may expand when the sample set is developed.
2. Modify the training network. This method will adjust the training network, one of which is to use a more nonlinear activation function in the last layer. Still, this method may lead to a decrease in training efficiency and effectiveness. The methods of modifying the training network are divided into complete resistance and detection-only. Complete resistance is to allow the model to identify adversarial samples as the correct classification, whereas detection-only is to discover such attack samples and thus refuse service.
3. Add network. This method uses an additional auxiliary network without changing the original model, which can keep the original network unchanged. The most effective method is the Generative Adversarial Network (GAN). GAN contains two essential parts, the generator and the discriminator, used to generate new data and judge whether the data is accurate. Similarly, this method is also divided into complete resistance and detection only.

References

1. Priyas, Singh S, Dandapat S K, et al. Identifying Infrastructure Damage During Earthquake Using Deep Active Learning[C] . Proceedings of the 2019 IEEE/ACM International Conference on Advances in Social Networks Analysis and Mining. 2019: 551–552.
2. Asghar N, Poupart P, Jiang X, et al. Deep active learning for dialogue generation[J]. 6th Joint Conference on Lexical and Computational Semantics (*SEM), 2017.
3. Hanbay K. Deep Neural Network based Approach for ECG Classification Using Hybrid Differential Features and Active Learning[J] . IET Signal Processing, 2019, 13(2) : 165–175.
4. Maldonado R, Harabagiusm. Active Deep Learning for the Identification of Concepts and Relations in Electroencephalography Reports[J] . Journal of Biomedical Informatics, 2019, 98: 103265. DOI: 10. 1016/j. jbi. 2019. 103265.
5. Taylor W L. “Cloze procedure”: A New Tool for Measuring Readability[J] . Journalism Quarterly, 1953, 30 (4) : 415–433.
6. Mikolov T, Sutskever I, chen K, et al. Distributed Representations of Words and Phrases and Their Compositionality[J] . Advances in Neural Information Processing Systems, 2013, 26: 3111–3119.

7. Bojanowskip, Grave E, Joulin A, et al. Enriching Word Vectors with Subword Information[J] . Transactions of the Association for Computational Linguistics, 2017, 5: 135–146.
8. Radford A, Wu J, Child R, et al. Language Models Are Unsupervised Multitask Learners[J] . OpenAI blog, 2019, 1(8) : 9.
9. Joshim , Chen D, Liu Y, et al. Spanbert: Improving Pre-trained by Representing and Predicting Spans[J] . Transactions of the Association for Computational Linguistics, 2020, 8: 64–77.
10. Zhang Z, Han X, Liu Z, et al. ERNIE: Enhanced Language Representation with Informative Entities[C] . Proceedings ofthe57th Annual Meeting of the Association for Computational Linguistics. 2019: 1441–1451.
11. Clark K, Luong M T, Le Q V, et al. ELECTRA: Pre-trained Text Encoders as Discriminators Rather than Generators[J] . ELECTRA, 2016, 85–90.
12. Wang Z, Wang X, Han X, et al. CLEVE: Contrastive Pre-trained for Event Extraction[C] . Proceedings of the59th Annual Meeting of the Association for Computational Linguistics and the11thInternationalJointConferenceon Natural Language Processing(Volume1: Long Papers) , 2021: 6283–6297.
13. Fu TJ, Li P H, Ma W Y. GraphRel: Modeling Text as Relational Graphs for Joint Entity and Relation Extraction[C]. Proceedings of the 57th Annual Meeting of the Association for Computational Linguistics. 2019: 1409–1418.
14. Kim D, Kim S, Kwak N. Textbook Question Answering with Multi-modal Context Graph Understanding and Self-supervised Open-set Comprehension [C]. Proceedings of the 57th Annual Meeting of the Association for Computational Linguistics. 2019: 3568–3584.
15. Vu M H,Löfstedt T,Nyholm T, et al. A Question-centric Model for Visual Question Answering in Medical Imaging[J]. IEEE Transactions on Medical Imaging,2020,39(9):2856–2868.
16. Qi P,Cao J,Yang T,etal.ExploitingMulti-domainVisualInformationforFakeNewsDetection[C].2019 IEEE International Conference on Data Mining(ICDM).2019:518–527.
17. Deng H,Kang P,Yang Z, et al. Dense Fusion Network with Multimodal Residual for Sentiment Classification[C].2021IEEEInternationalConferenceonMultimediaandExpo(ICME),2021:1–6.
18. Parisigi, Kemkerr, Part JL, et al. Continual Lifelong Learning with Neural Networks: A Review[J] . Neural Networks, 2019, 113: 54–71.

Chapter 4
Pre-trained Language Models

Pre-trained language models are integral to the field of natural language processing, continuously evolving as a prominent area of research. This chapter first introduces the concept and development history of pre-trained language models; then, it details the structure, features, and advantages of standard auto-encoding models and autoregressive models such as BERT, GPT-3, LSTM-based ELMo, and ERNIE, as well as the use of pre-trained language models; finally, it provides an outlook and analysis of the development trends and prospects of pre-trained language models.

4.1 Overview of Pre-trained Language Models

Everyday natural language processing tasks include Chinese word segmentation, word cloud portraits, part-of-speech analysis, automatic summarization, relationship mining, sentiment analysis, knowledge graphs, etc. Its real-life applications include chatbots, machine translation, voice recognition, etc. The difficulty of natural language processing tasks lies in how to make machines understand user's speech, which is the task that pre-trained language models need to complete.

4.1.1 Definition of Pre-trained Language Models

Pre-trained language models are an application of transfer learning, generally using self-supervised learning methods and large-scale data to learn the semantic representation related to the context of each member of the input sentence, not only implicitly learning general syntactic and semantic knowledge but also transferring the knowledge learned in the open field to downstream tasks to improve the shortcomings of low-resource task processing capabilities, which is very beneficial for

H. Zhang, J. Shang, *Natural Language Processing and Applications*,
https://doi.org/10.1007/978-981-97-9739-4_4

low-resource language processing tasks. Pre-trained language models have achieved the best results in almost all natural language processing tasks.

Transfer learning is a machine learning method that uses the model developed for one task as a starting point and reuses it for another task. Since all words are the same for natural language processing tasks, transfer learning methods are more suitable, saving time and computational resources. This is the origin of the pre-trained language model.

The use of pre-trained language models brings many benefits. It can train a universal language representation in large-scale corpora, provide a better model initialization, accelerate the convergence of the target task, and avoid overfitting on small datasets.

4.1.2 Development History of Pre-trained Language Models

In recent years, the emergence of pre-trained language models has played a milestone role in natural language processing. Natural language processing tasks no longer need to train models from scratch, and fine-tuning directly on pre-trained language models can achieve good results. The development of pre-trained language models can be divided into two generations.

1. First generation of Pre-trained Language Models

The first generation of pre-trained language models learns the vocabulary, context-independent, static word vectors. The practice of the first-generation pre-trained language model is to map each word in the vocabulary to a lookup table, and the training process is to obtain the lookup table. After obtaining the lookup table, the one-hot vector of each word is multiplied by the lookup table to get the word vector of this word.

The first generation of pre-trained language models has two apparent defects. One is that it cannot handle natural language inherent phenomena such as polysemy because it does not connect words with their context. For example, in the sentence "I hurt my back, while I backed my car," the former back is a noun, expressing the meaning of "back," and the latter back is a verb, describing the definition of "reverse." Therefore, these two-word vectors are different and should consider the context to determine a word's expression in a sentence. The second is the Out-of-Vocabulary (OOV) problem. If some words have yet to appear in the training data, then getting their word vector through the lookup table is possible. Words can be further divided into characters and other forms to solve the OOV problem to a certain extent.

The first generation of pre-trained language models is relatively shallow compared to the second generation. The classic structures are CBOW and Skip-Gram [1], and the most typical implementation is Word2Vec. There is also a classic structure called Glove [2], widely used to obtain word vectors.

At the same time, many works have studied sentence vectors, paragraph vectors, and even article vectors (such as skip-thought vectors, paragraph vectors, Context2Vec, etc.). These works are also classified as the first generation of pre-trained language models because they map inputs to fixed-dimensional vector representations.

2. Second Generation of Pre-trained Language Models

The second generation of pre-trained language models learns context-related, dynamic word vectors. Important representatives of the second-generation pre-trained language models are ELMo, OpenAIGPT, and BERT. Thanks to more substantial computing power, the development of deep models, the design of natural language processing pre-trained tasks, the use of large-scale training corpora, and the emergence of various training techniques, the second generation of pre-trained language models is flourishing and becoming increasingly powerful.

4.2 Introduction to Common Pre-trained Language Models

This section introduces the important representatives of the auto-encoding and auto-regressive models, the BERT model, and the GPT model. It then introduces the ELMo model based on LSTM and finally presents the Tsinghua ERNIE model and the Baidu ERNIE model.

4.2.1 BERT Model

BERT [3], short for Bidirectional Encoder Representations from Transformers, is a pre-trained language model composed of multiple Transformers. The BERT model is stacked from encoder layers. Benefiting from the powerful representation ability brought by the self-attention mechanism in Transformer, the BERT model can learn rich language information from a sizeable pre-trained corpus. After pre-training, fine-tuning can further improve its accuracy in specific subtasks. The following will detail the model architecture, pre-trained method, and fine-tuning process of BERT.

1. BERT Model Structure

Unlike standard language models such as recurrent neural networks and convolutional neural networks, BERT is inspired by Transformer, using the self-attention mechanism to learn long-distance dependencies in sentences, enabling the model to obtain accurate sentence representations. Moreover, BERT uses multiple encoder layers (12 layers) to enhance the model's capabilities further. BERT processes the text into various tokens and then uses a [CLS] token as the sentence's initial token to feed into the neural network. The neural network uses multiple encoder layers to process each token, ultimately obtaining the representation of each word. Figure 4.1 shows the principle of BERT.

2. BERT Pre-trained Method

For BERT to learn rich, large-scale language expression capabilities, researchers use a large amount of language data to pre-train the model. Through self-supervised learning, pre-trained language models unrelated to specific tasks are obtained from large-scale data. As an application of transfer learning, BERT learns the contextual representation of each member of the input sentence from a massive amount of text, implicitly learning general syntax and semantic knowledge. Through the learned syntax and semantic understanding, BERT can transfer the knowledge learned from the open field to downstream tasks to improve the accuracy of downstream tasks. Despite fine-tuning downstream tasks, BERT has performed best in almost all natural language processing tasks. The scalability of pre-trained BERT is also one of its advantages. Pre-trained BERT can adapt to most downstream tasks with slight changes to its structure, proving its strong generalization and scalability. However, it is worth noting that pre-training a BERT requires a large amount of computing resources (16 TPUs training for four days), so how to improve the training efficiency of pre-trained language models is still an open question.

3. Fine-Tuning of BERT

After obtaining a pre-trained language model, BERT can fine-tune the neural network for specific tasks. Unlike the self-supervised training of BERT pre-training, a supervised training method is usually used when fine-tuning BERT. BERT can cover many downstream tasks, such as text classification, text matching, sequence tagging, structure detection, machine translation, and text representation. Four typical downstream task fine-tuning methods include sentiment analysis, part-of-speech tagging, natural language inference, and sentence embedding.

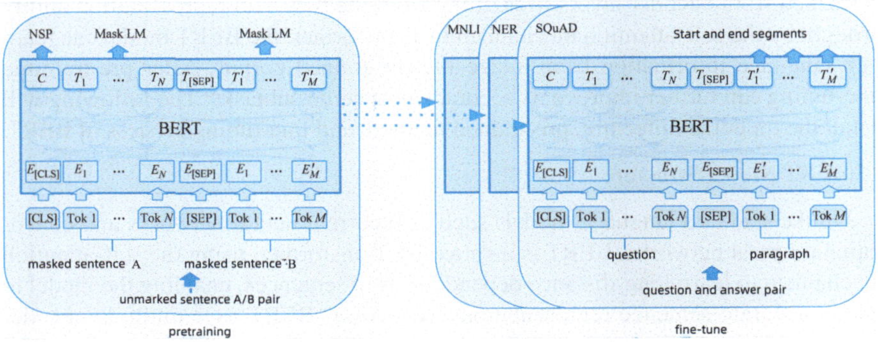

Fig. 4.1 Principle of BERT

4.2.2 GPT3 Model

The main innovation of the GPT series, as a pre-trained language model that appeared after BERT, is to make up for BERT's inability to handle sequence-to-sequence problems. The GPT series aims to develop a generative pre-trained language model to compensate for BERT's failure to complete generative tasks. Based on the Transformer's decoder layer, GPT-3 stacks the decoder layers, the core of which is the self-attention layer. The forward propagation method of GPT-3 is similar to recurrent neural networks, where the model only receives one character as input at the current moment and outputs the prediction of the next character.

GPT-3 also adopts the two-stage training method of "pre-trained + fine-tuning." In the pre-trained stage, GPT-3 uses a training method to predict the next character. The current character is used as input, and the model predicts the next character based on the current hidden state and input. Compared to BERT, GPT-3 has more model parameters, more pre-trained data, and consumes more computing resources.

Due to the large number of parameters in the GPT-3 model, fine-tuning the GPT-3 model also requires a large amount of computing resources. However, the excellent ability of GPT-3 itself is already sufficient for some natural language processing tasks. The un-fine-tuned GPT-3 has achieved surprising results in some downstream tasks, especially in tasks that do not require strong text logic. Additionally, GPT-3's powerful performance allows it to learn directly from several samples and output the correct results. In the few-shot learning defined by GPT-3, describing a small number of task patterns in the given samples can enable GPT-3 to learn task information and provide the correct results.

4.2.3 ELMo

ELMo stands for Embeddings from Language Model, and the paper proposing this model became the best paper at NAACL in 2018. Compared to the traditional form of word vectors, Word2Vec, ELMo is a dynamic model. In previous word vector representations, words are represented in a static form, using the exact vector representation in any context. This makes it challenging to represent polysemy. However, ELMo can dynamically generate word vectors based on context, is theoretically a better model, and has achieved state-of-the-art (SOTA) results in many tasks at the time. ELMo is undoubtedly a feature-based language model, using a pre-trained language model to generate better features. As shown in Fig. 4.2, ELMo uses a bidirectional LSTM language model consisting of a forward language model and a backward language model, and the objective function is the maximum likelihood of these two directional language models.

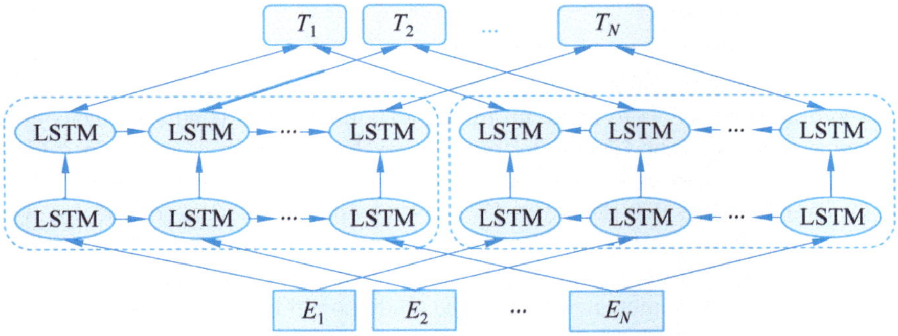

Fig. 4.2 ELMo structure

ELMo has two advantages: First, ELMo assumes that word vectors are not fixed, so in terms of polysemy, ELMo performs better than Word2Vec, and second, ELMo is based on learning from the entire corpus. The word vectors generated by the language model are equivalent to the word vectors learned from the whole corpus, which can more accurately represent the meaning of a word.

The main disadvantages of ELMo are in two aspects. First, regarding feature extractor selection, ELMo uses LSTM instead of a Transformer, and many studies have proven that the Transformer's feature extraction capability is far more robust than LSTM. Second, ELMo's bidirectional concatenation for feature fusion may need to be stronger than BERT's integrated feature fusion method. However, this is just a suspicion inferred, and there is currently no specific empirical evidence to prove this.

4.2.4 ERNIE

Tsinghua ERNIE, fully named Enhanced Language Representation with Informative Entities, proposed explicitly adding knowledge into BERT. Tsinghua ERNIE's model changes the model structure, fuses knowledge and language semantic information, enhances semantic representation, and has achieved significant success in knowledge-driven tasks. The model mainly consists of two types of Encoders: T-encoder and K-encoder. Among them, the T-encoder is primarily used for input text, extracting lexical and semantic information, with N layers, and the K-encoder mainly performs the embedding and knowledge fusion of knowledge entities (entity), with M layers.

Baidu ERNIE, fully named Enhanced Representation through Knowledge Integration, is a semantic representation model proposed by Baidu based on Transformer encoder. Compared to BERT, its pre-trained process uses richer semantic knowledge and more semantic tasks, achieving better results in multiple natural language processing tasks. The BERT model and others like it have shown excellent results. Baidu's ERNIE1.0 model uses words, entities, and their relationships in massive amounts of data to learn real-world semantic knowledge. Compared to BERT, which learns from raw language signals, Baidu's ERNIE1.0 can directly model prior semantic knowledge units, enhancing the model's semantic representation capabilities. Because Baidu's ERNIE1.0 learns at the entity level, it performs better in language inference tasks. Baidu's ERNIE1.0 has surpassed the BERT Chinese model in all Chinese tasks, including classification, semantic similarity, named entity recognition, question-answering matching, and others, with an average improvement of 1–2 percentage points. Baidu's ERNIE2.0 introduced continuous pre-training. Through continuous pre-training, the model can continuously learn various tasks, further improving the model's performance. Baidu claims continuous pre-training involves two steps: continuously constructing unsupervised pre-trained tasks with extensive corpus/prior knowledge; second, the ERNIE model will be gradually updated through multitask learning. Baidu's ERNIE2.0 is a sustainable semantic understanding framework that supports the incremental introduction of vocabulary, syntax, and semantics. It can fully capture lexical, grammatical, and semantic information in training materials. Baidu's ERNIE2.0 pre-training includes three major learning tasks: lexical layer tasks, learning to predict vocabulary in sentences; grammatical layer tasks, learning to reconstruct multiple sentence structures and reorder them; and semantic layer tasks, learning to judge logical relationships between sentences, such as cause and effect, contrast, parallelism, etc. Through these additional semantic tasks, Baidu's ERNIE2.0 semantic understanding pre-trained language model has obtained lexical, syntactic, and semantic information from training data in multiple dimensions, greatly enhancing its general semantic representation capabilities. Baidu's ERNIE2.0 model is almost universally superior to BERT and XLNet in English tasks, achieving the best results in seven GLUE tasks, and Baidu's ERNIE2.0 model is superior to BERT in all nine Chinese natural language processing tasks.

4.3 Use of Pre-trained Language Models

4.3.1 Transfer Learning

Transfer learning is usually the transfer of knowledge from one domain to another. There are many ways of transfer learning in natural language processing, such as domain adaptation, cross-language learning, and multitask learning. Applying pre-trained language models to downstream tasks is a sequential transfer task; these tasks need to be learned in sequence, and the target tasks have label information.

1. Choose Appropriate Pre-trained Tasks, Model Structures, and Corpora

Choosing the right pre-trained tasks, model structure, and corpus is essential. Different pre-trained tasks will have different effects on different downstream tasks. For example, the NSP task allows the pre-trained language model to better learn the connection between two sentences and have better results for downstream question-answering tasks and natural language understanding tasks. Pre-trained model: The structure of the language model is also essential for downstream tasks. For example, BERT can perform well on many downstream natural language understanding tasks but cannot be used for natural language generation. The data distribution of downstream tasks and the data distribution of pre-trained language models should be close, and a domain-related or specific pre-trained language model needs to be selected when applying.

2. Choose the Appropriate Layer

Research has found that pre-trained language models based on Transformers (such as BERT) represent the classic natural language processing: The first few layers learn syntactic information, and the higher layers learn higher-level semantic information. Depending on the specific task, different layers of the pre-trained language model are selected for representation, for example, only the static embedding layer is selected; the top layer can also be selected to represent the input to the target task, or a more flexible method can be used to choose the best layer representation.

4.3.2 Fine-Tuning

There are generally two ways to transfer the model: feature extraction (model parameters are fixed) and fine-tuning (model parameters participate in fine-tuning). When using the feature extraction method, the pre-trained language model needs to be regarded as a feature extractor, and the middle layer of the model needs to be exposed. Still, this method requires a complex network structure for the target task, so fine-tuning for the target task becomes a standard method. Figure 4.3 [4] shows the process of fine-tuning BERT on different downstream tasks.

The common fine-tuning methods include standard fine-tuning, two-stage fine-tuning, multitask fine-tuning, adopting additional adapters, and layer-by-layer freezing.

1. Standard Fine-Tuning

Standard fine-tuning refers to using the output of the top layer directly as the input of the downstream task and updating the parameters of the target task and the pre-trained language model simultaneously.

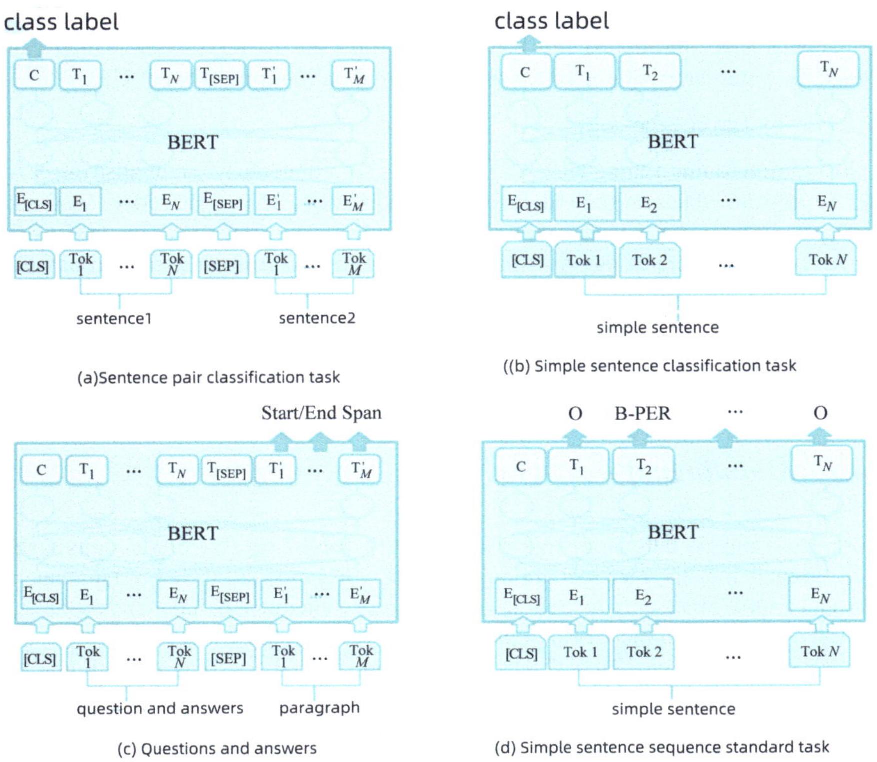

Fig. 4.3 BERT fine-tuning methods on different downstream tasks

2. Two-Stage Fine-Tuning

In two-stage fine-tuning, the first stage is to fine-tune or incrementally train the pre-trained language model using intermediate tasks or corpora, and the second stage is to fine-tune the target task using the model from the first stage. Research has shown that incremental pre-training on corpora related to the target task can achieve better results than BERT fine-tuning.

3. Multitask Fine-Tuning

Some researchers have fine-tuned BERT under the multitask learning framework, and the results show that multitask learning and pre-trained language models can complement each other.

4. Adopting Additional Adapters

The disadvantage of fine-tuning is that each downstream task must train its model parameters. A better method is introducing adaptive modules into the pre-trained language model while keeping the original parameters fixed. Some researchers have introduced the PAL module into the BERT model, which can reduce the parameter volume to 1/7 of the original on multiple tasks of GLUE.

5. Layer-by-Layer Freezing

In addition to fine-tuning all layer parameters simultaneously, gradually unfreezing the pre-trained language model from the top layer is also very effective. Method: Some researchers have proposed a layer-by-layer unfreezing method, first fine-tuning the parameters of the task-related layer, then fine-tuning the hidden parameters of the pre-trained language model, and finally fine-tuning the embedding layer.

4.4 Development Trend of Pre-trained Language Models

4.4.1 Multilingual

Most pre-trained language models are used only for a single language. Multilingualism is the idea that although people worldwide use different languages, they can express the same meaning. That is to say, semantics is independent of the symbol system, and a model can represent multiple language models.

mBERT (multilingual BERT) used a corpus and shared vocabulary composed of 104 languages on Wikipedia for pre-training through the MLM task. Each training sample is a single-language article without specific target tasks or data for cross-language. Even so, mBERT performs well on many cross-language tasks. Conneau and others [5] proposed the XLM model; XLM uses the CC-100 database, which contains 100 languages and also uses the BPE algorithm for tokenization, and the dictionary size is more significant than that of mBERT. The model training uses three objective functions: unsupervised CLM and MLM on single-language corpora. The goal used on parallel corpora is called TLM (Translation Language Modeling).

4.4.2 Multimodal

Pre-trained language models started in the text field, but the application of pre-trained language models in other fields has attracted more and more interest. Multimodal pre-training, also known as cross-modal pre-training, studies the use of unlabeled data from multiple modalities (such as vision, text, sound, etc.) for model pre-training, aiming to improve the performance of various multimodal downstream tasks (such as cross-modal search). The difficulty of multimodal pre-trained language models lies in integrating nontext information into BERT and other pre-trained language models. For example, in image–text, researchers hope that the model can connect the "dog" in the text with the "dog" in the image; in video–text, researchers hope that the model can match the "object/action" in the text with the "object/action" in the video. For example, multimodal models that integrate visual information tasks into pre-trained language models are designed like BERT and can be achieved through masked text prediction, masked object prediction, and video–text alignment pre-trained. Video text alignment refers to giving a picture–text relationship pair, letting the model judge whether the text describes the image.

The massive success of pre-trained language models in natural language processing has also inspired researchers to pay attention to pre-trained language models in the multimodal field. Many multimodal PTMs are designed to encode visual and language features and are pre-trained on many multimodal datasets. Typical tasks include vision-based MLM, masked visual feature modeling, and visual text matching, and the corresponding pre-trained language models include VideoBERT, VisualBERT, and ViLBERT.

4.4.3 Enlarging the Model

The current upper limit of pre-trained language models has yet to be reached, and many existing models can further improve their performance by increasing the number of training steps and data. Moreover, building a universal pre-trained language model usually requires a more complex network architecture, a large corpus, and more challenging pre-trained tasks. The training cost is also higher, and there are more complex distributed training and mixed precision training methods.

Better results can be achieved by merely increasing the scale of the pre-trained language model. In recent years, the parameter scale of pre-trained language models has advanced from the million to the billion and even the trillion levels. Figure 4.4 shows the changes in the parameters and the data scale of pre-trained language models from 2018 to 2021.

4.4.4 Replacement of Pre-trained Tasks

Pre-trained tasks often need to be challenging and require much training data. Pre-trained tasks can generally be summarized into three categories: supervised, unsupervised, and sub-supervised. The following are the main pre-trained tasks:

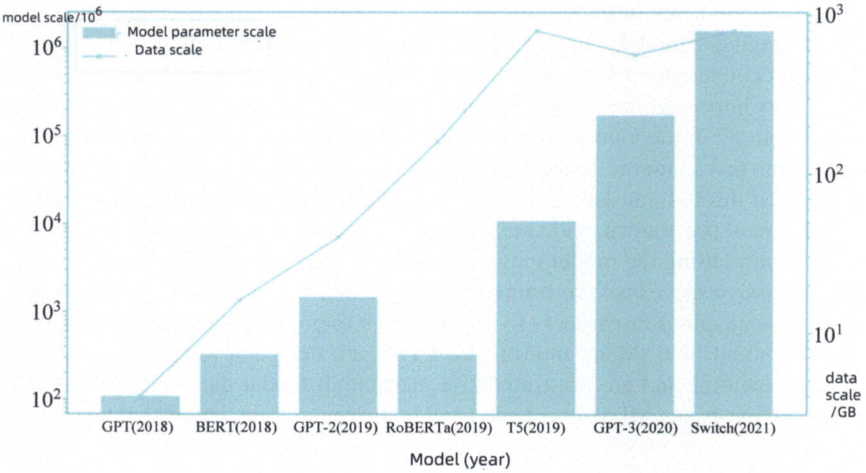

Fig. 4.4 Changes in the parameters and data scale of pre-trained language models from 2018 to 2021

1. LM

LM (Language Model) is the most classic way of modeling natural language processing. It includes the autoregressive LM in XLNet and the unidirectional LM used in GPT. The unidirectional LM has a clear disadvantage: For the representation of each token, it only encodes the context to the left of the token and the token itself. The bidirectional encoding BiLM was later introduced to solve this problem.

2. MLM

Taylor first proposed the MLM (Masked Language Model) task and then applied it to BERT's pre-trained tasks. MLM is like a cloze test, replacing some tokens in the input sentence with [MASK]. The goal of the pre-trained task is to predict the token replaced by [MASK] using both the left and right context. The feature encoder used in this is the Encoder part of the Transformer.

3. Seq2Seq MLM

Seq2Seq applies MLM to the encoder–decoder structure. The preprocessed text is sent to the encoder for feature encoding, and then the decoder generates the token replaced by [MASK] autoregressively. This method is beneficial for downstream tasks that use the Seq2Seq style (such as question-answering, summarization, machine translation, etc.). Pre-trained language models that use this method include MASS, T5, etc.

4. Translation Language Modeling (TLM)

Pre-trained language models can still be used when applied to translation tasks with parallel corpora. For example, the XLM model concatenates bilingual sentence pairs' original and target sentences and performs MLM operations. In this way, predicting the masked tokens in the original sentence can be associated with the content in the target sentence.

5. SBO

Span-BERT masks and replaces continuous characters (text fragments) instead of individual characters in MLM, which can incorporate strongly related word group information into training. Span-BERT further proposes a new target function, SBO (Span Boundary Objective), which only uses the tokens on the fragment's left and right boundaries to predict the fragment's information. Thus, word boundary information is introduced, and the relationship between continuous characters and their boundaries is better learned.

6. PLM

In PLM (Permuted Language Modeling), the permutation order is randomly drawn from all possible token permutations in the input text. Some target tokens are chosen, and the model is trained based on the characters preceding the target tokens in the permutation order, as well as the natural positions of the input text in its original order (this introduces a dualstream self-attention mechanism). In practical applications, due to slow convergence, only the last few tokens in the permutation sequence are selected as target tokens. They are currently used in XLNet.

7. RTD

In RTD (Replaced Token Detection), the ELECTRA model uses a generator (G) network and a discriminator (D) network and adopts a two-step strategy for training:

In the first step, only the G network is trained for MLM tasks.

In the second step, the D network is initialized with the weights of the G network, and the G network is freezed. The D network judges whether each token in the G network's output has been replaced. The downstream task uses the D network for fine-tuning. Here, the counterexamples generated by the G network can significantly increase the task's difficulty compared to randomly generated counterexamples.

8. NSP

The goal of the NSP (NextSentence Prediction) task is to judge whether the two input sentences are continuous, thereby learning the relationship between them. This benefits downstream tasks in the form of double sentences, such as question-answering and natural language inference tasks. When constructing the dataset, the second sentence has a 50% chance of being the following sentence of the first sentence and a 50% chance of being a random sentence.

4.4.5 Combining External Knowledge

Pre-trained language models can learn general language representations from large-scale general corpora but often need domain-related external knowledge. Introducing external knowledge into pre-trained language models has been proven to be effective. External knowledge includes language, semantics, common sense, facts, and domain knowledge.

External knowledge can participate in model training during the pre-trained stage. Early work mainly focused on learning knowledge graph embeddings and word vectors together. Researchers have designed some auxiliary pre-trained tasks, starting with BERT, to introduce external knowledge into pre-trained language models. LIBERT introduces external knowledge through additional language constraint tasks. Some researchers also added the emotional characteristics of each word to the MLM task, and the label-aware MLM achieved the best results in multiple sentiment classification tasks. Tsinghua ERNIE adds the entity and its corresponding pre-trained embedding into the model to enhance the representation of the text.

Similarly, Know BERT trains BERT and an entity-linking model end-to-end, thereby introducing entity representation information. These works indirectly introduce knowledge graph information through entity embeddings. In addition, K-BERT explicitly adds the triplet information corresponding to the entity from the knowledge graph into the sentence. Research also introduces the task of judging whether the entity has been replaced with the pre-trained language model, making the model pay more attention to learning factual knowledge. However, when most models introduce various external knowledge, the model update may lead to catastrophic forgetting of knowledge. Some researchers have proposed the K-Adapter model, which introduces different external knowledge by training different adapters. In addition, external knowledge can also be introduced without retraining. K-BERT also supports the introduction of external knowledge during the fine-tuning stage.

4.4.6 Pre-trained Language Model Compression

Transformer is the backbone architecture of many advanced pre-trained language models (such as BERT and GPT) in natural language processing. In practice, Transformer has been proven to be a powerful model, but it could be more efficient and requires a lot of computing resources. Researchers have proposed a series of solutions for the Transformer architecture, including parallel training, model compression, etc. Model compression methods mainly include model pruning, weight quantization, parameter sharing, and knowledge distillation.

1. Model pruning. Remove some unimportant parameters or structures.
2. Weight quantization. Use fewer bits to represent parameters.
3. Parameter sharing. The same modules in the model can share parameters. In ALBERT, cross-layer parameter sharing and embedding matrix decomposition are used to reduce the amount of model parameters.
4. Knowledge distillation. By training a smaller student model and learning from the intermediate representation of the original teacher model, the teacher model here can be one or more integrated models.

4.5 Applications and Analysis

GPT-3 has 175 billion parameters and was trained with 570 GB of data, but most of the corpus is in English, and the parameters of GPT-3 have not been published. To ensure ease of use and maximize the inclusion of Chinese corpus, we use a Chinese pre-trained language model that operates on a similar principle to GPT-3: the CPM (Chinese Pre-trained Language Model) developed by the Tsinghua University team.

4.5.1 Model Introduction

CPM contains 2.6 billion parameters and uses 100 GB of Chinese training data. It can be connected to downstream tasks, including dialogue, article generation, cloze test, and language understanding. As the parameter scale increases, CPM performs better on some datasets, indicating that larger models are more effective in language generation and understanding. In addition, it builds a new sub-word vocabulary based on the word segmentation corpus to adapt to the Chinese corpus and increases the batch size to 3072 to achieve more stable model training, and it still performs well in the small-sample or even zero-sample inference. The model architecture of Qingyuan CPM is similar to that of GPT-3, which is also a left-to-right Transformer encoder. Due to the uniqueness of Chinese, Qingyuan CPM has made improvements in the following two points: First, the character-level vocabulary structure in BERT-Chinese (2019) is not suitable for Chinese, so Qingyuan CPM uses a new sub-word vocabulary table, including characters and common vocabulary; second, because the distribution of Chinese is more discrete than English, the model uses a larger batch size during training than GPT-3's 1 million tokens, with the number of tokens reaching 3 million.

4.5.2 Model Usage

Since pre-trained language models are generally large and challenging to train, we downloaded the already trained Qingyuan CPM model for testing and observation. This model can use the official inference tool BMInf to complete the inference task directly. Users can also download the model themselves and customize the usage. In the paper on Qingyuan CPM, you can see several examples, such as the inference of capitals, cooking techniques, and disciplines in Fig. 4.5 and the dialogue inference in Fig. 4.6. It can be seen that the few-shot and supervised models of CPM-Large perform the best; the dialogue is smooth and natural; it will follow the conversation; and it will also be considered in reverse.

Relation:首都 (Capital)

Prompt: 美国 首都 华盛顿
America Capital Washington
中国 首都 北京
China Capital Beijing
日本 首都
Japan Capital

CPM: 东京
Toyko

Relation: 主要工艺(Main Process)

Prompt: 酱焖辣椒 主要工艺 焖
(Sauce Braised Chili) (Main Process)Stew
当归鸭肉煲 主要工艺 煲
(Duck with Angelica) (Main Process)Boil
韭菜煎蛋饼 主要工艺
(Leek Omelette) (Main Process)

CPM: 煎
Fried

Relation: 学科 (Subject)

Prompt: 恒星级黑洞 学科 宇宙论
(Stellar Black Hole) Subject Cosmology
品类需求强度 学科 品牌经济学
(Category Demand Intensity) Subject Economics
大地构造学 学科
(Tectonic Geology) Subject

CPM: 地质学
Geology

Fig. 4.5 Inference of the capital, cooking techniques, and disciplines

Post:	现在高中数学竞赛这么凶残了！ How crazy the math competition of high school is!
CDial-GPT: (Few-shot)	你还真信！ Do you really believe it?
CDial-GPT: (Supervised)	数学让我们学会了很多，好好学，明天回家。 Math lets us learn a lot,Study hard and go home, tomorrow.
CPM-Large: (Few-shot)	现在都是微积分，大学的东西比高中简单多了，还有什么呢? Now it's all calculus,The things in college art much simpler than those of high school, What else?
CPM-Large: (Supervised)	这是要逼死我吗?每道题都不会做啊。 Is this going to kill me? I can't answer any question.

Post:	每天嘻嘻哈哈地笑着，只有自己才知道活得有多累。 I laugh every day, but only I know how tired I am.
CDial-GPT: (Few-shot)	我也是。 Me too.
CDial-GPT: (Supervised)	只有自己才知道活得有多累。 Only I know how tired I am.
CPM-Large (Few-shot)	当想要哭泣的时候，就去那些熟悉的地方吧。 When you want to cry, to to those familiar places.
CPM-Large: (Supervised)	真的不知道生活怎么会这么累。 I really don't know how my life could be so tiring.

Fig. 4.6 Dialogue inference

References

1. Mikolov T, Sutskever I, Chen K, et al. Distributed Representations of Words and Phrases And Their Compositionality[J]. Advances in Neural Information Processing Systems, 2013: 3111-3119.
2. Pennington J, Socher R, Manning C D. Glove: Global Vectors for Word Representation[C]. Proceedings of the 2014 Conference on Empirical Methods in Natural Language Processing (EMNLP), 2014: 1532–1543.

3. Kenton J D M W C, Toutanova L K. BERT: Pre-trained of Deep Bidirectional Transformers for Language Understanding[C]. Proceedings of NAACL-HLT. 2019: 4171–4186.
4. Kenton J D M W C, Toutanova L K. BERT: Pre-trained of Deep Bidirectional Transformers for Language Understanding[C] . Proceedings of NAACL-HLT. 2019: 4171–4186.
5. Conneau A, Lample G. Cross-lingual Language Model Pretraining[C] . Proceedings of the 33rd International Conference on Neural Information Processing Systems. 2019: 7059–7069.

Part II
Information Processing

Chapter 5
Web Crawler Technology

This chapter first introduces the development history of web crawlers, the general process of crawling, the four main types and characteristics of web crawlers, and the Python library functions and frameworks needed to implement the crawling process. It then introduces the cutting-edge technology of web crawlers and the latest developments in website anti-crawling technology. Finally, this chapter presents an analysis conducted on job data crawled from Zhaopin.com, focusing on salary trends, work experience, and education requirements for positions in Beijing related to natural language processing, computer vision, backend, frontend, and big data engineering.

5.1 Overview

5.1.1 Conceptual Connotation of Web Crawlers

A web crawler, often called a crawler, is an integral part of a search engine. With the rapid advancement of information technology, web crawlers have always been a hot research topic as a component of search engines. Their development will directly determine the future of search engines. Research on web crawlers includes studying the characteristics of web search strategies and network analysis algorithms. Among them, the theme of web crawler network search is a research direction. According to some network analysis algorithms, irrelevant links are filtered out, connected to qualified web pages, and placed in a queue for the web crawler to crawl.

If the Internet is likened to a spider web, then the web crawler is the spider crawling around on this web. Web crawlers find web pages through the link addresses of web pages, starting from a specific page of a website (usually the homepage), reading the content of the web page, finding other link addresses in the web page, and then finding the following web page through these link addresses. This cycle

H. Zhang, J. Shang, *Natural Language Processing and Applications*,
https://doi.org/10.1007/978-981-97-9739-4_5

continues until all the pages of this website are crawled. If the entire Internet is regarded as a website, the web crawler can use the above principle to crawl all the web pages.

5.1.2 *Technological Development of Web Crawlers*

In 1989, the World Wide Web was born. Technically speaking, the World Wide Web and the Internet are different. The former refers to the information space, and the latter refers to the physical network connected by several computers. The World Wide Web has the following three main technological innovations:

1. Uniform Resource Locator (URL), which users use to access websites.
2. Embedded hyperlinks for navigation between web pages. For example, in the product details page, you can find product specifications and many other information, such as "customers who bought this product also bought certain goods"; this information is provided in the form of hyperlinks.
3. Web pages not only contain text but also images, audio, video, and software components.

In 1990, the first web browser was born, which was also invented by Tim Berners-Lee, the inventor of the World Wide Web.

In 1991, the first web server and the first HTTP protocol (starting with http://) web page appeared, and the number of web pages grew at a steady pace. By 1994, the number of HTTP servers exceeded 200.

In June 1993, the first web robot—the World Wide Web Wanderer was born. Although its function is the same as today's web robots, it was only used to measure the size of web pages. In December 1993, the first web search engine based on web crawlers—JumpStation was born. Because there were few websites on the network, website administrators often collected and edited web search engines. JumpStation brought a new leap; it was the first search engine to rely on web robots. Since then, people have started to use these automated web crawler programs to collect and organize the Internet. From Infoseek, AltaVista, and Excite to today's Bing and Google, the core of search engine robots remains the same: Find a web page, download (retrieve) it, scrape all the information displayed on the web page, and then add it to the database of the web search engine. Since web pages are designed for human users, even with the development of web robots, it is still difficult for computer engineers, scientists, and ordinary people to scrape web data. Therefore, people have been striving to make web crawlers easier to use.

In 2000, Web APIs and API crawlers were invented. API stands for Application Program Interface. It is an interface that makes program development more convenient by providing pre-built modules. In 2000, Salesforce and eBay launched their APIs, which programmers could use to access and download some public data. Since then, many websites have provided web APIs for people to access their public databases. Users send HTTP requests and receive feedback in JSON or XML. Web

APIs, by collecting data provided by websites, offer programmers a more user-friendly way to run web crawlers.

Not all websites provide APIs. Even if they do, they may offer only some data users want. Therefore, programmers still need to develop methods to improve web crawlers. In 2004, BeautifulSoup was released. It is a library designed for Python. In computer programming, a library is a collection of script modules, like commonly used algorithms, allowing programmers to use it directly, simplifying the programming process. With simple commands, BeautifulSoup can understand the structure of a website and help users parse content from HTML containers. It is considered the most complex and advanced library for web crawling and one of the most common and popular methods today.

In 2006, Kapow Software released Kapow Web Integration Platform 6.0, a visual web crawler software that allows users to quickly select web content and construct these data into usable Excel files or databases. Ultimately, visual web data scraping software can allow many nonprogrammer users to run web crawlers themselves. Since then, web scraping has become mainstream. Now, nonprogrammer users can find over 80 data collection software that provides a visual process.

5.1.3 Web Crawler Crawling Process

The crawling process of a web crawler is shown in Fig. 5.1 and mainly consists of the following five steps:

1. Analyze the target website. Clarify the structure of the target website and identify the location of key data.
2. Send requests. Use HTTP libraries or browser simulation tools to send requests to the target website.
3. Get responses. If a response is received, it means that the target website is in a normal state.
4. Parse the response content. Parse HTML data and JSON data to obtain key data information or the next URL to be crawled.
5. Save data.

5.2 Classification of Crawlers

5.2.1 General Web Crawler

The general web crawler, also known as a full-web crawler, expands its crawling objects from some seed URLs to the entire web, mainly collecting data for portal site search engines and large web service providers. Due to commercial reasons, their technical details are rarely disclosed. This type of web crawler has a vast range

Fig. 5.1 Crawling process of a web crawler

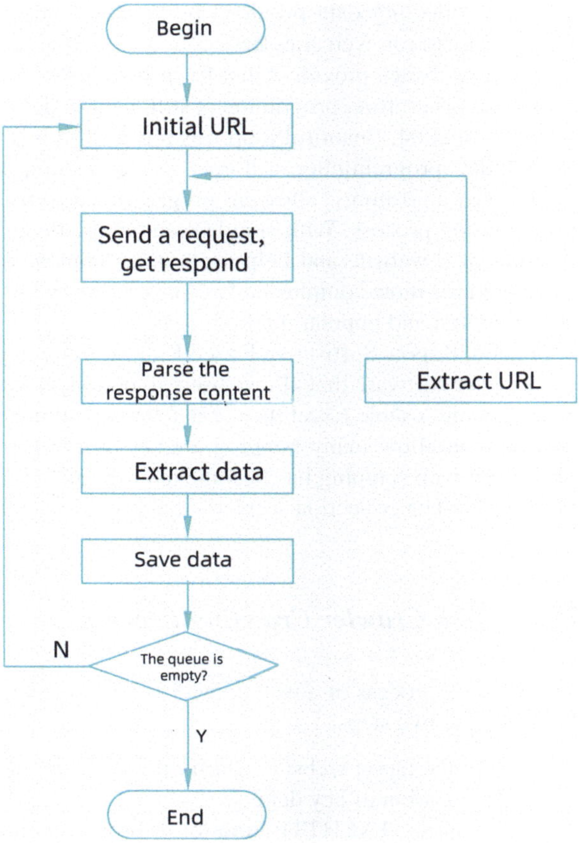

and number of web pages to crawl, requiring high speed and storage space but lower requirements for the order of crawling web pages. At the same time, because there are too many web pages to be refreshed, this web crawler usually works in parallel, but it takes a long time to refresh a web page once. Despite certain shortcomings, the general web crawler can search for a wide range of topics for search engines and has substantial application values.

The general web crawler has two strategies for web page crawling:

1. Breadth-first strategy. It crawls web pages in the order of the depth of the web page content directory, and web pages at a shallower directory level are crawled first. When the web pages at the same level are crawled, the web crawler goes deeper into the next level to continue crawling. The advantage of this strategy is that it can effectively control the crawling depth of the web page, avoid the problem of being unable to end the crawl when encountering an infinitely deep branch, and is easy to implement. It can store a large number of intermediate

nodes for subsequent information extraction. However, this strategy requires a long time to crawl to web pages with deeper directory levels.
2. Depth-first strategy. In the order of depth from shallow to deep, visit the next-level web page links until you can no longer go deeper. After completing a crawl branch, the web crawler returns to the previous link node to search other links further. When all links have been traversed, the crawling task ends. This strategy is very suitable for vertical search or site search. However, due to the crawling order, this strategy will cause a massive waste of resources when crawling websites with more web page content levels.

5.2.2 Deep Web Crawler

Web pages can be divided into surface and deep web pages according to their existence. Surface web pages are pages that traditional search engines can directly discover, mainly composed of static web pages that hyperlinks can reach. Deep web pages are those where most of the content cannot be obtained through static links, hidden behind search forms, and can only be obtained by users submitting some keywords.

The crawling steps of the deep web crawler are as follows:

1. Automatically find the entry points of deep web pages. To crawl deep web pages, you need to fill in search forms (these forms constitute the entry points of deep web pages) and access the information behind these search forms. The entry point is the search form on the website that allows access to deep web pages.
2. Form modeling. Since search forms are designed for human–computer interaction, accessing deep web pages requires an automated process to simulate this interaction. Any agent interacting with the search form must complete form modeling and query selection. Form modeling involves understanding the search form so that the semantics related to each field, the type and value of expected values in each field, and the relationship between fields can be identified. Query selection involves generating combinations of values of the appropriate type and domain to fill each field. Both of these tasks can be automated using unsupervised or semi-supervised techniques.
3. Learning crawl paths. The work of deep web crawlers is not limited to filling out forms and returning result pages. In some cases, they need to delve further into the web, finding subsets of web pages related to the user or process using the web crawler. Depending on the purpose of crawling these subsets of web pages, these web crawlers can be divided into general web crawlers, focused web crawlers, and hotspot web crawlers.

5.2.3 *Focused Web Crawler*

5.2.3.1 System Structure of the Focused Web Crawler

General crawlers can only provide rough information. The focused web crawler has a clear theme and can accurately obtain valid information. When storing web page URLs, the focused web crawler needs to judge the relevance of the URL to the theme and filter out web pages related to the theme as much as possible. The work of the focused web crawler includes web page acquisition, web page filtering, web page storage, and web page analysis.

1. Web page acquisition. Simulate the client to send HTTP requests, download web pages after getting the server's response, and complete the crawling work of the focused web crawler system.
2. Web page filtering. Filter URLs related to the theme, grab web pages related to the theme, and ensure the accuracy of the focused web crawler system.
3. Web page storage. Store the data parsed by the web page parsing module in the form of files or databases.
4. Web page analysis. It includes two parts: The first part is the judgment of theme relevance, and the second part is the prediction of theme relevance.

5.2.3.2 Similarity Discrimination Algorithm of Focused Web Crawler

The similarity discrimination algorithm of the focused web crawler is divided into two types, namely, the vector space model algorithm and the semantic similarity algorithm. The vector space model algorithm transforms text processing into vector operations in vector space, represents each document as a vector in vector space, and measures the similarity between documents by calculating the closeness of vectors in vector space. The semantic similarity algorithm is based on the two quantities of word frequency and document frequency, enabling the computer to judge the similarity of documents from a semantic perspective.

5.2.3.3 Search Strategy of Focused Web Crawler

The search strategy of the focused web crawler is divided into two types: static search strategy and dynamic search strategy.

The static search strategy searches according to specific rules and will not change due to web page structure or text information changes. It mainly includes breadth-first search, depth-first search, and best-first search.

The dynamic search strategy aims to efficiently and quickly complete the crawling task as its first principle and adjusts the search route in real time. Dynamic search strategies adjust in real time according to the relevance of the URL's topic, mainly including text-based Fish-Search and Shark-Search and link analysis-based PageRank, HITS, and HillTop.

Below, we take Fish-Search as an example, and its algorithm description is as follows:

First, put the seed links related to the topic into the URL queue to be crawled. If the current webpage is related to the topic, the first 'xwidth' links from this webpage are placed at the top of the URL queue to increase the crawling width, where the parameters 'x' and 'width' are given initial values. If the current page is unrelated to the topic, put the first width links of this web page in the middle of the UR L queue to be crawled, that is, behind the links related to the topic; if it is other situations, put the sublinks of this web page into the tail of the URL queue to be crawled, and only crawl these links when there is enough time. A variable potential score 1 can be defined to describe topic relevance. When the web page is related to the topic, the potential_ score is set to 1; when the web page is unrelated to the topic, the potential_ score is set to 0.5; in other cases, the potential_ score is set to 0. Finally, sort the URLs in the URL queue to be crawled according to the value of the potential_ score.

5.2.3.4 Research Direction of Focused Web Crawler

Focused web crawlers currently have two types based on web page content and link analysis.

Currently, the methods for calculating text similarity of focused web crawlers based on web page content are roughly divided into two categories: One is based on word statistics models, such as vector space models, and the other is based on semantic understanding models. Researchers hope to use semantic relevance to make web crawlers get more accurate results. In the entire text similarity judgment process, first, determine the topic of the web crawler; then, calculate the web page content and structure information to calculate the relevance of the web page topic and the relevance of the URL to be crawled, and determine the links to be crawled and the priority of crawling based on the relevance of the web page topic. This type of web crawler can achieve a high accuracy rate.

Tens of billions of web pages on the Internet are linked together through links on the World Wide Web; researchers are trying to obtain the meaning of the link context in a practical way, to parse and extract the link context, or to improve the traditional link selection algorithm based on web page content, to improve the accuracy of the web crawler's collection. This type of algorithm judges the importance of a web page by analyzing web page links, emphasizes the authority of page links to user needs, and, at the same time, solves the "topic drift" problem by combining web page text, link anchor text, and web page content analysis and link analysis of the anchor text context, improving the accuracy of topic crawling.

5.2.3.5 Application of Focused Web Crawler

In order to make up for the shortcomings of general search engines and achieve the retrieval of specific topic information, vertical search engines have emerged, which produce more accurate results, dig deeper into information, have less invalid information, and are more suitable for services in vertical fields. Vertical search engines are aimed at specific fields, serving particular users. A kind of search engine is a deep excavation of professional field information; it integrates information after filtering, screening, and sorting, providing users with searches for professional knowledge. The working principle of vertical search engines is similar to that of full-text search engines; the difference lies in the spider program in the crawling module and the subject thesaurus.

5.2.4 Incremental Web Crawler

5.2.4.1 Crawling Steps of the Incremental Web Crawler

The crawling process of the incremental web crawler is shown in Fig. 5.2, which can be divided into the following three steps:

1. Before sending a request, determine whether this URL has been crawled before (suitable for websites that constantly have new pages).
2. After parsing the content, determine whether this part of the content has been crawled before (suitable for websites where the content is updated regularly).
3. When writing to the storage medium, determine whether the content already exists in the medium (to achieve the purpose of deduplication to the greatest extent).

5.2.4.2 Deduplication Method of Incremental Web Crawler

There are two methods for deduplication of incremental web crawlers:

1. Store the URLs generated during the crawling process in the Redis set. When crawling next time, first determine whether the URL corresponding to the upcoming request already exists in the URL set. If it exists, do not send the request; otherwise, send the request.
2. Assign a unique identifier (data fingerprint) to the crawled web page content, and then store this unique identifier in the Redis set. When the web page data is crawled next time, before persistent storage, determine whether the unique identifier of this data exists in the Redis set, thereby deciding whether to persistently store it.

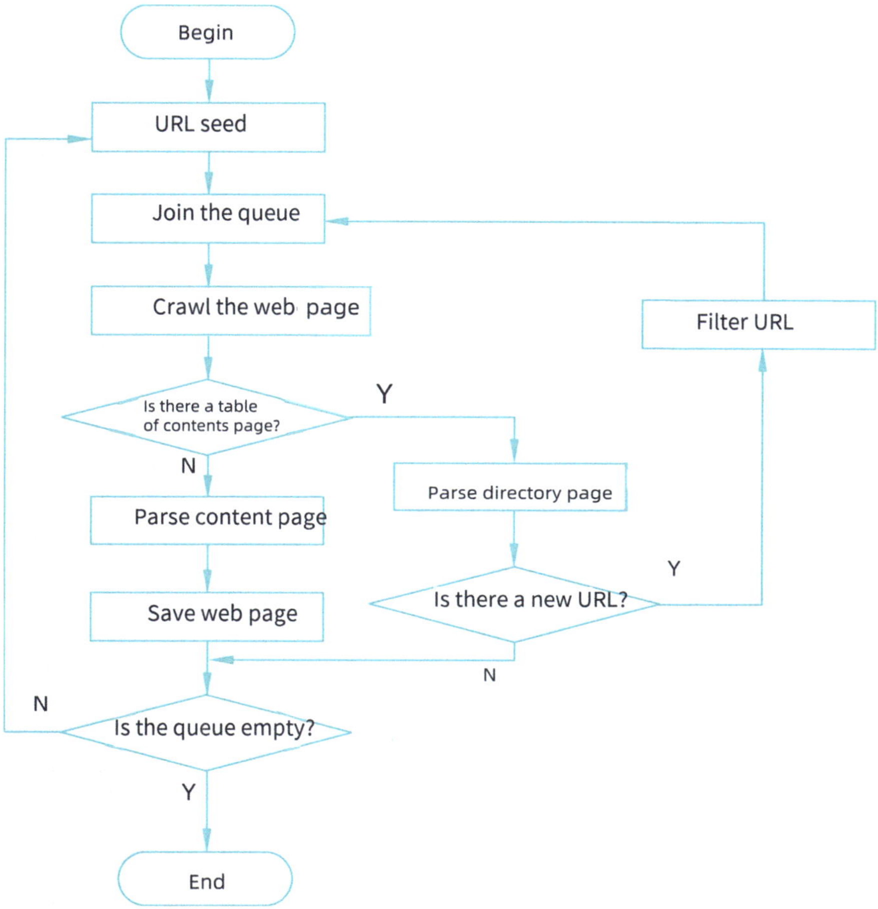

Fig. 5.2 Crawling process of incremental crawler

The deduplication method based on Redis's Bloomfilter not only brings into play the massive deduplication ability of Bloomfilter but also utilizes Redis' persistence ability. Bloomfilter is a very long binary vector and a series of random mapping hash functions. The standard method to identify whether an element is in a set is to compare this element with the elements in the set one by one. Bloomfilter can check whether a component is in the set in a short time.

5.3 Crawler Libraries and Frameworks

Common web crawler libraries can be divided into three categories according to the crawling link: web page crawling library, web page analysis library, and data storage library. The commonly used web crawler frameworks mainly include three frameworks: Scrapy framework, PySpider framework, and feapder framework.

5.3.1 Web Crawler Library

5.3.1.1 Web Page Crawling Library

The primary purpose of the web page crawling link is to implement HTTP requests, read the network resources specified in the URL address, and save them locally. The commonly used web page crawling libraries include urllib, requests, selenium, pyppeteer, aiohttp, etc. Selenium was originally an automated testing tool, but web crawlers mainly use it to solve the problem that requests cannot. Executing JavaScript code often involves dealing with information hidden within the JavaScript code itself. Selenium essentially drives the browser, fully simulating browser operations such as navigation, input, clicking, dropdown, etc., thereby obtaining the results after the web page is rendered. Pyppeteer uses the Python asynchronous coroutine library asyncio, which can be integrated with Scrapy for distributed crawling. Although Pyppeteer supports a relatively single browser, it far surpasses Selenium in terms of installation convenience and operational efficiency. Both urllib and requests block HTTP request libraries, i.e., after sending a request, the program will wait for the server response, and only after the server responds will the program proceed to the next step, whereas aiohttp can implement asynchronous web requests, significantly improving the efficiency of web page crawling. The following mainly introduces urllib and requests:

urllib

urllib is a built-in HTTP request library in Python, which includes the following four modules:

1. request. This is the HTTP request module, which can be used to simulate sending requests. Simply pass the URL and corresponding parameters to the library method, and it can simulate the operation of entering a URL in the browser address bar and pressing the Enter key.
2. error. This is the exception handling module. When a request error occurs, these exceptions can be caught and then retried, or other operations can be performed to ensure that the program does not terminate unexpectedly.

3. parse. This is a utility module that provides many URL processing methods, such as splitting, parsing, merging, etc.
4. robotparser. This is a website recognition module, mainly used to identify the robots.txt file of a website and then determine which websites can be crawled.

requests

Requests is a third-party library in Python. It further encapsulates urllib, making it more convenient to use. In practical applications, requests are used the most. Its functional features mainly include p-Alive and connection pool, international domain name and URL, session with persistent cookies, browser-style SSL certification, automatic content decoding, basic/digest authentication, key/value pair cookies, automatic decompression, Unicode response body, HTTP/HTTPS proxy support, file chunk upload, stream download, connection timeout, chunked request, and support for .netrc.

5.3.1.2 Web Page Analysis Library

The web page analysis link is a key step in the crawler, which requires locating and extracting the required information from XML or HTML. Common web page analysis libraries include lxml, Beautiful Soup, re, and PyQuery. PyQuery is a Python implementation of JQuery, which can parse HTML documents in a syntax similar to JQuery, and its ease of use and parsing speed are very good. The following mainly introduces lxml and BeautifulSoup:

lxml

lxml is a parser for XML and HTML; its main function is to parse and extract data from XML and HTML. Similar to regular expressions, lxml can also use XPath syntax to locate specific elements and node information. Moreover, it is compatible with the well-known ElementTree API. XPath, the full name of which is XML Path Language, was originally used to search XML documents, but it is also suitable for searching HTML documents, so it can be used in web crawlers to extract corresponding information.

BeautifulSoup

BeautifulSoup can convert complex HTML or XML documents into a complex tree structure (document tree), where each node on the tree is a Python object. All objects can be categorized into four types: Tag, NavigableString, BeautifulSoup, and Comment.

1. Tag: Tags in HTML documents. You can use soup plus the tag name to get the content of these tags, and the type of these objects is Bs4, Element.Tag.
2. NavigableString: The content in the tags.
3. BeautifulSoup: Represents the entire content of a document; in most cases, it can be treated as a Tag object; it supports most of the methods described in traversing and searching the document tree.
4. Comment: Comments in the document. The Comment object is a special type of NavigableString object.

5.3.1.3 Data Storage

The data storage section is the database storing the information that has been extracted, such as MySQL, MongoDB, Redis, etc. It can persistently store data. Commonly used data storages include PyMysql, PyMongo, Redis, etc.

5.3.2 Web Crawler Framework

5.3.2.1 Scrapy Framework

The Scrapy framework is a fast, high-level screen scraping and web scraping framework, used for scraping websites and extracting structured data from web pages. The appeal of Scrapy lies in it being a framework; users can conveniently modify it as needed. It also provides base classes for various types of web crawlers (such as BaseSpider, Sitemap, etc.). The architecture of the Scrapy framework is shown in Fig. 5.3.

1. Scrapy engine. The core of the Scrapy framework, used to handle the entire system's data flow and trigger transactions.
2. Scheduler. Receives requests sent by the Scrapy engine and adds them to the queue, providing them to the Scrapy engine when it needs to send the request again.
3. Downloader. The component with the highest load among all, used for high-speed downloading of resources on the network.
4. Web crawler. This module helps users customize their own web crawlers (using regular expressions and other syntax) to extract the information users need from specific web pages.
5. Item pipeline. Used to process the entities extracted by the web crawler. The main functions are to persist entities, validate the entities, and clean up unnecessary information.

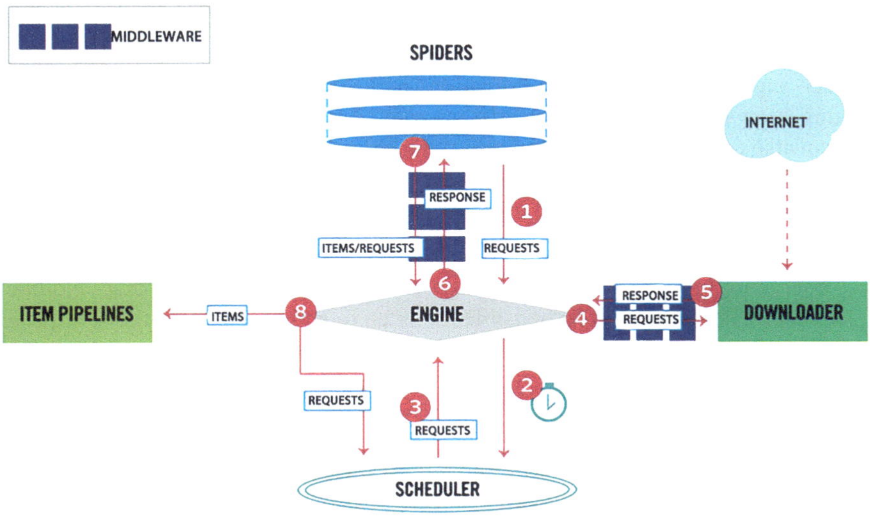

Fig. 5.3 Architecture of the Scrapy framework

5.3.2.2 PySpider Framework

The PySpider framework is a framework with a powerful WebUI, script editor, task monitor, project manager, and result processor. It supports multiple database backends, multiple message queues, and JavaScript rendering page collection functions.

Compared with Scrapy, PySpider has the following features:

1. Provides WebUI; the writing and debugging of web crawlers are all done in WebUI.
2. It has PyQuery built in as a selector.
3. It supports PhantomJS for JavaScript rendering page collection.

5.3.2.3 feapder Framework

feapder is a simple-to-use, powerful Python crawler framework, similar to Scrapy, convenient for those switching from the Scrapy framework. The feapder framework has three built-in web crawlers:

1. AirSpider. It is a lightweight crawler with a low learning cost. For some data with less volume, no need for breakpoint continuation, and no need for distributed collection, this web crawler can be used.
2. Spider. It is a Redis-based distributed web crawler suitable for massive data collection, supporting breakpoint continuation, alerting, automatic data entry, and other functions.

3. BatchSpider. It is a distributed batch web crawler. For data that needs to be collected periodically, consider using this web crawler first.

feapder supports breakpoint continuation, data loss prevention, monitoring and alerting, browser rendering downloads, automatic data entry (MySQL or MongoDB), and can also connect to other storage libraries by writing pipelines. The structure of the feapder framework is shown in Fig. 5.4.

The modules of the feapder framework are as follows:

1. spider. The crawler, which is the core of the framework scheduling.
2. parser_control. Parsing controller, responsible for scheduling the parser module.
3. collector. Task collector, responsible for batch fetching tasks from the task queue into memory, reducing the access frequency and concurrency of the web crawler to the task queue database.
4. parser. Data parser.
5. start_request. Initial task dispatch function.
6. item_buffer. Data buffer queue, used to batch store data into the database.
7. request_buffer. Request task buffer queue, used to batch store request tasks into the task queue.
8. request. Data downloader, encapsulates the request, used to download data from the Internet.
9. response. Request response module, encapsulates the response, supports XPath, CSS, re and other parsing methods, and automatically handles Chinese garbled text.

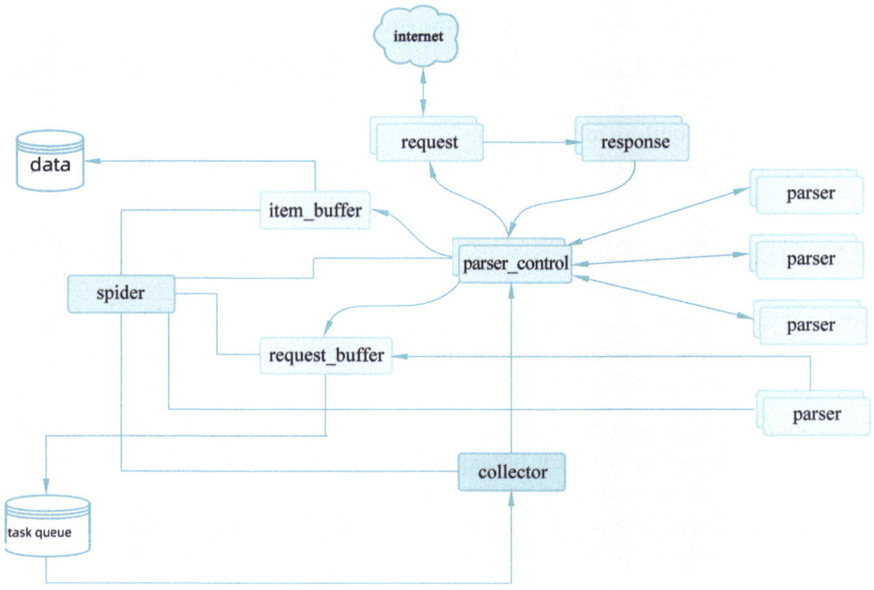

Fig. 5.4 Structure of the feapder framework

The flow of the feapder framework is as follows:

1. The spider schedules the start_request to generate initial tasks.
2. The start_request dispatches tasks to the request_buffer.
3. The spider schedules the request_buffer to batch store tasks into the task queue.
4. The spider schedules the collector to batch fetch tasks from the task queue into memory.
5. The spider schedules the parser_control to fetch tasks from the collector's memory queue.
6. The parser_control schedules the request to fetch data.
7. The request requests and downloads data.
8. The request sends the downloaded data to the response, further encapsulates it, and returns the encapsulated response to the parser_control (multiple parser_controls in Fig. 5.4 represent multithreading).
9. The parser_control schedules the corresponding parser to parse the returned response (multiple parsers in Fig. 5.4 represent different website parsers).
10. The parser_control distributes the data (entities) and new requests obtained from the parser analysis to the item_buffer and request_buffer.
11. The spider schedules the item_buffer and request_buffer to batch the data into the database.

5.4 Cutting-Edge of Web Crawler Technology

5.4.1 Latest Developments in Web Crawler Technology

As an emerging automated data collection technology, web crawlers have rapidly evolved in the era of big data, becoming an integral part of various industries and offering significant convenience. However, their rapid development has also led to some challenges. The diversity of web crawler users and the lack of Internet regulations have allowed malicious web crawlers to seriously compromise network security. Yet, the potential of web crawlers to revolutionize data collection and analysis is a fascinating area for exploration.

Next, we will introduce three examples of combining web crawlers with practical technology.

The first is the Google Translate crawler based on the Gecko browser kernel. This method simulates the browser loading web pages, completes user input, triggers the execution of scripts, and finally obtains the target data. Using the above method, a dedicated crawler for Google Translate is designed and implemented, which can solve the problem of automatic translation of large-scale corpora in a "multiple less take" way.

The second is to use web crawlers to predict the network's public opinion. Using web crawler technology to obtain data from a hot event from the Baidu Index, preprocess these data, and then establish a logistic differential equation model for network public opinion. Combining existing data, intelligent algorithms are used to determine the three key parameters in the solution of the differential equation, and finally, they are applied to network public opinion prediction.

The third is to use web crawlers to conduct government information disclosure audits. According to the traditional mode, conducting government information spot checks at a 5% rate cannot achieve full coverage, and frequent manual website visits and data verification are ineffective. However, with a distributed cloud computing platform at its core and a custom approach to dealing with different formats of web page data sources, the problem of government information disclosure audit can be solved.

5.4.2 Cutting-Edge Anti-crawling Technology

Although web crawler technology brings great convenience, if used improperly or maliciously, it can also cause harm to normally operating servers, so anti-crawling technology is needed.

5.4.2.1 User Agent Control

The user agent can carry a string used to represent user device information, including browser, operating system, CPU, etc. You can set a user agent whitelist on the server; only user agents that meet the conditions can access the server. The disadvantage of this technology is that it is easy for web crawler programs to forge header information and then be cracked.

5.4.2.2 Session Limitation

A session is a credential for a user to request a server response; web crawlers often simulate regular user requests to the server by carrying everyday user-session information. Therefore, it is also possible to determine whether there is a web crawler program based on the size of the access volume in a short period of time and add the user session of the suspected web crawler program to the blacklist.

5.4.2.3 Spider Traps

Spider traps lead web crawler programs into infinite loop traps, consuming their resources and leading to their crash and inability to continue crawling data. This method has the disadvantage of adding many resource-wasting files and directories and preventing the search engine's crawler program from crawling information, thereby causing the website to rank behind.

5.4.2.4 IP Address Restriction

You can set a threshold on the server, add IP addresses with a large amount of access in a short period of time to the blacklist, and prohibit its access to achieve the purpose of anti-crawling.

5.4.2.5 Captchas

When a user logs in or accesses essential information, a captcha can block web crawler programs. Captchas include image captchas, SMS captchas, numerical calculation captchas, sliding captchas, and pattern marking captchas.

5.4.2.6 Data Encryption

Before the front end requests the server, it encrypts the request parameters, user agents, Cookies, etc., and requests the server with the encrypted data. In this way, the web crawler program does not know the set encryption rules and cannot perform simulation request operations. However, this technology's encryption algorithm is written in JavaScript code, which is easy for users to find and crack.

5.4.2.7 Restricting Cookies

When a user sends a request to visit a website, the data will contain specific Cookie data. The website will verify the value of the Cookie to determine whether the operation comes from a web crawler script or a real user. When the "user" opens the web page for the second and third times without Cookie data, it indicates that the operation comes from a web crawler script.

The automation technology of web crawlers indeed brings many benefits to the Internet, but the abuse of web crawler technology also has many disadvantages. "Water can carry a boat, and it can overturn it"; therefore, we should use web crawler technology correctly under laws and regulations and good social behavior norms in order to maximize its advantages and avoid causing harm to the Internet environment.

5.5 Application and Analysis

Zhaopin.com is one of the commonly used platforms for job seekers to find jobs. This example uses web crawler technology to crawl and perform a simple visual analysis of several job information in Beijing on Zhaopin.com. First, use natural language processing, computer vision, C++ engineering, backend engineering, frontend engineering, and big data as job titles to crawl job titles, salaries, work locations, work experience requirements, education requirements, job descriptions, etc., and then analyze the average salary, work experience requirements, and education requirements of each job.

5.5.1 Data Crawling

Implement data crawling based on Python's requests library and Selenium, using the lxml library to parse HTML code to obtain the required data, and converting it into JSON format for storage, while introducing multithreading to speed up the crawling speed. To avoid crawling duplicate data, a Bloom filter is used for data filtering.

Log in to the Zhaopin recruitment website, enter the job title you want to query, such as "Natural Language Processing," perform a search, get the list page search results, then analyze the HTML code, and you can extract the URL link of the details page of each job information through XPath rules. The job information list page of the Zhaopin recruitment website is shown in Fig. 5.5.

The details page contains information such as job title, salary, work location, work experience requirements, educational requirements, job description, and job requirements. This crawl collects all field information, and finally, the "Natural Language Processing" position collected 277 pieces of information, the "Computer Vision" position collected 254 pieces of information, the "Backend Engineer" position collected 336 pieces of information, the "Frontend Engineer" position collected 922 pieces of information, and the "Big Data" position collected 675 pieces of information. The format of the job information data collection results on the Zhaopin recruitment website is shown in Fig. 5.6.

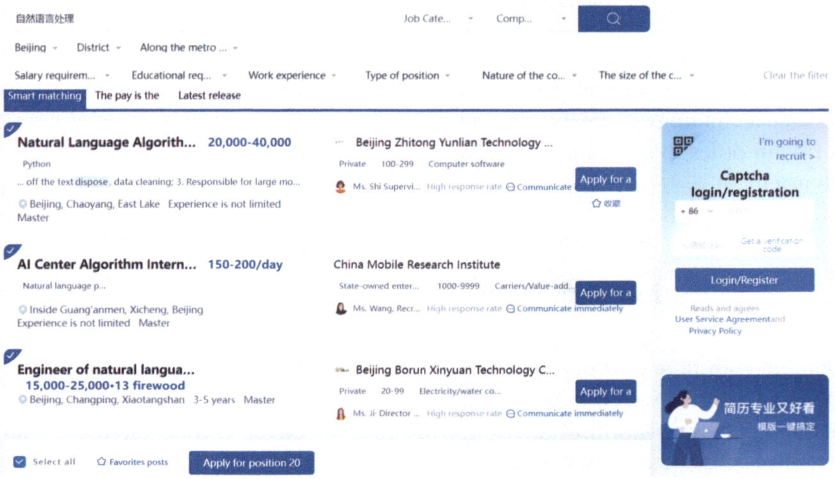

Fig. 5.5 Zhaopin recruitment website job information list page

```
{
        "Title": "Natural Language Processing,"
        "Salary": "15000-250000",
        "Place": "Beijing Haidian District",
        "Experience": "1-3 years",
         "Education": "Undergraduate",
        "Description": [
                "Job Responsibilities."
                "1. Responsible for NLP modeling work applicable to medical/financial big data."
                "2. Track the latest MP algorithms. "
                " Job Requirements:
                "1. Bachelor's degree or above in computer science, mathematics, statistics, communication, and other related fields."
                "2. Familiar with commonly used LP algorithms and master the use of tensorflow or pytorch."
                "3. Proficient in Python and conventional computer science solutions.
                "4. Proficient in reading professional foreign literature. "
                "5. Likes learning, enjoys collaboration, and is willing to challenge.
        ]
        "descriptionskill": [
                "Python".
                "Speech algorithms",
                "Knowledge graph",
                "Text classification",
                "Chapter analysis" ,
                "OpenNLP,"
                Stanford NLP "
        ]
}
```

Fig. 5.6 Zhaopin recruitment website job information data result format

5.5.2 *Data Analysis*

After crawling the recruitment data of several positions such as Natural Language Processing, Computer Vision, C++ Engineer, Backend Engineer, Frontend Engineer, and Big Data, simple data analysis can be carried out from various aspects, such as the average salary of the position, the number of recruitments for each position, and the requirements for education and work experience for each position.

For the analysis of average salary, calculate using the lowest salary (monthly salary) and draw it into a bar chart for visualization, as shown in Fig. 5.7. It can be found that the salaries of Computer Vision and Natural Language Processing positions are higher, and the average salary is over 20 k. At the same time, analyze the work experience requirements of different positions and visualize them, as shown in Fig. 5.8. It can be found that almost all positions require some level of work experience.

People with 3–5 years of work experience are more recognized, and those with less than 1 year of work experience and no experience account for the least; the proportion of recruiters who do not limit experience in computer vision, natural language processing, and big data positions is relatively high. An analysis and visualization of the educational requirements for different positions is shown in Fig. 5.9. It can be found that the proportion of positions requiring at least a bachelor's degree is the highest, the proportion of those with a master's degree or above in natural language processing and computer vision is relatively high, while big data, backend engineers, and frontend engineer positions have lower educational requirements compared to natural language processing and computer vision positions.

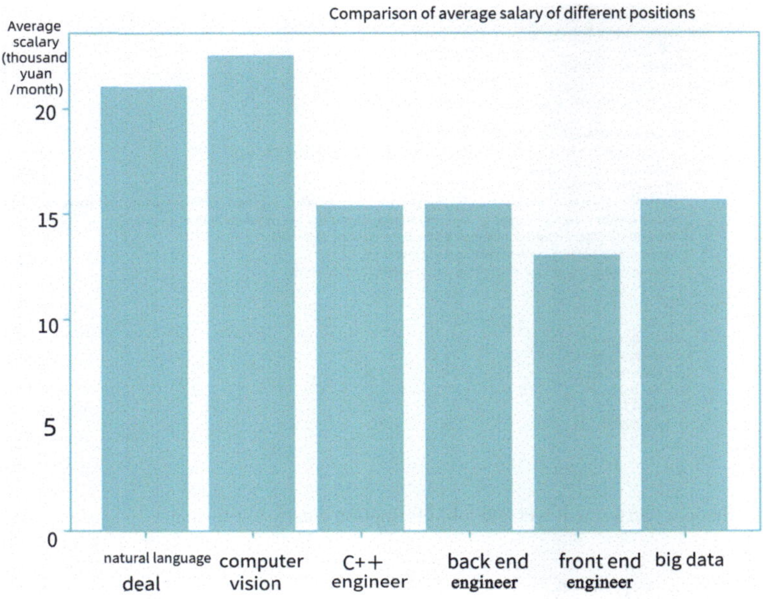

Fig. 5.7 Comparison of average salaries for different positions

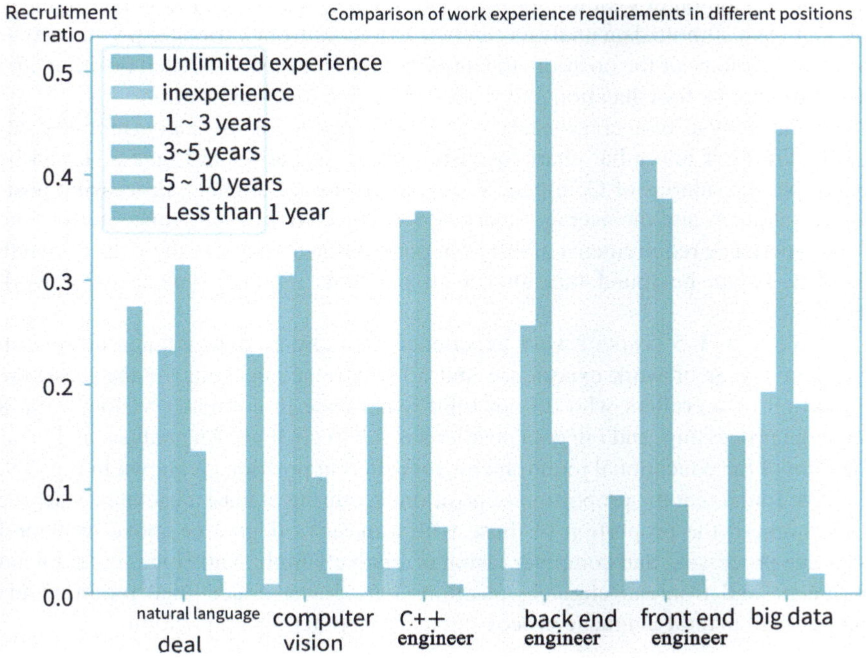

Fig. 5.8 Comparison of work experience requirements for different positions

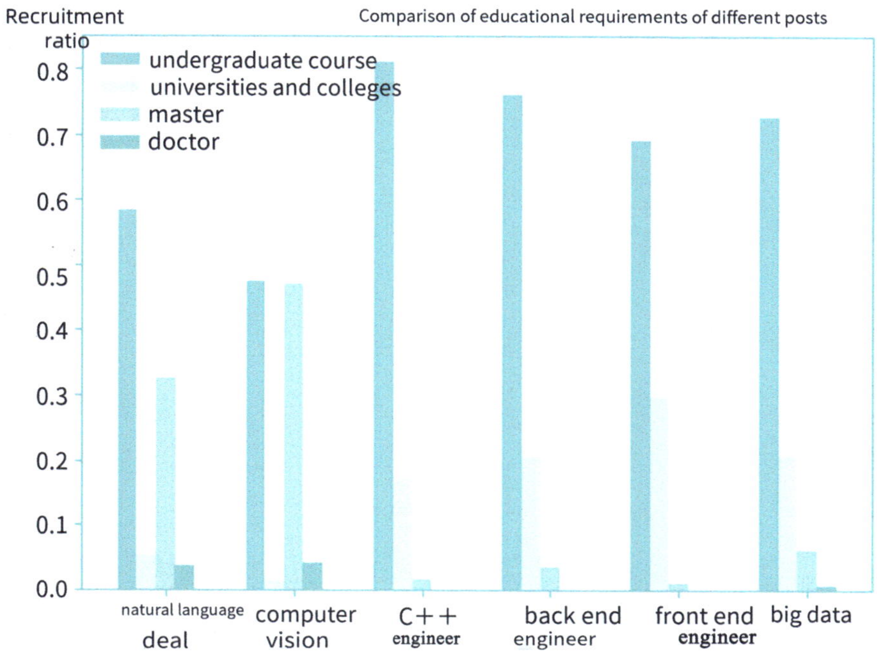

Fig. 5.9 Comparison of educational requirements for different positions

Chapter 6
Multiformat Document Parsing and Management

This chapter explores the principles and applications of multiformat document parsing and management, beginning with an introduction to document formats and standards. It then delves into the underlying mechanisms of multiformat document parsing, focusing on widely used formats such as Word, PDF, and PPT. Key parsing techniques are covered using various Python libraries and frameworks, with practical examples provided for parsing Office documents. This chapter further provides application examples of widely used Office document parsing.

6.1 Overview

6.1.1 Document Format

Document format refers to computers' particular encoding method to store information and recognize internally stored data. Typical documents include PNG, GIF, and other images, MP4, FLV, and other videos, Word, PowerPoint, Excel, PDF, and other office documents. Office documents are widely used daily, so this chapter mainly introduces the parsing and management of Office documents. For different types of documents, the computer will call different parsers to parse the document information according to the document's suffix name. If the suffix of an MP4 video document is forcibly changed to pdf, then the computer will call the PDF parser to parse the document, leading to document parsing failure.

Document management refers to viewing, storing, classifying, and retrieving documents, spreadsheets, graphics, and scanned image documents. Each document has a record similar to an index card, which includes information such as the author, document description, creation date, and type of application used. Traditional document management mainly manages archived materials, which are not frequently searched and reused. But in modern enterprise scenarios, the main management is

H. Zhang, J. Shang, *Natural Language Processing and Applications*,
https://doi.org/10.1007/978-981-97-9739-4_6

processing documents; the frequency of document generation and modification is high, and the requirements for the time of completion are also high; users are required to see the latest version of the information at any time, and anywhere, and can search for the necessary information from a large number of documents.

6.1.2 Development History of Document Standards

At the beginning of the twenty-first century, most Office software used closed document formats, and it was impossible to obtain complete and accurate description information of the document format through legal means. The closed document format Office software makes it very difficult for other Office software to be compatible with it, gradually becoming a shackle that hinders user information exchange. Using a closed document format, even the document owner and government departments cannot truly own the document, so they may lose the ability to access, modify, and save these documents at some point in the future, posing many security risks that we must be aware of and address.

The traditional problem that documents based on binary closed formats may face is that previous documents can no longer be used due to Office software upgrades or the closure of the original Office software company. Obviously, this means that users will face a terrible risk of data loss. Government agencies worldwide with high file security requirements are particularly worried about this.

More and more government agencies are realizing the urgency of adopting open formats to store electronic versions of national policy documents. This shift brings a promising future, where data loss risks are minimized and document compatibility is enhanced.

In order to change the terrible situation of Office software being mutually closed and document formats being incompatible, the ODF (OpenDocumentFormat) alliance, created by 36 members including SUN and IBM, is promoting the ODF document format globally, making it the standard for Office documents. In May 2006, ODF was established as an international standard. In 2002, Red Flag Chinese 2000, Kingsoft, Yongzhong, other companies, and the Software Research Institute of the Chinese Academy of Sciences launched the Chinese document standard UOF (Unified Office document format) compatible with ODF. UOF was established as a national standard in China in May 2007. At the same time, Microsoft developed a new document standard, OOXML (Office Open XML), based on the Office document format, which contains many of Microsoft's private standards and technologies. Its document is as long as 6000 pages, and only Microsoft's products can implement all functions.

6.2 Multiformat Document Parsing

6.2.1 Word Document Analysis

6.2.1.1 Introduction to OpenXML

OpenXML is a new document format proposed by Microsoft in Office 2007. In Office 2007 and later versions, Word, Excel, and PowerPoint all use the OpenXML format by default. Under the OpenXML form, the original dox, xls, and ppt formats have been transformed into docx, xlsx, and pptx formats. These files are essentially compressed files containing a series of XML files. Using common compression software, you can decompress it. Compared with the original format, the OpenXML format takes up less space. The following will take the docx format document of Word as an example to explain its structure, analysis method, and application scenarios of document analysis.

6.2.1.2 Word Document Structure

After decompressing a docx format Word document, you can get a series of related files. Further analysis shows that a docx format Word document is composed of many different XML files, each storing corresponding information. The core XML files include app.xml, core.xml, .rels.xml, etc.

The explanation of some XML files is shown in Table 6.1. The docProps directory is used to record property information, where app.xml stores application-specific properties, and core.xml stores common file properties of the document format. The _rels subdirectory under the Word directory stores all XML file information and index ids of the parent directory, and the XML files under the Word directory store various information about the Word document.

Table 6.1 Explanation of some XML files

Filename	Illustrate
app.xml	Application-specific properties
core.xml	Common file properties for document formats
.rels.xml	Stores the information and index ids of all XML files in the parent directory
document.xml	The content and properties of all text in the document
fontTable.xml	Font information used by the document
setting.xml	Document general settings information
style.xml	The overall style information of the document

6.2.1.3 Word Document Parsing

Word document parsing can be divided into two methods: The first is to call APIs written by others to parse docx format Word files directly; these APIs are generally encapsulated in function libraries, such as Java's POI component's xwpf and Python's python-docx library; the second is to obtain Word document information by parsing XML files. As mentioned earlier, docx format Word files are composed of a series of XML files, so DOM, SAX, JDOM, and DOM4J methods can be used to parse XML files to achieve the purpose of Word document parsing.

Compared with this, using APIs to parse Word documents directly is more concise, but obtaining Word document information by parsing XML files is more widely used. As shown in Fig. 6.1, XML files can be seen as a tree in structure, and by traversing the tree nodes, you can obtain relevant information about the XML file and then achieve the purpose of parsing the Word document.

6.2.1.4 Application Scenarios

In addition to obtaining simple text information, Word document parsing has two higher-level applications. One is Word desensitization technology [1], which parses XML files using BMHS and Word2Vec keyword matching algorithms and other technologies to process sensitive words. The second is malicious document detection. Attackers embed malicious code, images, etc., into XML files and then package them into docx format Word documents to send to users. Once the user opens this document, they will suffer a malicious attack.

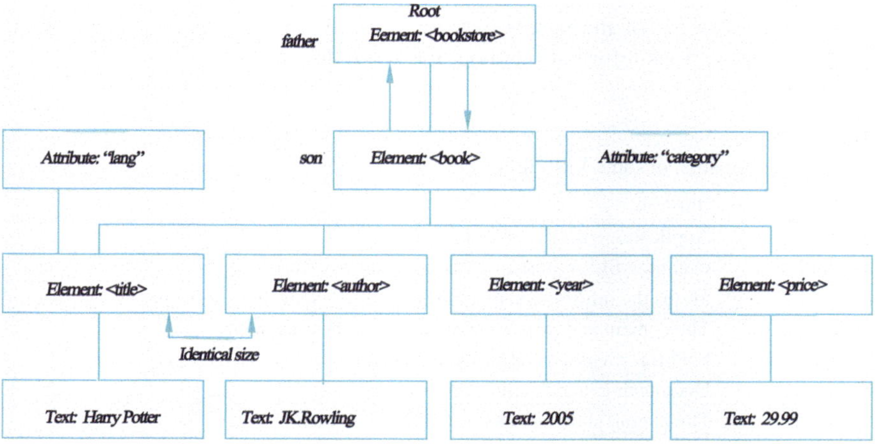

Fig. 6.1 Tree structure of XML file

By using XML parsing combined with machine learning, deep learning, and other related technologies, the document can be detected, and malicious attacks can be intercepted.

6.2.2 PDF Document Parsing

6.2.2.1 Overview of PDF Documents

PDF (Portable Document Format) was first proposed by Adobe, a US typesetting and image-processing software company, in 1993. This format allows document archiving and exchange on any operating system [2]. PDF is a text format composed of human-readable identifiers, but it also contains binary data, such as images, embedded files, or encrypted data. PDF documents were initially mainly used in the electronic printing industry, but now they are popular everywhere. PDF documents, with their cross-platform nature, independence of display and document information, excellent security, etc. Good features are widely used in daily office scenarios such as scientific research, enterprises, and government departments.

The basic elements of PDF documents are objects. PDF objects can be divided into two types: direct and indirect. PDF is developed from PostScript, so its data information type is similar to the data type of ordinary programming languages. The basic object types of PDF [3] are shown in Table 6.2.

Table 6.2 Basic object types of PDF

PDF object type	Description	Example
Array	Ordered list	[1 2 3 4 5]
Boolean value	Logical true/false value	False
Numerical value	There are two types of numeric objects: integers and real numbers	1.2
Name object	An atomic symbol, uniquely defined by a sequence of characters	Type
String	String	(abc) 或<Aabb>
Empty object	Represented by the keyword null	null
Dictionary object	A list consisting of many objects, enclosed in a pair of double angle brackets	<</Type/Catalog/Page30R/Outline20R>>
Stream object	Usually represents a compressed data stream	130obj<</Type/Xobject>>stream 030004040404040 end stream

6.2.2.2 PDF Document Structure

The structure of a PDF document often takes the/Catalog tag as the root node, pointing to the child object node according to the referenced object number and progressing layer by layer until there is no referenced object number, that is, reaching the leaf node. The logical structure of PDF is more complex than the original content of the document due to the reference relationship between the objects, but it also expresses a lot of document semantics, such as whether certain objects exist and the reference relationship between objects. This semantic information plays a vital role in comprehensively judging the behavior of the document.

The structure of a PDF document mainly includes four parts: document header, document body, cross-reference table, and document tail. The document header usually appears on the first line of the PDF document, indicating the version number of the PDF document. The document body is the main part of the PDF document, mainly including various objects that make up the PDF document. The cross-reference table contains information about indirect objects in the PDF document and stores an index table for indirect access to object addresses. The document tail provides the address of the cross-reference table and some special objects (such as Catalog).

6.2.2.3 PDF Document Parsing

The process of parsing a PDF document is also the process of reproducing the structure of a PDF document. Access to the content of a PDF document starts from the catalog object and further accesses the content stream object corresponding to the page. Each object has a digital number as a reference value, and objects are connected through references. In this way, related objects can be referenced by corresponding objects. PDF document parsing follows the document structure and its rules, visiting and processing each page object layer by layer. If some PDF documents are encrypted, it is necessary to use third-party software to decrypt them and then parse the PDF document. As shown in Fig. 6.2, the specific process of PDF document parsing is as follows:

1. Find the keyword Root from Trailer, Root points to Catalog (dictionary); Catalog is the total entrance of a PDF document, and it contains Page tree, Outline hierarchy, etc.
2. Find the keyword Pages from Catalog; Pages is the total entrance of all pages of the PDF document, that is, Page Tree Root.
3. Find the keywords Kids and Count from Pages. Kids contains Page nodes, and Count lists the total number of pages in the document. At this point, we already know how many pages the PDF document has.
4. Obtain information such as MediaBox, Contents, Resources, etc., from Page. MediaBox contains page width and height information, Contents contains page content, and Resources contains the resources required by the page.
5. Obtain page content from the content stream pointed to by Contents.

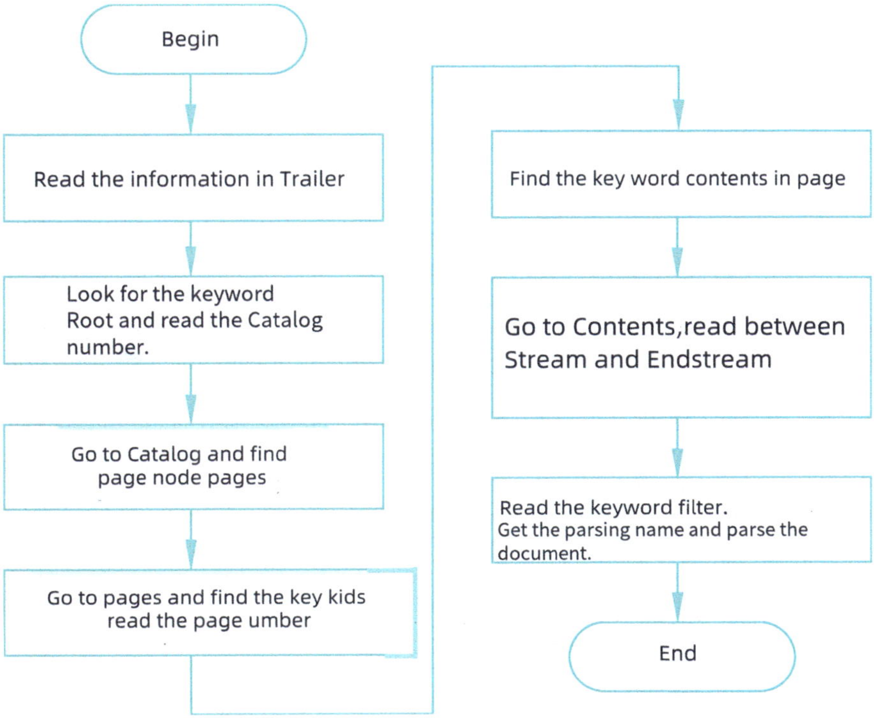

Fig. 6.2 PDF document parsing process

6.2.2.4 Application Scenarios

In addition to obtaining simple text information, PDF document parsing has more advanced applications. In the aspect of graph report, using PDF document parsing technology can convert various graph reports into customized data, achieve effective integration of laboratory resources, and create conditions for data mining and artificial intelligence applications of graph reports [4]. For academic papers, using PDF document parsing technology can effectively extract the structural information of PDF documents, thereby enabling automatic extraction and processing of academic resources, which is beneficial for the further use of PDF documents in the field of academic papers and holds important significance for the current research on knowledge mining of academic paper resources [5]. In addition, using the PDF document parsing technology, combined with desensitization processing technology, we can recognize and desensitize online PDF files, which have certain application values [6].

6.3 Multiformat Document Management

6.3.1 Online Document Management

6.3.1.1 Collaborative Editing

This document explores the significance and challenges of collaborative editing in distributed work environments, particularly in the context of the rapid development of Internet-distributed technology. More and more work is now being completed by users in different geographical locations, necessitating the development of collaborative systems in various fields. These systems support scientific research and help corporate teams complete collaborative work. As a typical representative of the collaborative field, collaborative editing has always been one of the primary research directions in the CSCW (Computer-Supported Cooperative Work) field. It provides people with a fast and efficient cooperation mode, ensuring smooth document collaboration editing among team members. It is significant within universities and corporate teams in information exchange, collaborative office, Wiki writing, etc.

At present, collaborative editing is mainly divided into asynchronous collaborative editing and synchronous collaborative editing.

Asynchronous collaborative editing allows multiple people to edit the same document separately in time and ensures that the document data is consistent through mechanisms such as locking and version control tools (such as Git, SVN), but this method has major defects in user perception, concurrency, and conflict resolution. If the user does not submit the local updated version during the editing process, other users cannot know the editing result of this user. At this time, if other users submit changes, conflicts may occur. To avoid this problem, you need to manually check the document conflict content and then modify and merge, increasing the complexity and latency of the collaborative process.

Synchronous collaborative editing allows collaborative users to perceive other members' editing operations in real time, increasing user concurrency. The efficiency of collaborative editing has been improved, achieving a what-you-see-is-what-you-get effect. The real-time perception characteristics of collaborative editing will inevitably lead to the problem of data consistency among users. Therefore, the research focus of collaborative editing mainly focuses on how to maintain the data consistency model.

6.3.1.2 Data Consistency Model

The data consistency model in collaborative editing includes Causality, Convergence, and Intention. Hence, it is abbreviated as the CCI model. Although most current collaborative editing algorithms can ensure the first two consistencies, they are inefficient and lack flexibility in maintaining operation intention consistency.

Causality Consistency

For any two operations, A and B, if A→B exists, then the execution order on all nodes should also be A before B. If the order of arrival of the two operations due to network delay or other issues is contrary to their causal order, it results in a problem of inconsistency in causal relationships.

Data Convergence Consistency

In collaborative editing, when all nodes perform the same operations in the same order and do not edit, the shared document data of all nodes is consistent; when the execution order of operations on each node is different, it may lead to inconsistent results on each node, resulting in a problem of data nonconvergence.

Operation Intention Consistency

Sun et al. defined the consistency of operation intention from the execution effect, that is, for any operation, its effect after execution on any node in the collaborative editing system should be consistent with the desired effect. Li et al. pointed out that the definition of execution effect is somewhat vague and proposed the CA model. They believe that as long as each operation does not violate the established operation effect relationship when it is executed, it can maintain operation intention consistency.

The design purpose of all concurrency control algorithms is to maintain the data consistency model of the collaborative editing system, that is, to keep the consistency of causality, data convergence, and operation intention of each node through the design of concurrency control process and transformation functions.

6.3.1.3 Concurrency Control Algorithm

Traditional concurrency control algorithms mainly use serialization, token passing, and locking. Serialization executes operations in the same order on each node, usually implementing concurrency control in two ways: One is a pessimistic delay operation, waiting until all previous operations are completed before execution; the other one is an optimistic execution operation, which eliminates the impact of the early execution of this operation on the results through the undo/redo mechanism. The serialization method cannot guarantee each node's causal relationship consistency and operation intention consistency. Token passing and locking methods actually control the fact that only one user can edit the document at any time in the system to ensure data consistency. Still, this method limits the user's operation and does not reflect the advantages of multiple concurrency in real-time collaborative editing systems.

Below are two commonly used concurrency control algorithms.

dOPT Algorithm

The dOPT algorithm was proposed by Ellis and Gibbs and is the earliest concurrency control algorithm based on OT (Operational Transformation). It introduced a state vector to solve the problem of causal relationship consistency. The basic idea is to perform concurrent transformations of concurrent operations and then execute the operations of each node in order.

adOPTed Algorithm

Ressel and others proposed the adOPTed algorithm at the CSCW conference. The algorithm's innovation lies on the introduction of the L transformation function and the use of a multi-dimensional interaction graph, which stores all possible forms of operations.

The multidimensional interaction graph of the adOPTed algorithm is shown in Fig. 6.3.

In the multidimensional interaction graph, the vertices represent the document's current state, and the directed edges' transformation properties are defined as TP1 and TP2. TP1 represents a simple concurrent relationship, and TP2 represents a partial concurrent relationship. Satisfying properties TP1 and TP2 are sufficient and necessary conditions for achieving data convergence. The adOPTed algorithm

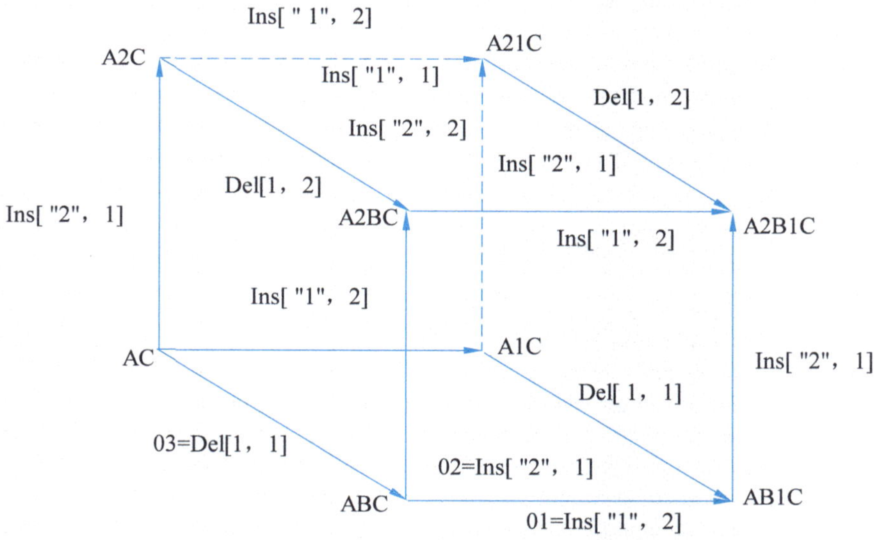

Fig. 6.3 Multidimensional interaction graph of the adOPTed algorithm

stores all operations' generation and intermediate forms in a linear log, ensuring the consistency of user operation intentions and causal relationships. However, implementing this algorithm is relatively complex, leading to a longer time required for transformation and reducing the system response speed.

6.3.2 Blockchain Document Management

6.3.2.1 Blockchain

Blockchain is a new technology for sharing data among untrusted peers [7]. Since the advent of Bitcoin in 2008, blockchain has gradually entered the public eye and has become a hot topic at home and abroad. Before 2014, blockchain-related research was limited to theory, mainly focusing on blockchain principles, the blockchain structure, and blockchain-related technologies. From 2015 to 2017, some suggestions have appeared on how to apply blockchain to the industry. Since 2018, while theoretical research has made new progress, blockchain has also begun to be applied to some industries.

There is no universally accepted definition of blockchain in the industry. In a narrow sense, a blockchain is a special data structure that links all data blocks in order using a chain structure. In this process, a decentralized public ledger that provides antitampering and anticounterfeiting functions is used with cryptographic technology. Generally speaking, blockchain is a new type of decentralized computing paradigm that combines cryptography, consensus mechanisms, and smart contracts. Some encryption algorithms in cryptography are used to verify and store data; consensus mechanisms are used to achieve block formation and data updates; and smart contracts are used for programming and operating data.

6.3.2.2 Document Blockchain

At present, domestic and foreign parties are comprehensively discussing and solving the specific methods of trustworthy document management problems from multiple approaches such as standard specifications, system architecture, process systems, and technical models. Among them, the most prominent is the blockchain technology that has received much attention in recent years. In 2017, the National Archives of the United Kingdom led the ARCHANGEL project, using distributed ledger technology to ensure the sustainability, accessibility, and authenticity of digital archives in long-term preservation [8]. In 2019, the National Archives and Records Administration of the United States issued the "Blockchain White Paper" to guide how to apply blockchain to document management. At the same time, domestic attention to blockchain technology is also increasing day by day. For example, China Petrochemical Corporation integrates the archive system with the blockchain platform, and the "Netcom Law Chain" of the Guangzhou Internet Court has realized digital deposit and electronic file management [9].

At present, document blockchain systems are mostly oriented toward a single business, serving a single entity, and have not fully reflected the advantages of blockchain's distributed and multiparty participation. With the deep application of blockchain technology, future document blockchain systems will inevitably interact and exchange data with more systems. For example, the State Grid Corporation will provide transaction evidence verification services to the "Tianping Chain" it participates in. Given that the current cross-chain technology is still immature, each blockchain is still a value island, and different types of blockchain systems due to programming languages, data dictionaries, smart contracts, etc., are inconsistent, and it is still relatively difficult to realize data value intercommunication. Therefore, in the future, we need to further explore cross-chain integration technology to achieve more flexible and efficient cross-chain business and data interaction. The effective integration of blockchain in document science management still requires people to further in-depth research and comprehensive thinking.

6.4 Application and Analysis

This section uses Python-based related modules to parse documents of different formats, perform deformatting and normalization processing on various types of documents, and prepare conditions for subsequent flexible document management work.

6.4.1 Multiformat Document Reading Algorithm

6.4.1.1 Word Document Reading

Python-docx is a dedicated library for Python to create and update docx format Word documents. It supports editing the content, font style, paragraph style, and headers/footers of Word documents and can also insert new content such as tables and pictures, thus better supporting the reading of the main content, format, and other useful information of the document. If necessary, you can also use this module to create a new Word document and write updates in the font format you want. However, it is worth noting that this module only supports reading the docx format of Office 2007 and higher version Word documents, because the docx format uses the new format of OpenXML, and this module is essentially a convenient editing interface for various elements of the OpenXML format. So, here we choose to use it first. The win32com library directly calls the Word program to first save the doc format document to be read as a docx format and then uses python-docx for parsing.

In the python-docx library, the document object represents a Word document, the paragraph object is used to read paragraphs, and the run object is used to divide the paragraph into individual sections. Figure 6.4 shows the basic content organization of a Word document in this library.

Further, you can use the sections object to view chapter information, including headers/footers, page margins, page orientation, etc. Entering a specific chapter, you can use the paragraph object to view the specific paragraph's indentation method, alignment method, etc. At this level, you can directly use the paragraph. Text method to get the text in the paragraph. Here, the reading of the text content of the paragraph is also done through the smaller run object, so if necessary, you can also operate on the finer-grained run object to get the corresponding section text, font, style, and other more specific information. In addition, this library also supports using the tables object to operate on tables in the document and supports using the add picture method to add pictures.

6.4.1.2 PDF Document Reading

Currently, there are many Python extension packages that can read and parse PDF documents, including PyPDF2, pdfplumber, Camelot, pdfminer, etc. PyPDF2 provides extraction, splitting, merging, encryption, and decryption methods, but it is not very friendly to Chinese, and reading Chinese may cause errors. Pdfplumber includes various libraries for handling PDF format information and can parse the text and tables of PDF documents well. Camelot is mainly used to extract table information from PDF documents, but it has functional limitations, such as not being able to automatically detect tables when using stream. pdfminer focuses more on obtaining and analyzing text data in PDF documents, allowing the extraction of exact text positions, fonts, line numbers, etc., providing richer text content reading and writing functions, suitable for analyzing and processing the overall content of

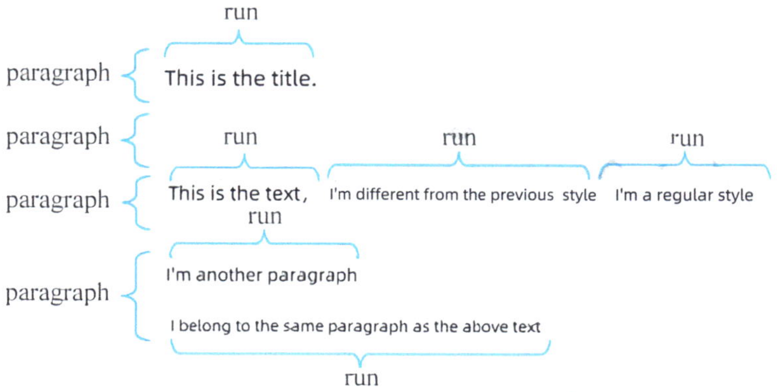

Fig. 6.4 Basic content organization of a Word document in the python-docx library

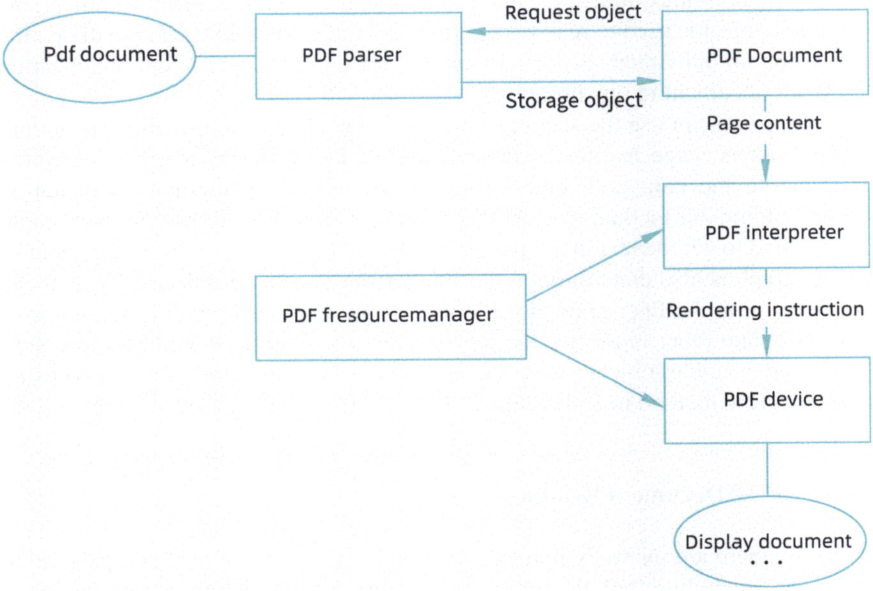

Fig. 6.5 Relationship and data flow between the modules of pdfminer3k

the document. Pdfminer3k is the Python3 version of pdfminer. The following mainly uses pdfminer3k to complete the related functions of the PDF document reading module.

When using the pdfminer3k module, you need first to use the PDFParser module to create a PDF document analyzer for the file to be read. Then, use the PDFDocument module to create a virtual PDF document through this analyzer for subsequent document object operations. You can use the PDFResourceManager module to create a resource manager object to manage the parsed content, and you can use the PDFPageAggregator and PDFPageInterpreter modules to create a PDF document page aggregator object and interpreter object, respectively. The LTPage object in the page aggregator object contains most of the document content that users can see; for objects containing main text information, such as horizontal text boxes, you can conveniently use the get_text() method to obtain the text inside. So far, the PDF document format has been unfolded layer by layer, which is convenient for the extraction and normalization of the information inside. Figure 6.5 shows the relationship and data flow between the modules of pdfminer3k.

6.4.1.3 Reading PPT Documents

Python-pptx is similar to python-docx; it is Python-specific for reading and parsing PPT documents (pptx format).

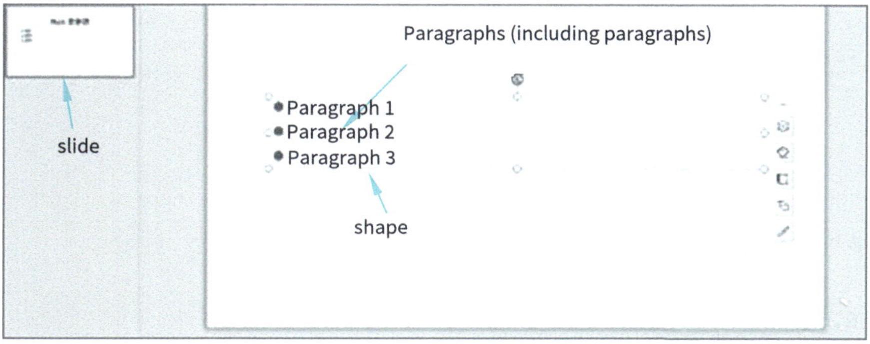

Fig. 6.6 Basic content organization structure of the PPT document in the python-pptx library

The library also has a rich set of document element parsing functions and supports reading slides, text boxes, text paragraphs, and their shape structures. Similarly, for versions before Office 2007, the win32com library needs to be used for storage conversion, changing to pptx format, to achieve more convenient and flexible document parsing.

In the python-pptx library, the Presentation object represents a PPT document, the slides object represents one of the slides, and the shape object is the shape object. The shape should include paragraphs and segmented objects if it is a text box. Figure 6.6 shows the basic content organization structure of the PPT document in this library.

Through the slides object, you can get the content of each slide in the PPT document and then use the related methods in the shape, paragraph, and other objects to extract the main content, and the overall function organization style is similar to python-docx.

6.4.2 *Multiformat Document Parsing Example*

6.4.2.1 Overall Framework

This example extracts the core content of various different format documents uniformly and converts it into a pure text file that is easy to read. Using the format parsing technology mentioned in Sect. 6.2, first detect the document format; then, call the related functions to extract the main content of the document; further use simple word frequency statistics and keyword extraction and other natural language processing technologies to tag the document content, and combine the original file name, path, modification time, original format, and other information of the document itself; write in a pure text file in a unified format, and complete the normalization process. The overall framework of the multiformat document parsing example is shown in Fig. 6.7.

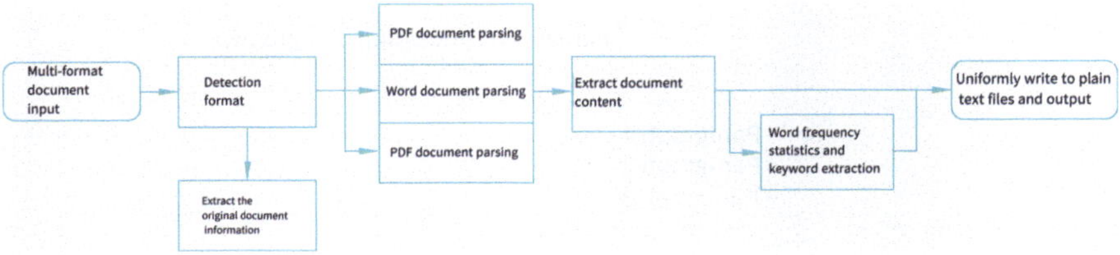

Fig. 6.7 Overall framework of the multiformat document parsing example

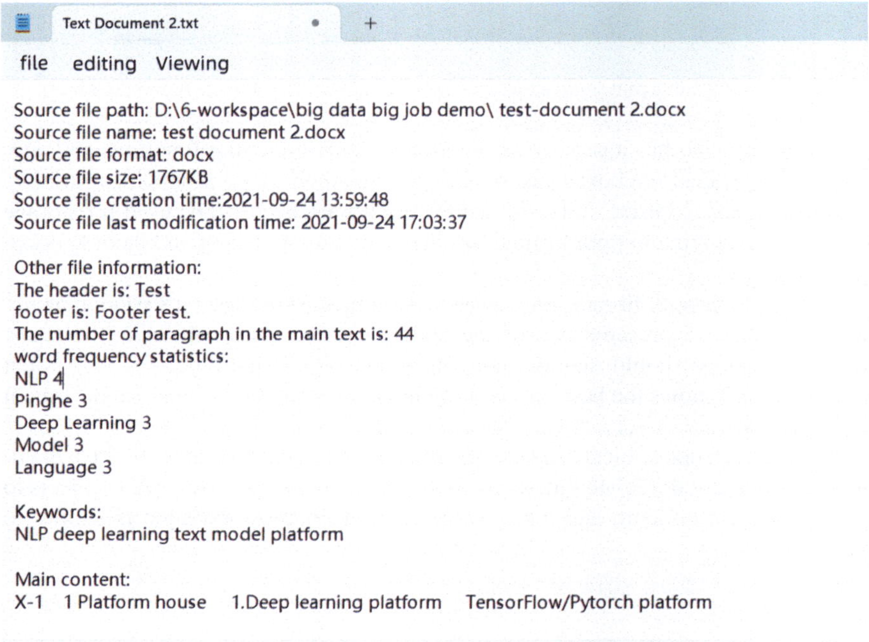

Fig. 6.8 Word document parsing

6.4.2.2 Main Programs

The main program has the following three parts:

1. Word document parsing. Use Python's docx library and win32com library to parse Word documents. The code is as shown in Fig. 6.8.
2. PDF document parsing. Use pdfminer to parse PDF documents. The code is as shown in Fig. 6.9:
3. PPT document parsing. Use pptx library and win32com library to parse PPT documents. The code is as shown in Fig. 6.10:

```
Source file path:D: \6-workspace \big data big job demo\ test-documents5.pdf
Source file name: test documents5.pdf
pdf Source file format: pdf
Source file size: 263.45KB
Source file creation time: 2021-09-24 13:59:48
Source file last modified time: 2021-09-24 17:08:37

Other file information:
Number of pages: 1
Number of pictures: 0
Number of curves: 59
Number of charts: 20
Number of rectangles: 58
Number of horizontal text boxes: 32

Word frequency statistics:
detection 20
novel coronavirus 5
Report 5
2019ncov 4

Keywords:
detection   novel coronavirus   2019ncov   results   Sample

text content:
1-1      1      novel coronavirus      2019 nCov      Nucleic Acid Test Report
1-2      2      NM Sample No:
1-3      3      Sample Type:
```

Fig. 6.9 PDF document parsing

The vast majority of existing data exists in the form of documents, such as PDF documents, Word format papers, PPT format lectures, web format websites, etc. However, people have not fully mined and utilized these data. Using multiformat document parsing and management technology, massive document data can be transformed into a more easily utilized knowledge base through extraction, analysis, and unified management. Multiformat document parsing is the first step in all knowledge mining tasks, the first process in managing massive knowledge data, and a job that is easily overlooked. Especially for libraries and electronic journal databases with a massive demand for electronic document management, the research and application of document parsing technology will become increasingly important.

```
Source file path: D:\6-workspace\bigdatajobdemo\test_document 3. pptx
Source file name: test_document 3. pptx
File format: pptx
File size: 21512.8KB
Source file creation time: 2021-09-26 09:04:17
Source file last modification time: 2021-09-27 22:50:11

Other file information:
number of slides: 55

words Statistics:
document 58
file 47
Object 32
Management: 28
Blockchain 27

Key words:
document    Blockchain    file    PDF    Object
body content:
1-1       1       Document object management
1-2       2       Blockchain technology
```

Fig. 6.10 PPT document parsing

References

1. Liao Yuanting. Research on Word Document Parsing and Desensitization Technology [D]. Xi'an: Xi'an Jiaotong University, 2018.
2. SCHMITT F, GASSENJ, GERHARDS-PADILLA E. PDF Scrutinizer: Detecting JavaScript-based Attacks in PDF Documents. Tenth Annual International Conference on Privacy, Security and Trust, 2012: 104–111.
3. Liu Xianying. Design and Implementation of PDF Text Content Extraction System for Medical Knowledge [D]. Harbin: Harbin Institute of Technology, 2018.
4. Liu Yu, Wang Hui, Wang He. Analysis, Examples and Application Prospects of Atlas Report PDF Files [J]. Computer Knowledge and Technology, 2021, 17(34): 134–140. DOI: https://doi.org/10.14004/j.cnki.ckt.2021.3616.
5. Zhou Yilian. Research on PDF Structure Analysis Technology of Academic Papers [D]. Changsha: Hunan University, 2020. DOI: https://doi.org/10.27135/d.cnki.ghudu.2020.003225.
6. Zhu Lingyu. Research on PDF Document Analysis and Content Desensitization Technology [D]. Chengdu: Southwest Jiaotong University, 2018.
7. KAN L, WEI Y, MUHAMMAD A H, et al. A Multiple Blockchains Architecture on Inter-Blockchain Communication[C]. International Conference on Software Quality, Reliability and Security Companion, IEEE, 2018: 139–145.
8. COLLOMOSSEJ, BUIT, BROWN A, et al. ARCHANGEL: Trusted Archives of Digital Public Documents [C/OL]. 2018[2023-07-05]. DOI: https://doi.org/10.1145/3209280.3229120.
9. Jin Yang Network. Guangzhou Internet Court "Netcom Law Chain" intelligent credit ecosystem online [EB/OL]. (2019-03-30). https://news.ycwb.com/2019-03/30/content_30229720.htm

Chapter 7
Speech Text Recognition

Speech recognition, with speech as the research object, realizes natural language interaction between humans and machines. It is a branch of pattern recognition and involves many fields, such as physiology, psychology, linguistics, computer science, and signal processing. This chapter covers the evolution of speech recognition technologies, from early isolated word recognition systems to modern continuous speech recognition models. Classic algorithms such as Hidden Markov Models (HMM), Gaussian Mixture Models (GMM), and recent deep learning methods like DNN, LSTM, and BiLSTM are introduced, alongside cutting-edge end-to-end models such as CTC and Seq2Seq. This chapter also discusses challenges like robustness, cocktail party effects, and personalized recognition, and explores the latest developments such as hybrid network conformers and DFCNN models. Through practical applications and examples, this chapter illustrates the current state and future potential of speech recognition technology.

7.1 Overview

The goal of speech recognition is to convert speech into text; hence, speech recognition systems are also known as STT (Speech to Text) systems. Speech recognition is a very important first step in achieving natural language interaction between humans and machines. After converting speech into text, the natural language understanding system performs semantic calculations.

Speech recognition tasks can be classified according to the following four dimensions:

1. Based on the vocabulary size, it can be divided into small vocabulary and large vocabulary speech recognition.

H. Zhang, J. Shang, *Natural Language Processing and Applications*,
https://doi.org/10.1007/978-981-97-9739-4_7

2. Based on the speaker, it can be divided into speaker-dependent and speaker-independent speech recognition.
3. Based on the acoustic environment, speech recognition can be divided into studio and different degrees of noise environment.
4. Based on the speaking style, it can be divided into continuous speech or isolated word speech recognition, and it can also be divided into planned or spontaneous speech recognition, for example, "Uh, this thing, no, what is that?"

The combination of these dimensions determines the difficulty of different tasks, for example, the earliest speech recognition systems could only recognize isolated words (there are pauses between words, so it is easy to segment), and the vocabulary was very small (e.g., it could only recognize numbers 0–9). But now, speech recognition systems can complete the task of recognizing large vocabulary in a noisy environment, and the speaking style is continuous; it can handle the differences between different speakers and even handle nonstandard pronunciation (e.g., Mandarin with an accent).

7.1.1 Development History

In 1952, Bell Labs first implemented the Audrey English number recognition system [1], which could recognize the pronunciation of individual numbers 0–9, pronunciation, and the accuracy rate for acquaintances was over 90%. At this time, speech-to-text recognition could effectively recognize isolated words. In the 1970s, Carnegie Mellon University developed the Harpy speech recognition system, which could recognize 1011 words, greatly expanding the previously isolated word recognition. In the 1980s, important technologies such as the Hidden Markov Model (HMM), n-gram language model, etc., which are still in use today, appeared, and speech-to-text recognition entered the continuous word recognition stage. Subsequently, discriminative model training methods MCE and MMI and model self-adaptation methods MAP and MLLR appeared, further increasing the vocabulary of continuous word recognition. In 2006, Hinton proposed the deep belief network (DBN) [2]. In 2009, Hinton and his student Mohammed applied deep neural networks to speech recognition, achieving success on the small vocabulary continuous speech recognition task TIMIT. Since then, deep learning models have been applied to the field of speech-to-text recognition. In recent years, classic end-to-end speech-to-text recognition models represented by CTC and Seq2Seq have continued to develop. The development history of speech-to-text recognition is shown in Fig. 7.1.

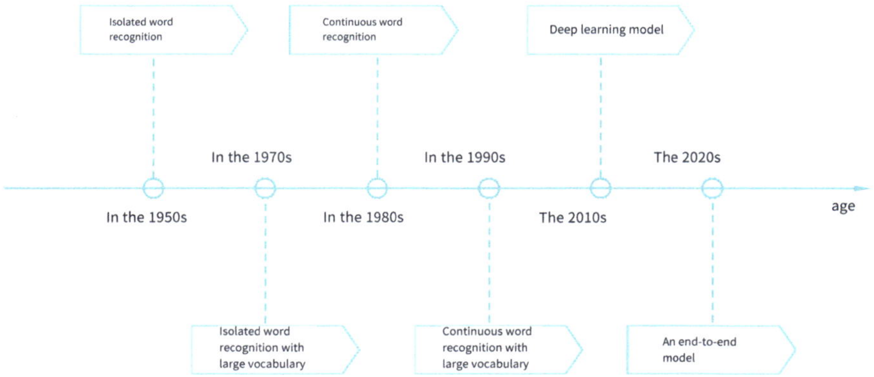

Fig. 7.1 Development history of speech-to-text recognition

7.1.2 Basic Principles

Sound is a wave. The first step in speech-to-text recognition is to record the sound and obtain the corresponding sound waveform. To analyze the sound waveform, it needs to be framed first. That is, the sound is cut into equal-length segments, and each segment is called a frame. Figure 7.2 shows the framing result of the sound waveform of "Hi, everyone."

It is worth noting that the framing here is not directly cut; there is a certain overlap between frames. After framing, the waveform of the small segment needs to be transformed so that the computer can process it; this process is called acoustic feature extraction. There are generally two ways of acoustic feature extraction, namely, MFCC (MelFrequency CepstralCoefficient) feature extraction and deep learning feature extraction. After framing, both need to perform discrete Fourier transform and square, perform Mel filtering, take logarithm and calculate cepstral coefficients, then perform Fourier transform, use a neural network to recognize characters, and finally get the mapping diagram. After the acoustic feature extraction is completed, the sound becomes an M-row N-column matrix called the observation sequence, where M is the feature dimension, and N is the total number of frames. The MFCC feature map obtained by extracting the sound features of the previously obtained framed sound waveform is shown in Fig. 7.3.

Next, the matrix is converted into the corresponding text. The specific operation can be summarized as follows: recognizing frames as states, combining states, combining them into phonemes, and combining phonemes into words. Among them, a phoneme (phone) is the smallest unit in speech, and based on the pronunciation action analysis in the syllable, one action constitutes one phoneme. In Chinese, the phoneme set is generally directly composed of the initials and finals of Chinese Pinyin, and the commonly used phoneme set in English is a set of 39 phonemes from Carnegie Mellon University.

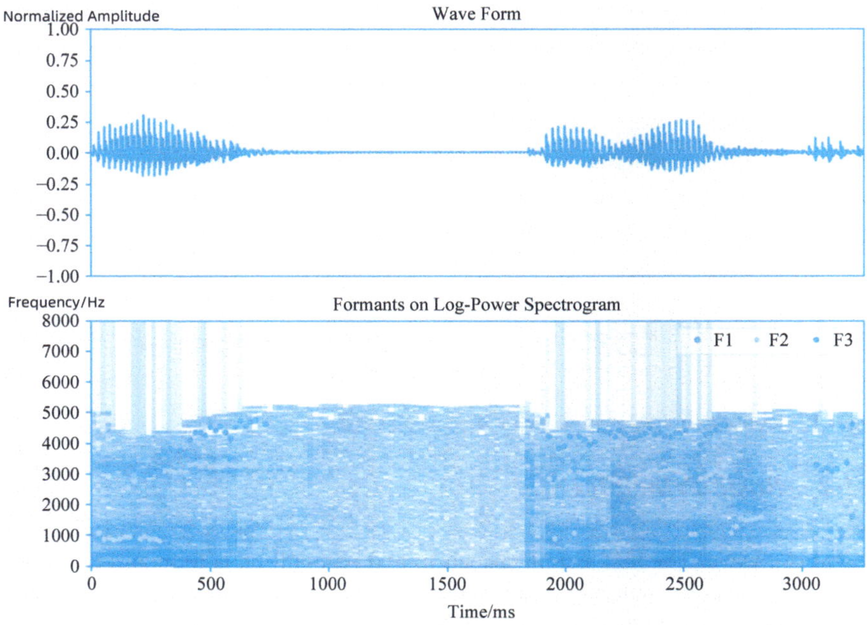

Fig. 7.2 Framing result of the sound waveform of "Hi, everyone"

To associate frames with states, it is necessary to calculate the probability of the frame under each state. The overall structure of speech-to-text recognition is shown in Fig. 7.4.

The state with the highest probability is the state to which the frame belongs. This probability is the probability corresponding to each frame and each state, also known as observation probabilities. In addition to observation probabilities, the recognition process also requires the probability of each state transitioning to itself or to the next state, known as transition probabilities. The reading and calculation of observation and transition probabilities come from the acoustic model; the specific content is introduced in Sect. 7.2.2. In order to match the text corresponding to the speech, it is also necessary to obtain the probability obtained from the language model based on language statistics. These three probabilities are combined to get the final cumulative probability, and the entire speech recognition process is completed. The above process is expressed in mathematical formulas as follows:

$$W^{*} = \arg\max_{W} P(W|Y) = \arg\max_{W} \frac{P(Y|W)P(W)}{P(Y)} \approx \arg\max_{W} P(Y|W)P(W) \tag{7.1}$$

In Eq. (7.1), W represents the text sequence, Y represents the speech input, and arg max represents the function to find the parameter (set). Equation (7.1) indicates that the goal of speech recognition is to find the text sequence with the highest

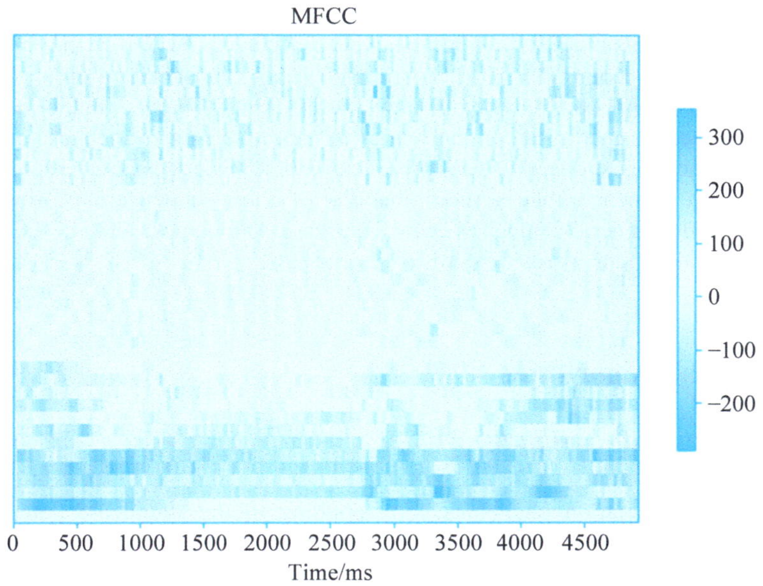

Fig. 7.3 MFCC feature map obtained from sound feature extraction

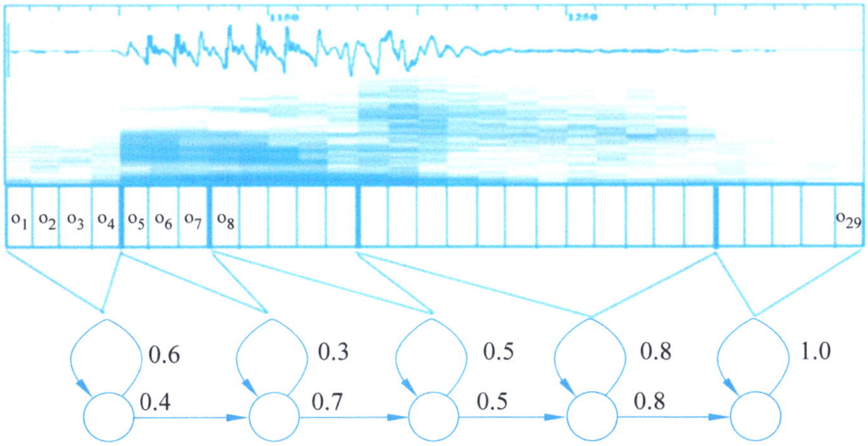

Fig. 7.4 Overall structure of speech-to-text recognition

probability given the speech input. According to Bayes' theorem, the second expression in Eq. (7.1) can be obtained, where the denominator represents the probability of this speech occurring, which has no parameter relationship with the text sequence to be solved, and can be ignored when solving, thus obtaining the third expression. In the third expression, $P(Y|W)$ represents the probability of this speech occurring given a text sequence, which is the acoustic model in speech recognition, and $P(W)$ represents the probability of this text sequence occurring, which is the language model in speech recognition. The acoustic model is a representation of knowledge about acoustics, phonetics, environmental variables, and differences in the speaker's gender and accent, that is, the probability of this speech being emitted after a given text, while the language model is a representation of knowledge about the composition of a text sequence, that is, the probability of a text sequence appearing. The acoustic model generally maps speech features to phonemes, and the language model generally maps words to words and sentences.

7.2 Classic Algorithms

7.2.1 Classical Language Model

Once the pronunciation phonemes corresponding to the sound are determined through the acoustic model, the language model needs to consider what the text with the highest probability corresponding to such a pronunciation phoneme is. Taking the "Hi, everyone" speech in Sect. 7.1.2 as an example, after extracting the sound features of "hai," it generally corresponds to the word "hi," not "harm," "still," and "sea" because in the language model, the probability of "hai" corresponding to the word "hi" is the highest. More complex homophonic word problems also need to be handled through the language model.

7.2.1.1 n-gram Language Model

When a piece of speech is processed to the sound feature or phoneme stage, it will be transformed into a sequence in the language model. It is mathematically described as follows: T is composed of word sequences $A1, A2\ldots An$, and the text probability $P_{(T)}$ should be maximized, where $P_{(T)}$ is

$$P(T) = P(A1A2\ldots An) = P(A1)P(A2|A1)\ldots P(An|A1, A2, \ldots, An-1) \tag{7.2}$$

Obviously, such calculation is very tedious. To simplify the calculation of the language model, the Markov assumption is introduced:

In a series of events, the probability of an event occurring is only related to its previous m events. When $m = 2$, we can get

$$P(T) = P(A1A2\ldots An) = P(A1)P(A2|A1)P(A3|A2)\ldots P(An|An-1) \tag{7.3}$$

7.2.1.2 RNN Language Model

Another type of language model is built using recurrent neural networks (RNN) [3]. The task of the language model is to predict a sequence, and RNN is naturally used to solve sequence problems. In the RNN model, theoretically, all previous words will affect the prediction of the current word. In practical implementation, phoneme information is input, and the most likely text sequence is output. In the process of inputting words into the recurrent neural network one by one, each time a word is input, the recurrent neural network outputs the most likely next word so far.

7.2.2 Classic Acoustic Model

The acoustic model can be understood as modeling speech, which can convert speech input into acoustic representation output; more accurately, it gives the probability that speech belongs to a certain acoustic symbol. The acoustic model is more complex overall, and the traditional model, deep learning-based model, and end-to-end model are explained separately below.

7.2.2.1 Traditional Model

The traditional acoustic model is generally based on GMM (Gaussian Mixture Model) and HMM (Hidden Markov Model) [4]. Figure 7.5 is the overall processing flow of the acoustic GMM–HMM.

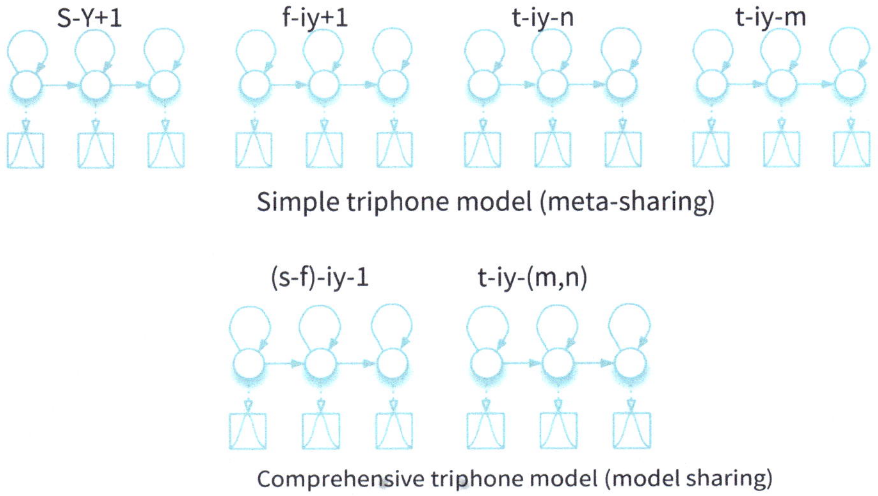

Fig. 7.5 Overall processing flow of the acoustic GMM–HMM

GMM is a probability model that mixes data from multiple normal distributions (Gaussian distributions). In speech recognition, the previously obtained sound features can be further transformed into the state of the sound through the Gaussian mixture model. In actual GMM training, the EM (Expectation-Maximization) algorithm is usually used for iterative optimization to obtain the weighting coefficients and the mean and variance parameters of each Gaussian function in the GMM. As a statistical model based on Fourier spectrum speech features, GMM has played an important role in traditional speech recognition systems. Its disadvantage is that it cannot consider speech sequence information, and Gaussian mixture distribution is also difficult to fit nonlinear or approximately nonlinear data features. Therefore, when the concept of state is introduced into the acoustic model, a new type of acoustic model is created—HMM.

HMM is also a statistical model, which is used to describe a Markov process with hidden unknown parameters. The Markov process is that the current state is only related to the previous one or a few states, a process that is unrelated to the previous state. The previously obtained state needs to be determined by the HMM whether to retain or jump. When a Markov process contains hidden unknown parameters, such a model is called a hidden Markov model. The core concept of HMM is the state; the state itself is a discrete random variable, and each state of the Markov chain has added uncertainty or statistical distribution, making the HMM a double stochastic process. The main content of HMM includes parameter characteristics, simulation methods, maximum likelihood estimation of parameters, EM estimation algorithm, Viterbi state decoding algorithm, and other detailed knowledge. In HMM, the Viterbi algorithm is used to find the shortest path in a directed acyclic graph. The state obtained by further processing the sound features will form a huge state network, and the Viterbi algorithm can easily find the shortest path in the state network, which is the final text sequence.

7.2.2.2 Deep Learning-Based Models

In 2006, Hinton proposed the deep belief network (DBN), which revived the research of deep neural networks (DNN) [5]. In 2009, Hinton applied DNN to the acoustic modeling of speech, achieving the best results on TIMIT at the time. At the end of 2011, Dong Yu and others from Microsoft Research applied DNN technology to large vocabulary continuous speech recognition tasks, greatly reducing the error rate of speech recognition. Since then, speech recognition has entered the DNN–HMM era.

DNN–HMM mainly uses the DNN model to replace the original GMM to model each state. Simply put, DNN gives the state probability corresponding to a series of input features. The advantage it brings is that it no longer needs to make assumptions about the distribution of speech data; the concatenation of adjacent speech frames also contains the temporal structure information of speech, making the model's classification probability for states significantly improved. At the same time,

DNN also has a strong environmental learning ability, which effectively improves the robustness of the model to noise and accents. The structure of the DNN model is shown in Fig. 7.6. This model includes one input layer, three hidden layers, and one Softmax layer.

Another deep learning-based language model is Long Short-Term Memory (LSTM) [6], and its structure is shown in Fig. 7.7. Since speech signals are continuous, not only are there no clear boundaries between phonemes, syllables, and words but each pronunciation unit is also affected by the context. Although frame concatenation can increase context information, it is still not enough for speech. RNN can remember more historical information, which is beneficial for modeling the context information of speech signals, but simple RNN exists.

The problems of gradient explosion and gradient vanishing make it impossible to directly apply to speech signal modeling. The solution to these problems is LSTM. LSTM can better control the flow and transmission of information through the input gate, output gate, and forget gate and has the ability to remember for long

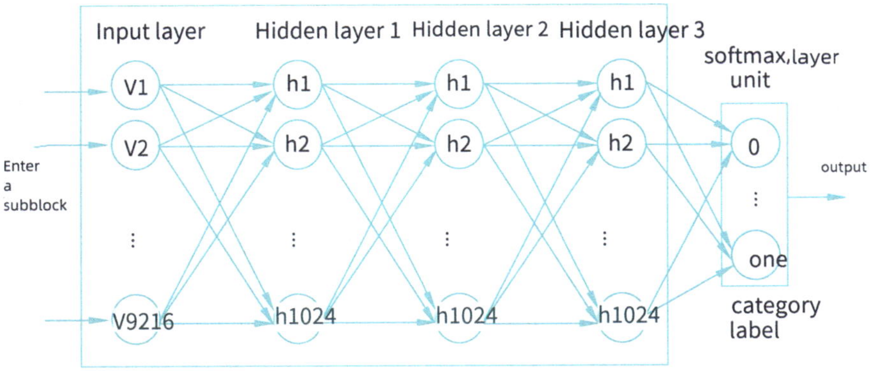

Fig. 7.6 Structure of the DNN model

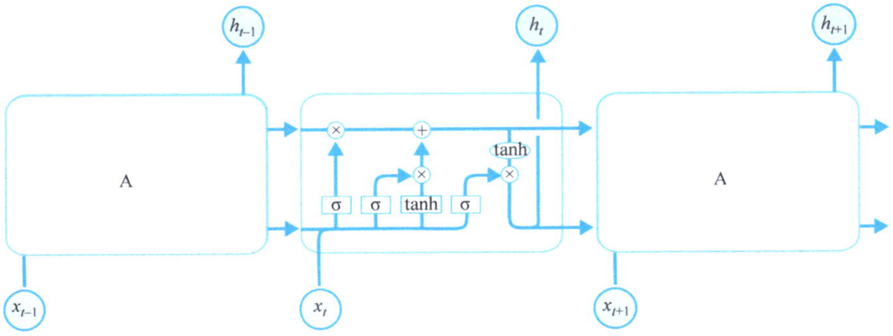

Fig. 7.7 Structure of the LSTM model

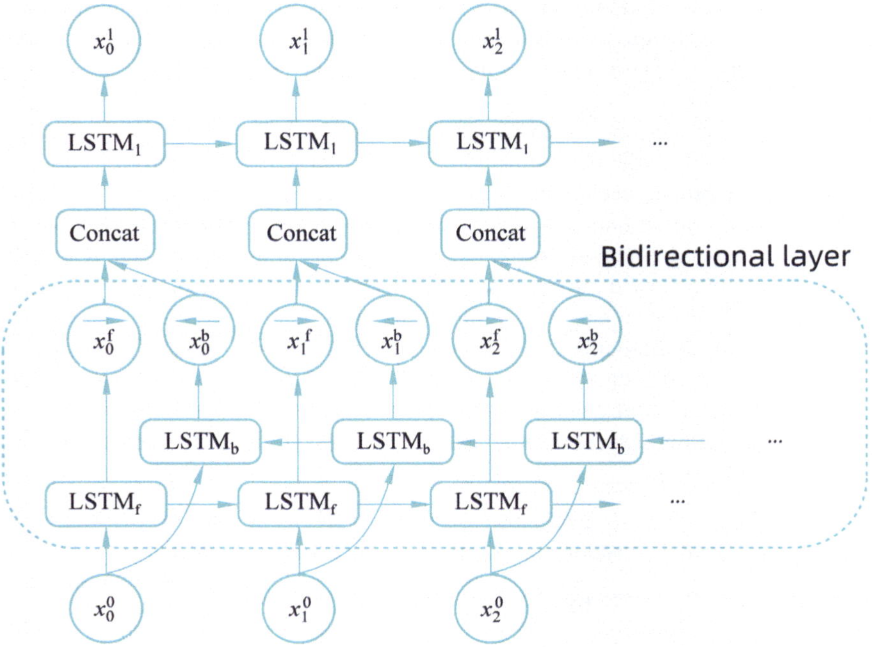

Fig. 7.8 Structure of the BiLSTM model

and short times. Although the computational complexity of LSTM is higher than that of DNN, its overall performance is about 20% more stable than DNN.

Based on LSTM, further improvements are made, not only considering the impact of the historical information of the speech signal on the current frame but also considering the impact of future information on the current frame, thus constructing BiLSTM (Bi-Long Short-Term Memory). The structure of BiLSTM [7] is shown in Fig. 7.8. In BiLSTM, there are two information transmission processes along the time axis, which can more fully consider the impact of the context on the current speech frame, greatly improving the accuracy of speech state classification. The cost of BiLSTM considering future information is that sentence-level updates are required, and the convergence speed of model training is relatively slow.

7.2.2.3 End-to-End Model

The breakthrough of end-to-end technology makes speech-to-text recognition no longer need HMM to describe the changes in the internal state of phonemes but unifies all modules of speech recognition into a neural network model, making the entire process simpler, more efficient, and accurate.

The end-to-end method of speech recognition mainly changes the cost function, but the structure of the neural network model has not changed much. In general, end-to-end technology solves the problem that the length of the input sequence is much greater than the length of the output sequence.

End-to-end technology is mainly divided into two categories: one is the CTC (Connectionist Temporal Classification) method [8], and the other is the Seq2Seq method [9]. In the traditional speech recognition DNN–HMM structure, each frame input corresponds to a label category, and the labels need to be iteratively adjusted to ensure a more accurate alignment. When using CTC as the loss function for the acoustic model sequence, there is no need to pre-align the data; only an input sequence and an output sequence are needed for training. CTC is concerned with whether the predicted output sequence is close to the real sequence and does not care whether each result in the predicted output sequence aligns exactly with the input sequence in time. The CTC modeling unit is a phoneme or a word, so it introduces the concept of blank. For a speech, the final output of CTC is a peak sequence, the position of the peak corresponds to the label of the modeling unit, and other positions are blank.

The Seq2Seq method was originally mainly used in the field of machine translation. In 2017, Google applied it to the field of speech recognition and achieved very good results, reducing the word error rate to 5.6%. As shown in Fig. 7.9, the Seq2Seq model consists of three modules. The Encoder module, similar to the standard acoustic model, inputs the time–frequency features of the speech signal, maps it into high-level features through a series of neural networks, and then passes it to the Attention module. The Attention module uses the henc feature to learn the alignment between the input x and the predicted subunit, and the subunit can be a phoneme or a word. Finally, the output of the Attention module is passed to the Decoder

Fig. 7.9 Sequence-to-sequence model diagram

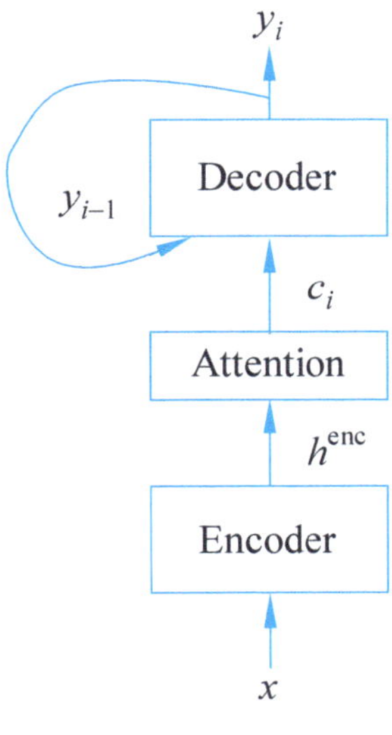

module to generate a series of hypothetical word probability distributions, similar to traditional language models. This is the structure of the Seq2Seq model.

7.3 Latest Progress

In fact, the speech recognition products on the market do not meet expectations in many real application scenarios, especially in far-field, Chinese mixed with English, and when there are people talking nearby. Therefore, there are still many problems to be researched in the field of speech recognition.

1. Robustness problem

   The robustness of current speech recognition systems is not high, and it depends on increasing data (including synthesized simulated data) to improve robustness. In a quiet environment or a scene that matches the training set, most manufacturers can achieve a CER (Character Error Rate) of less than 5%, but in far-field, low signal-to-noise ratio, out-domain, heavy accent, and other situations, the accuracy rate drops significantly. This is particularly evident for systems based on deep learning. The limitation of these methods is that they do not perform well in uncovered situations. In the history of speech recognition research, people have long recognized this problem and developed many adaptive algorithms, trying to adapt according to changes in the scene and environment. At present, adaptive algorithms have played a certain role, but they cannot completely solve the problem of robustness.

2. Problems with the complete end-to-end method

   Some related articles on ICASSP suggest that with a large amount of training data, end-to-end models can potentially outperform hybrid models on specific speech search tasks. However, the performance is still lacking in real-world scenarios where the tail-end search words have not been encountered. This indicates that these models have strong memory but have yet to develop their generalization capabilities fully. The progress made is significant, as the gap between previous end-to-end systems and hybrid models was substantial. Now, this gap is narrowing, and in some scenarios, the end-to-end model can even surpass the hybrid model, marking significant progress. The potential for the end-to-end system to replace the hybrid model is an intriguing question that remains to be answered.

   The current end-to-end system is based on the CTC framework, and the Seq2Seq framework is based on the attention mechanism. Google's paper uses the attention-based framework, less used in practice. The CTC model is used more. Tencent's products include both the CTC and the hybrid models, and there is not much difference in performance.

   The advantage of CTC is that it can use larger modeling units, and the disadvantage is that there is a problem of random delay, that is, the time when the result is produced is not known in advance. The consequence of random delay is that it is difficult to break sentences, leading to increased delay. Therefore, most interactive systems, such as voice assistant systems, still use hybrid systems. For products that do not require real time, such as YouTube's subtitle generator, because it can be offline, it does not matter if there is a delay.

3. Cocktail party problem
   Current speech recognition technology can recognize what a person is saying with high accuracy, but when the number of speakers is two or more, the speech recognition rate will significantly decrease. This difficult problem is known as the cocktail party problem. One method to solve the cocktail party problem is permutation-invariant training. In addition, other important methods have been proposed, such as MERL's deep clustering method and Columbia University's deep attractor network.

4. Low-resource speech recognition problem
   The current commercial speech recognition system requires tens of thousands or even tens of thousands of hours of annotated audio. The annotation cost is roughly 400–600 yuan/hour, and 10,000 hours of annotated audio requires about 5 million yuan. Therefore, solving the problem of low-resource speech recognition algorithm modeling can greatly reduce the cost of landing speech recognition algorithms.

5. Personalized speech recognition problem
   Model adaptation is based on user behavior habits, specifically improving individual recognition effects. Of course, this requires a prerequisite—the effective protection of user data and privacy.

7.3.1 DFCNN Model

Currently, mainstream speech recognition frameworks consist of three parts: acoustic model, language model, and decoder; some frameworks also include front-end processing and post-processing. Due to the complexity of Chinese speech recognition, domestic research on acoustic models is progressing faster, with the mainstream direction being deeper and more complex neural network technologies integrated with end-to-end technologies. In 2018, Wang Hai kun and others [10] proposed the Deep Fully Convolutional Neural Network (DFCNN). DFCNN uses a large number of convolutional layers to directly model the entire sentence of speech signals, mainly drawing on the network configuration of image recognition; each convolutional layer uses a small convolution kernel and adds a pooling layer after multiple convolutional layers, thus accumulating a large number of convolutional layer–pooling layer pairs, so that more historical information can be seen.

In 2021, Bo Caitong and others [11] first applied the distilled federated learning technology to the training of robust speech recognition models. On the basis of balancing performance and communication consumption, they locally distilled the downlink global model parameters and global average Logits and local model parameters, further overcoming the distribution problem of nonindependent and identically distributed data. They proposed an improved distilled federated learning algorithm with personalized local distillation to solve the problems of nonindependent and identically distributed data and lack of personalization in the model encountered when applying distilled federated learning technology in robust speech recognition. The trained robust speech recognition model with high robustness was applied in military equipment control tasks, with an accuracy rate of up to 92%, and it can complete equipment control tasks.

Due to the problems of traditional language model n-gram, such as ignoring word semantic similarity and large parameters, Hu Zhangfang and others [12] proposed a new type of speech recognition system, using Chinese syllables (pinyin) as intermediate characters and deep feedforward sequence memory.

Deep feedforward sequential memory network (DFSMN) as an acoustic model performs the task of speech to Chinese syllable conversion and then interprets the conversion of Pinyin to Chinese characters as a translation task, introducing Transformer as a language model; at the same time, it proposes a simple method to reduce the computational complexity of Transformer, introducing Hadamard matrix for filtering when calculating attention weights, discarding parameters below the threshold, making the model decoding speed faster. Experiments on datasets such as Aishell-1 and Thchs30 show that compared with the DFSMN combined with the 3-gram model, the character error rate of the speech recognition system based on DFSMN and improved Transformer dropped by 3.2% on the optimal model, reaching a character error rate of 11.8%, compared with the BLSTM model speech recognition system, whose character error rate relatively decreased by 7.1%.

7.3.2 *Hybrid Network Conformer*

Since the introduction of Transformer, it has been brilliant in the field of natural language processing. At the same time, convolution also plays a vital role in the visual field. However, the characteristics of these two models are different: for the Transformer, the cascaded self-attention mechanism can capture long-distance feature information, but it weakens local feature information, while the convolution operation of CNN is very good at capturing local feature information, but it is very difficult to capture global feature information in the image, as shown in Fig. 7.10.

The problem with Transformer is that it blurs the local feature information of the foreground and background, for example, the edge information of the peacock (foreground) and leaves (background) in Fig. 7.10 has been lost. Similarly, the global feature information obtained by CNN is also very small, for example, whether it is the shallow or deep layer of the network, the outline of the peacock is not fully displayed.

The emergence of Conformer is to solve this problem: the CNN branch is based on ResNet, and the Transformer branch is based on ViT. The combination of local feature information from the convolution branch and global feature information from the Transformer branch can obtain better feature representation. The structure of the Conformer model is shown in Fig. 7.11.

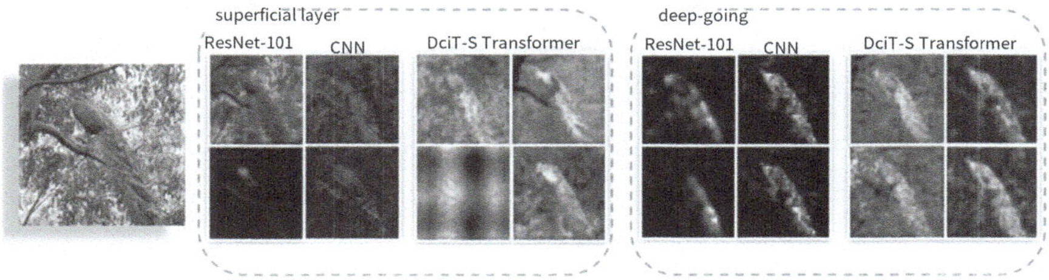

Fig. 7.10 Feature information diagram of different models

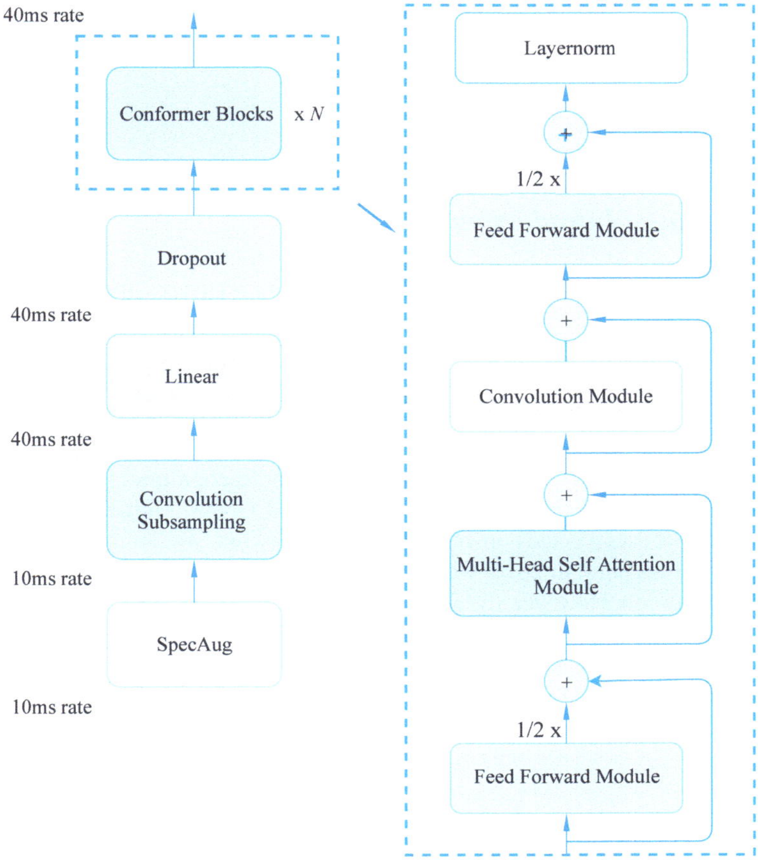

Fig. 7.11 Conformer model structure

The calculation formula of the Conformer block is shown in Eq. (7.4), where FFN is the feedforward module, MHSA is the multihead self-attention module, and Conv is the 2 × 2 convolution module. This is actually like a sandwich structure, because the front and back are feedforward modules, which is inspired by MacaronNet, that is, using two FFNs, each FFN contributing half of the value.

$$\tilde{x}_i = x_i + \frac{1}{2}\mathrm{FFN}(x_i)$$

$$x_i^{'} = \tilde{x}_i + \mathrm{MHSA}(\tilde{x}_i)$$

$$x_i^{''} = x_i^{'} + \mathrm{Conv}(x_i^{'})$$

$$y_i = \mathrm{Layernorm}\left(x_i^{''} + \frac{1}{2}\mathrm{FFN}(x_i^{''})\right) \tag{7.4}$$

7.4 Application and Analysis

7.4.1 Dataset Introduction

The data used in this sample program comes from the THCHS-30 dataset.

7.4.2 Model Introduction

The model uses MASR (Magical Automatic Speech Recognition), which is simple and practical, with an internal structure of stacked CNNs. Unlike other network models, MASR uses a gated convolutional neural network, similar to the Wav2Letter proposed by Facebook in 2016, which uses only convolutional neural networks for speech recognition. However, the activation function used by MASR is not ReLU or HardTanh, but GLU (Gated Linear Unit), hence it is called a gated convolutional network. Experimental results show that the convergence speed of using GLU is faster than that of HardTanh.

After training, the final word error rate of the model is around 0.06751.

7.4.3 Code Sample

```
self.conv = ConvStack(feat_size = feat_size, conv_out_channels=cnn_size)
self.rnn = RNNStack(i_size= self.conv.output_dim, h_size= rnn_size, num_rnn_
layers = num_rnn_layers)
```

Consumption time :70ms; Identification results: In recent years, not only do I use books to give my daughter pressure, but also persuade relatives and friends not to give her lucky money and send pressure books, score :94

Fig. 7.12 Model output result example

7.4.4 Model Effect

Input speech (wav format) file, after the recognition program extracts its MFCC feature values, is then fed into the model, and the output is decoded by the CTC decoder. The final recognition result is obtained, and the time required from the input speech file to the output result and the score of the CTC decoding are calculated. The model output result example is shown in Fig. 7.12.

References

1. BREMSDJ. Automatic Speech Recognition(ASR)Processing Using Confidence measures[J]. Journal of the Acoustical Society of America, 1995, 102(1):25–32
2. HINTON G E, OSINDERO S, TEH Y W. A Fast Learning Algorithm for Deep Belief Nets[J]. Neural Computation, 2006, 18(7): 1527–1554.
3. GRAVES A, MOHAMED A, HINTON G. Speech Recognition with Deep Recurrent Neural Networks[C]. 2013 IEEE International Conference on Acoustics, Speech and Signal Processing. IEEE, 2013.
4. XUAN G, ZHANG W, CHAIP. EM Algorithms of Gaussian Mixture Model and Hidden Markov Model[C]. Proceedings 2001 International Conference on Image Processing (Cat. No. 01CH37205). IEEE, 2001.
5. SELTZER M L, YU D, WANG Y. An Investigation of Deep Neural Networks for Noise Robust Speech Recognition[C]. 2013 IEEE International Conference on Acoustics, Speech and Signal Processing. IEEE, 2013.
6. SøNDERBY S K, SøNDERBY C K, NIELSEN H, et al. Convolutional LSTM Networks for Subcellular Localization of Proteins[C]. International Conference on Algorithms for Computational Biology. Cham: Springer, 2015.
7. GRAVES A, JAITLY N, MOHAMED A. Hybrid Speech Recognition with Deep Bidirectional LSTM[C]. 2013 IEEE Workshop on Automatic Speech Recognition and Understanding. IEEE, 2013.
8. CHIU C C, SAINATH T N, WU Y, et al. State-of-the-art Speech Recognition with Sequence-to-Sequence models[C]. 2018IEEE International Conference on Acoustics, Speech and Signal Processing (ICASSP). IEEE, 2018.
9. GRAVES A, FERNÁNDEZ S, GOMEZ F, et al. Connectionist Temporal Classification: Labelling Unsegmented Sequence Data with Recurrent Neural Networks[C]. Proceedings ofthe23rd International Conference on Machine Learning. 2006.
10. Wang Haikun, Pan Jia, Liu Cong. Research development and forcast of automatic speech recognition technologies [J]. Telecommunications Science, 2018, 34(2): 1–11.
11. Bo Caitong, Cui Xiaolong, Li Ai. Robust Speech Recognition Technology Based on Distilled Federated Learning[J]. Computer Engineering, 2022(10): 48–53.
12. Hu Zhang fang, Jian Fang, Tang Shanshan, et al. DFSMN-t: Chinese Speech Recognition Combining Strong Language Model Transformer[J]. Computer Engineering and Applications, 2022, 58(9): 187–194.

Chapter 8
Image Semantic Representation and Character Recognition

Image semantic representation focuses on how computer systems interpret visual information to achieve a level of understanding comparable to human perception. It addresses the extraction of relevant information from images to perform specific tasks and convert this data into meaningful interpretations. Character recognition involves the identification of printed or handwritten text, transforming it into a digital format for storage and further processing. Both fields encompass a variety of methods, devices, and implementations crucial for understanding and processing visual and textual data.

8.1 Picture Caption

8.1.1 Background

Image captioning describes images in natural language, generates meaningful syntax, and corrects visual content using a visual understanding system and language model. Neuroscience research has only established the connection between human vision and language generation in the past few years. Similarly, in the field of artificial intelligence, designing an architecture capable of handling images and generating language is a very new problem. The goal of these research works is to find the most effective process to handle input images, represent their content, and transform them into a string of words by generating connections between visual and textual elements while maintaining the fluency of the language. In standard situations, image captioning is a problem from image to sequence, where the input is pixels. These pixels are encoded into one or more feature vectors, called visual encoding. Then, a sequence of words or subwords decoded according to a given vocabulary is generated using a language model [1].

H. Zhang, J. Shang, *Natural Language Processing and Applications*,
https://doi.org/10.1007/978-981-97-9739-4_8

Traditional image captioning methods are of two types: one is a template-based image description method, which detects objects, actions, scenes, etc., in the image based on visual dependency tables and then generates text using fixed sentence templates, and the second is a retrieval-based image description method, which collects a large number of images from the network and annotates titles, descriptions, etc., to build a database and finds the most similar matching image by calculating the global similarity between the image to be described and the network database image.

8.1.2 *Technical Analysis*

Current visual encoding methods can be divided into four categories: non-attention methods based on global CNN features, additional attention methods based on grid or regional embedding, graph-based visual encoding methods, and self-attention methods based on the Transformer paradigm.

8.1.2.1 Non-attention Methods Based on Global CNN Features

Global CNN features refer to directly inputting the entire image into the CNN. In the simplest method, the activation function of the last layer of the CNN is used to extract high-level and fixed-size feature representations, which are then used as conditional elements of the language model. The advantage of this type of method lies in its simplicity and compactness, including the ability to extract and condense information from the entire input and consider all contexts of the image. However, its disadvantages are also obvious. This type of method leads to excessive compression of information, and all objects and regions are fused into one vector.

This makes it difficult for the subtitle model to generate specific and detailed descriptions.

8.1.2.2 Additional Attention Methods Based on Grid or Region Embedding

Additional Attention Methods

Additional attention methods can intuitively be seen as a weighted average method. The attention mechanism was originally used to model the relationship between two element sequences (i.e., the relationship between the hidden states of the cyclic encoder and decoder), and later it was found that it could also be used to link a set of visual representations with the hidden states of the language model.

Additional attention methods based on grid or region embedding are shown in Fig. 8.1.

Fig. 8.1 Based on grid or region embedding

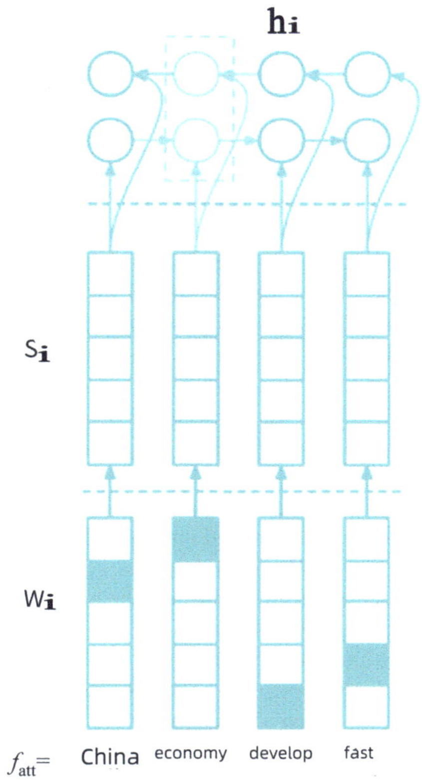

The formal definition of these methods is given below. Given two sets of vectors:

Input vectors $\{x_1, x_2\ldots x_n\}$ and hidden sequence vectors $\{h_1, h_2,\ldots h_m\}$, the additional attention score between h_i and x_j is calculated as follows:

$$f_{att}\left(h_i,\ x_j\right) = W_3^T \tan h\left(W_1 h_i + W_2 x_j\right) \tag{8.1}$$

Participating in Convolutional Activation

Xu et al. [2] first proposed the method of using additional attention on the spatial output grid of the convolutional layer, allowing the model to selectively focus on certain elements of the grid by choosing a subset of features for each generated word. The specific process is as follows:

The model first extracts the activation values of the last convolutional layer of the VGG network and then uses additional attention to calculate the weight of each grid element. The attention mechanism enhances and suppresses the extracted feature map, interpreted as the relative importance of this element for generating the next word, serving as the input data for the subsequent LSTM model.

Review Network

The main idea of the review network proposed by Yang et al. [3] is as follows: The attention mechanism only pays attention to local factors each time, without considering the impact of global factors on prediction. Therefore, they use the feature map as the global information of the image and then obtain a thought vector that can better represent the global information of the image than the feature map through the LSTM unit, which is more compact and abstract.

Utilizing Human Attention Characteristics

Attention is generally divided into two types. One is top-down conscious attention, referred to as focused attention. Focused attention refers to the attention that is purposefully task-dependent and consciously focused on a certain object. Another is the unconscious attention from bottom to top, referred to as saliency-based attention or stimulus-based attention. Therefore, saliency information (i.e., what humans pay more attention to in a scene) can be integrated to guide subtitle generation. This idea was first proposed by Sugano and Bulling, who used the information of human eye gaze for image description.

8.1.2.3 Graph-Based Visual Encoding Method

In order to further encode the relationship between image regions, some researchers consider using graphs to model image regions, enriching the representation of images by introducing semantic and spatial connections. Graph encoding provides a mechanism for utilizing the relationships between detected targets, allowing information exchange between adjacent nodes, thus completing local exchange. It can seamlessly integrate external semantic information, and the manually constructed graph structure can limit the interaction of visual features.

Semantic Relationship Graph and Spatial Relationship Graph

Semantic relationship graphs and spatial relationship graphs use graph convolutional networks (GCN) to integrate the semantic and spatial relationships between targets. Semantic relationship graphs are obtained using a classifier pre-trained on Visual Genome, which predicts actions or interactions between target pairs. Spatial relationship graphs are inferred from the geometric measures between the bounding boxes of target pairs (i.e., intersection over union, relative distance, and angle).

Hierarchical Tree

Yao et al. [4] first represented the image as a tree-like hierarchical structure, where the root node represents the whole image, the intermediate nodes represent image regions and their contained subregions, and the leaf nodes represent segmented objects within the region. Then, the image tree is fed into Tree LSTM to obtain image encoding.

8.1.2.4 Self-Attention Method Based on Transformer

The self-attention mechanism uses methods based on regions, image patches, or early fusion of image text and can be seen as a complete graph representation that connects all elements. The self-attention mechanism was first proposed by Vaswani in 2017. The Transformer proposed by him has dominated the field of natural language processing and has been widely applied to computer vision in recent years. It is a special attention mechanism where each element in a set is connected to all other elements, and a new representation of the elements within the set is calculated through residual connections.

Early Self-Attention Models

The first self-attention model applied to image captioning was proposed by Yang et al. [5], using a self-attention model to encode the relationships between features generated by the target detector. Later, Li et al. [6] proposed a Transformer model, which includes a visual encoder that encodes region features and a semantic encoder that obtains information from an external marker. Both of these encoders are based on self-attention and FFN layers. The decoder fuses the outputs of the two encoders through a gate mechanism that controls the propagation of visual and semantic information. The bidirectional encoder maps the original input to a highly abstract representation, and then the decoder merges multimodal information to generate subtitles word by word.

Variants of the Self-Attention Model

Here are some variants of the self-attention model.

1. Geometry-aware encoding. Herdade et al. [7] introduced a modified version of self-attention that takes into account the spatial relationships between regions. Specifically, additional geometric weights are calculated between objects and used to adjust self-attention weights. They proposed a normalized version of geometry-aware self-attention based on the relative geometric relationships between input objects.

2. Attention on attention. Huang et al. [8] proposed an extension of the attention operator, where the final attention information is weighted by context-guided gates. Specifically, the output of self-attention is connected with the query, and then the information vector and gate vector are multiplied to get the final information. This mechanism is used in their designed visual encoder to improve visual features.
3. Memory-augmented attention. Cornia et al. [9] proposed a new structure based on Transformer, where the self-attention calculation of each encoder layer is enhanced with a set of memory vectors. It expands the key–value pair collection with additional slots, which are learned during training and can encode multi-level visual representations using prior knowledge.

Visual Transformer

In addition to using object regions as input sequences for attention mechanisms like the above methods, the Transformer structure can also be directly applied to image blocks, thereby reducing or completely discarding convolution operations.

Language models are mainly used to predict the probability of a given word sequence appearing in a sentence; a key component of many natural language is processing tasks, giving computers the ability to understand and process language problems in a random process.

Typically, for a sequence containing n words, the language model of image captioning will assign a probability to this sequence, as shown in Eq. (8.2):

$$P(y_1, y_2, \ldots y_n | X) = \prod P(y_i | y_1, y_2, \ldots, y_{i-1}, X) \tag{8.2}$$

Here, X represents the visual encoding with specific conditions on the language model. This means that the entire language model is autoregressive when predicting the next word of a given word, that is, each word is predicted based on known words, and the next word is predicted on this basis. In image captioning, a sequence end marker is also output to terminate the generation of image captions.

8.1.3 Modeling Methods

The main language modeling methods currently used for image captioning are divided into four categories: LSTM-based methods, convolutional methods, Transformer-based methods, and BERT-like methods.

8.1.3.1 LSTM-Based Methods

LSTM is a variant of RNN, often used in language modeling.

Single-Layer LSTM

The single-layer LSTM was first used in image captioning by Vinyals et al. [10] in 2015. This method uses visual encoding as the initial hidden state of the LSTM and then generates image captions. At each time step, the hidden state is projected onto a vector of the same size as the vocabulary, and then the Softmax function is used to predict words. The output at each step is the probability of all words in the vocabulary. During training, the input is the sentence that serves as the ground truth, that is, word embeddings. During inference, the input is the words that have already been generated.

On this basis, Xu et al. [2] proposed a new attention mechanism that no longer uses a unified semantic feature but instead uses dynamic and time-varying image representations to replace static global vectors, while also enhancing the alignment of text and image content. After extracting the positional features of the image, the attention mechanism allows the decoder to have the ability to select from the features. The early hidden state will guide the attention calculation of the visual features to produce a context vector, which is then put into the MLP to predict the output word.

Liu et al. [11] used a visual sentinel to capture spatial image features. At each time step, the visual sentinel calculates and generates words based on the previous hidden state. Then, the model generates a context vector, which is a combination of image features and the visual sentinel, the importance of which is reflected by a learnable gate through weighting.

Chen et al. [12] proposed a dynamic transformation method for standardizing language models based on the previous hidden state of the current state using a second LSTM. Ge et al. [13] proposed to better capture context information by using a bidirectional LSTM with an auxiliary module. The auxiliary module of the bidirectional LSTM in one direction approximates the hidden state of the LSTM in the other direction. Finally, a cross-modal attention mechanism combines grid visual features with two sentences in the bidirectional LSTM to get the final image caption.

Wang et al. [14] proposed to decompose the image caption generation process into two stages, both of which use a single-layer LSTM. Gu et al. [15] designed a coarse-to-fine multistage framework using a sequence of LSTM decoders, each performing decoding. Each decoder operates on the output of the previous one to produce increasingly refined image captions.

Dual-Layer LSTM

LSTM can be extended to a dual-layer or even multilayer structure to enhance the ability to capture high-order relationships.

For a dual-layer LSTM, these two layers can be specialized to perform visual attention and actual language modeling. The first layer LSTM acts as a top-down visual attention model, which can obtain previously generated words, previous hidden states, and average pooled image features, which then use the current hidden state to calculate the probability distribution on the image area of the additional attention mechanism. The obtained image features are put into the second layer LSTM, combined with the hidden state of the first layer LSTM, to generate the probability distribution of vocabulary.

Dual-layer LSTM and its internal attention mechanism represented the best language modeling method before the emergence of methods based on the Transformer architecture, so many variants of dual-layer LSTM have been proposed to improve its performance.

1. Put text into image areas and integrate a pointer network to regulate the content-based attention mechanism. During the generation process, the pointer network predicts slots in the image caption and then fills in the image area classes. This method uses an object detector as a feature area extractor and visual word prompter for the language model.
2. Introduce two reconstruction modules. The first reconstruction module calculates the relevance between the hidden state of all past predicted words and the current word, thereby modeling longer dependencies and improving historical relevance. The second reconstruction module improves the syntactic structure of the sentence through word co-location information (e.g., the subject usually appears at the beginning but is predicted to appear in the middle).
3. Use two modules, of which the review module considers the vectors involved before when calculating the next vector and the prediction module predicts two new words at the same time.
4. Adopt an adaptive attention mechanism. In this, the decoder can perform any number of attention operations for each generated word, and the output is determined by the confidence network on the second layer LSTM.
5. Introduce a recall mechanism modeled by a text retrieval system. It provides useful words for each image of the model. Auxiliary word distribution is obtained from the recalled words and used as a semantic guide.

Enhancement of LSTM with Self-Attention

Using self-attention operation instead of additional attention in LSTM-based language models can enhance LSTM with attention operators and calculate another step of attention based on visual self-attention. Cornia et al. [9] introduced the

X-linear attention block, which enhanced self-attention through second-order interaction and improved visual encoding and language models.

8.1.3.2 Convolutional Methods

Aneja et al. [16] used CNN as a language model, combining vectors and word embeddings and inputting them into CNN. During the training process, all words are operated in parallel and operated in sequence during inference. Convolution is right-covering to prevent the language model from using the information of the word tokens behind. Although parallel training has certain advantages, due to its performance limitations and the emergence of Transformer, the method of performing convolution operations in language models has not been widely used.

8.1.3.3 Transformer-Based Methods

When dealing with image regions, image captions can be transformed into a set sequence problem, so Transformer can also be used for image caption generation. The standard Transformer decoder performs hidden attention operations, applies them to words, and then performs cross-self-attention operations, where words serve as questions. The output of the last encoder layer serves as keys and values and then feeds into the feed-forward neural network. During the training process, the masking mechanism is used for known words, the purpose of which is to constrain the unidirectional generation process. Initially, when the Transformer was used for image captioning, its structure had a few modifications. Later, many variants based on Transformer methods were produced to improve the ability of language models and visual feature encoding.

Li et al. [6] proposed a gating mechanism for cross-attention operators, which controls the flow of visual and semantic information by combining and modulating the image region representation and semantic attributes from external markers.

Ji et al. [17] integrated a context gating mechanism to regulate the impact of global image representation on each generated word, modeled through a multihead attention mechanism.

Cornia et al. [9] considered all encoding layers, instead of only performing cross-attention operations on the last encoding layer.

8.1.3.4 BERT-Like Methods

BERT-like architecture was used to fuse text and visual features early on as the language model for image captioning. The advantage of this method is that it can use the parameters learned from pre-training in large-scale text corpora to initialize the parameters for processing the text layer.

Zhou et al. [18] established a unified model, integrating both visual and text into a BERT-like architecture to solve the image captioning problem. The model consists of a shared multilayer encoding and decoding transformer, pre-trained on a large image captioning corpus, and then fine-tunes the image captions by masking the sequence to the right to simulate a unidirectional generation process.

Li et al. [19] proposed a method to better learn the alignment in the joint representation of visual encoding and language models by using the object tags detected in the image as anchors. Their proposed model represents the input image–text pair as a triple feature of word tag, object tag, and region, where the object tag is the text class extracted by the object detector.

8.1.4 Applications and Analysis

8.1.4.1 Model Structure

StyleNet is a novel framework for generating attractive captions for images and videos with different styles. In StyleNet, a new model component called FactoredLSTM is used, which can automatically extract style factors from monolingual text corpora.

8.1.4.2 Code Demonstration

The model building code for the CNN encoder is as follows:

```
  class EncoderCNN(nn.Module):
 def __init__(self, emb_dim):
super(EncoderCNN, self).__init__()
resnet = models.resnet152(pretrained=True)
modules = list(resnet.children())[:-1]
self.resnet = nn.Sequential(*modules)
self.A = nn.Linear(resnet.fc.in_features, emb_dim)
 def forward(self, images)
features = self.resnet(images)
features = Variable(features.data)
features = features.view(features.size(0), -1)
features = self.A(features)
return features
Part of the code for the FactoredLSTM framework i
Part of the code for the FactoredLSTM framework is as follows:
def forward(self, captions, features=None, mode="factual"):
batch_size = captions.size(0)
embedded = self.B(captions) # [batch, max_len, emb_dim]
```

```
if mode == "factual":
    if features is None:
sys.stderr.write("features is None!")
    embedded = torch.cat((features.unsqueeze(1), embedded), 1)
# initialize hidden state
h_t = Variable(torch.Tensor(batch_size, self.hidden_dim))
c_t = Variable(torch.Tensor(batch_size, self.hidden_dim))
nn.init.uniform(h_t)
nn.init.uniform(c_t)
if torch.cuda.is_available():
    h_t = h_t.cuda()
    c_t = c_t.cuda()
all_outputs = []
# iterate
for ix in range(embedded.size(1) - 1):
    emb = embedded[:, ix, :]
    outputs, h_t, c_t = self.forward_step(emb, h_t, c_t,
mode=mode)
    all_outputs.append(outputs)
all_outputs = torch.stack(all_outputs, 1)
return all_outputs
```

8.2 OCR and Domain Optimization

8.2.1 Problem Background

According to research, 91% of the information humans use to understand the world comes from vision. Similarly, computer vision is the basis for computers to perceive the world and a hotspot in artificial intelligence research, while text recognition is an integral part of computer vision. Text is ubiquitous in daily life, and without text, people would find it very inconvenient in all aspects of food, clothing, housing, and transportation. How to quickly extract text information from various images is also a technology that people urgently need. Optical Character Recognition (OCR) refers to analyzing, recognizing, and processing images containing text to obtain text and layout information, that is, recognizing the text in the image and returning the result in the form of text. The principle of OCR recognition of text is that the computer preprocesses, segments, and locates the text in the image, detects the brightness and darkness, enlarges the image to determine its shape features, and finally matches the image of the black and white dot matrix with the text character encoding. The text is converted into text based on the degree of match in the image [20]. The basic steps of OCR are shown in Fig. 8.2.

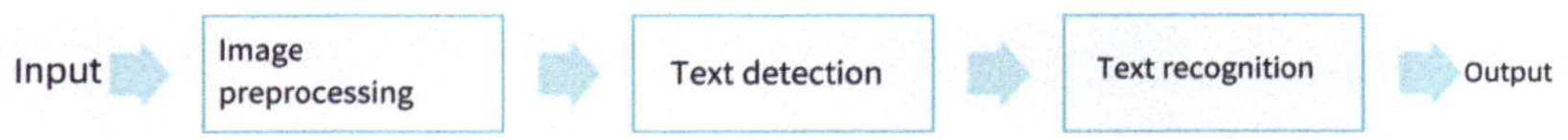

Fig. 8.2 Basic steps of OCR

In Fig. 8.2, image preprocessing is usually used to correct the imaging problems of the image. Common preprocessing processes include geometric transformations (perspective, distortion, rotation, etc.), distortion correction, defocusing, image enhancement, light line correction, etc. Text detection and text recognition are the technical bottlenecks that affect recognition accuracy and are also the focus of OCR technology. Text detection detects the text's position, range, and layout, which usually includes layout analysis, text line detection, etc. The main problem that text detection solves is determining the position and range of the text. Text recognition is the recognition of the content of the text based on text detection and the conversion of the text in the image into text information. The main problem that text recognition solves is to determine what each text is.

8.2.2 Technical Analysis

8.2.2.1 Image Preprocessing

This mainly introduces binarization, denoising, and tilt angle detection and correction.

Binarization

Binarization is to convert the grayscale image signal into a binary image signal with only black (1) and white (0) by setting the grayscale value of the pixel to 0 or 1. In traditional methods and current popular methods, high-quality binary images can significantly improve the effect of OCR, not only reducing the data dimension but also excluding noise to highlight the effective area. The comparison of images before and after binarization is shown in Fig. 8.3.

Denoising

Image noise refers to unnecessary or redundant interference information in image data. Image noise is generated during the acquisition, quantization, or transmission of the image, which will have a great impact on the subsequent processing and analysis of the image.

Fig. 8.3 Comparison of images before and after binarization

The traditional denoising (or noise reduction) method is to estimate the noise using the image prior and noise model; for example, NLM and BM3D estimate the noise using the local similarity and noise independence of the image. The wavelet denoising method estimates the noise using the sparsity of the image in the transformation domain, and NBNet [21] performs noise reduction through adaptive projection of the image. The projection can better retain the structure information of the image and is also a way to capture global relevance. The projection can train an A network that separates information from noise. The specific method is generating a series of image basis vectors from the input image and then reconstructing the denoised image in the subspace composed of these basis vectors. NBNet is generally a UNet-style network, where the key is the Subspace Attention Module (SAM). NBNet includes two main steps, namely basis vector generation and subspace projection. The innovation of NBNet lies in subspace projection.

Tilt Angle Detection and Correction

Images can easily experience medium rotation and displacement during the scanning process. Common tilt angle detection and correction methods include the Hough transform, the Radon transform, and methods based on PCA.

The most commonly used method is the Hough transform, which generally consists of three steps: ① Detect all straight lines in the image. ② Calculate the tilt angle of each line and take their average. ③ Rotate the image according to the tilt angle to correct it. The Hough transform first dilates the image, connecting discontinuous text into a straight line and facilitating line detection. After calculating the angle of the line, you can use the rotation algorithm to correct the tilted image in the correct direction. An example of tilt angle detection and correction is shown in Fig. 8.4.

Fig. 8.4 Example of tilt angle detection and correction

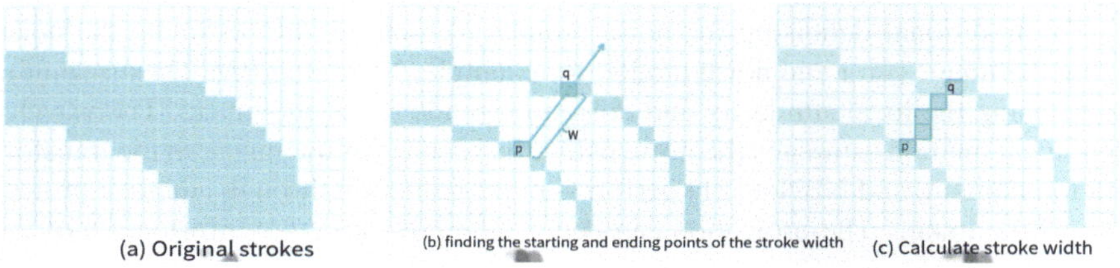

Fig. 8.5 Example of the specific process of the SWT algorithm

8.2.2.2 Text Detection

Traditional Text Detection Methods

Traditional text detection methods generally use manual feature extraction methods, such as SWT and MSER algorithms, and then use template matching or model training methods to recognize the detected text, while current text detection mainly uses deep learning methods which use convolutional neural networks to replace manual feature extraction methods [22]. The following briefly introduces traditional text detection methods using the SWT algorithm as an example.

The steps of the SWT (Stroke Width Transform) algorithm are as follows: First perform stroke width transformation, output SWT image, then obtain multiple connected domains through the SWT image, then use custom rules to filter some connected domains, get candidate connected domains, and finally, merge the connected domains to get the text. The specific process of the SWT algorithm is shown in Fig. 8.5.

Text Detection Method Based on Candidate Boxes

The idea of text detection methods based on candidate boxes is as follows: According to the set anchor, a series of text candidate boxes are generated, and then a series of adjustments and screenings are carried out. Finally, the text boundary is obtained through Non-Maximum Suppression (NMS).

The text detection based on CTPN (Connectionist Text Proposal Network) has been improved based on Faster RCNN, and it is one of the most widely used text detection models at present [23]. Its key points are as follows:

1. The vertical anchor regression mechanism is used to detect small-scale text candidate boxes.
2. The method of vertical anchor is used to solve the problem of uncertain text length changes. Only the position of the text in the vertical direction is predicted, not its position in the horizontal direction. When determining the horizontal position, it is only necessary to detect small fixed-width text segments one by one, predict their corresponding heights accurately, and finally connect them together to get the text line.
3. RNN is used to connect the detected small-scale text to get the text line.
4. The end-to-end training method of CNN+RNN is adopted, which supports multiscale and multilanguage, and does not need post-processing.

8.2.2.3 Text Recognition

Traditional Text Recognition Method

The traditional text recognition method uses template matching for classification. However, the content of text lines can only be determined by recognizing each character. Therefore, character segmentation can be performed on the text line to obtain single characters.

Over-segmentation-dynamic programming is the most common segmentation method. First, the candidate characters are over-segmented to make them sufficiently fragmented, and then the fragments are merged through dynamic programming.

Another method is to match each possible character through a sliding window. Because a single character may have multiple recognition results due to the cutting position, such as the word "like" will be cut into "female mouth" when the cut is improper, it is necessary to segment the candidate characters to make them sufficiently fragmented and then merge the fragments through dynamic programming to get the optimal combination. This process requires manual design of the loss function. The accuracy of this method depends on the size of the sliding window. If the sliding window is too large, it will cause information loss; if it is too small, it will significantly increase the computational demand.

CNN+RNN+CTC Method

Similar to the problem of speech recognition, the OCR task can be modeled as a sequence-dependent vocabulary or recognition problem. Some scholars have tried to borrow the CTC loss function to OCR recognition, and CRNN is a representative algorithm among them.

Using CNN features as input and bidirectional LSTM for sequence processing, the efficiency of text recognition has been greatly improved, and the generalization ability of the model has been improved. This method first obtains the feature map by the classification method. Then, the results are translated through CTC to get the output. This method is currently the most widely used text recognition framework, but it requires users to build their own word library (including common words, various characters, etc.).

The algorithm of training RNN with CTC significantly surpasses traditional speech recognition algorithms in the field of speech recognition phrases [24].

This method combines the potential of CNN in image feature engineering with the potential of LSTM in serialization recognition, extracting robust features and avoiding the extremely difficult single character segmentation and single character recognition in traditional algorithms. At the same time, serialization recognition can also embed time dependence (implicitly using corpus).

CNN+RNN+Attention Method

The convolutional recurrent neural network based on the attention model mainly consists of three parts: the convolutional neural network, the recurrent neural network, and the attention model. The convolutional neural network extracts features from the image, and the attention model, with its precision, calculates the attention weight. The combination of these two after decoding results in the probability distribution of the character set, with the character with the highest probability selected as the recognition result.

This recognition framework is an encoder–decoder structure, with the bottom layer extracting the original image and its enhanced feature map using the CNN structure.

The convolutional neural network first automatically extracts features from the input image; then, the attention model calculates the attention weight based on the hidden state of the recurrent neural network neuron and the output of the previous moment and finally combines the feature map output by the convolutional neural network with the attention weight, inputs it into the recurrent neural network for encoding and decoding, obtains the probability distribution of the entire character set, and finally directly selects the character with the highest probability as the recognition result.

8.2.2.4 Optimization in the OCR Field

Text Detection Optimization

In practical applications, the sources and quality of images are complex, with sources including screen captures, photos, and handwriting, and the text may be blurred, broken, or folded. Current solutions include enhancing contrast, adding blur algorithms to handle image blur problems, and preprocessing to solve problems such as tilt and noise.

The PP-OCR model currently mainly uses the following methods for optimization:

1. Choice of ultra-lightweight backbone network. A major factor affecting the size of the detector model is the choice of the backbone network structure. An ultra-lightweight text detector should use an ultra-lightweight backbone network.
2. Lightweight head. The head of DB text detection is a fusion of feature maps of multiple resolutions, which can improve the detection effect of targets of different scales.
3. Cosine learning rate. The cosine learning rate is used instead of the segmented learning rate, and the learning rate is relatively large throughout the process, because this convergence is slow, but the final convergence effect is better. Figure 8.6 shows a comparison of different learning rate methods.
4. Warm-up learning rate. Many studies have shown that using a too large learning rate at the beginning of training can lead to numerical instability in the learning

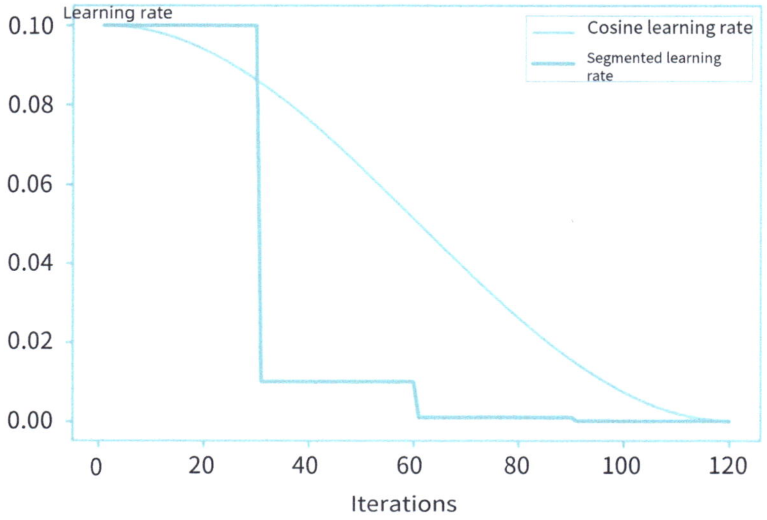

Fig. 8.6 Comparison of different learning rate methods

process. It is recommended to start with a smaller learning rate and gradually increase to the initial learning rate, which can help improve performance.
5. Removal of the SE module. There are multiple SE modules in the MobileNetV3 structure. For text detection tasks, images often contain multiple text targets and noisy background information, and the input resolution is generally quite large, such as 640 × 640. It is difficult to mine information between features through SE, and it also increases the computational load.
6. FPGM model pruning method. The FPGM model pruner is used to further reduce the model size.

Text Recognition Optimization

The main optimization strategies for text recognition are as follows:

1. Data augmentation. Also known as data expansion, it refers to the practice of making limited data equivalent to more data without substantially increasing the amount of data.
2. Increase the resolution of the feature map. In PP-OCR, in order to retain more information in the horizontal and vertical directions of the image, all stride transformations are changed from (2, 2) to (1, 1), and the resolution of the feature map of the entire network is increased. The results show that increasing the resolution of the feature map improves the recognition accuracy to a certain extent.
3. Regularization parameters. Regularization is generally used to prevent overfitting. Adding regularization parameters to the loss function makes the entire network's weight values tend to be smaller, thereby improving the model's generalization ability.
4. Pre-train large models. In image classification, object detection, and image segmentation, using ImageNet1000-trained image classification pre-trained language models helps the model converge quickly and improve performance.
5. PACT quantization. This method can reduce redundancy in the model, reduce the model size, shorten the recognition time, and thus improve the model's performance.

8.2.3 Application and Analysis

In response to the shortcomings of the spatial transformation grid in existing scene text recognition models, this section proposes a scene text recognition method based on a linearly constrained correction grid. The novelty of this method lies in adding linear constraints to the correction information of the spatial transformation grid, making its output shape conform to the polynomial curve and, at the same time, sharing the primary features extracted by the spatial transformation grid to achieve scene text recognition.

8.2.3.1 Model Structure

In order to better recognize irregular scene text, this method corrects severely deformed text images into horizontal text images. The model structure is shown in Fig. 8.7. During correction, the convolutional network extracts features and predicts position information, which serves as the parameters of the spatial transformation grid.

After the transformation, a correction control point grid is obtained with a shape of M × N. The predicted output points are all distributed on the curve determined by the polynomial. This transformation process ensures a smoother transition between the grid points of the spatial transformation grid. The corrected image extracts features through the convolution network. The convolution operation here can be the same as the convolution operation in the correction, using the same residual network.

The sequence encoder consists of two layers of bidirectional LSTM, which encodes the features obtained from feature extraction into an intermediate state. This state contains contextual information. The sequence decoder includes word

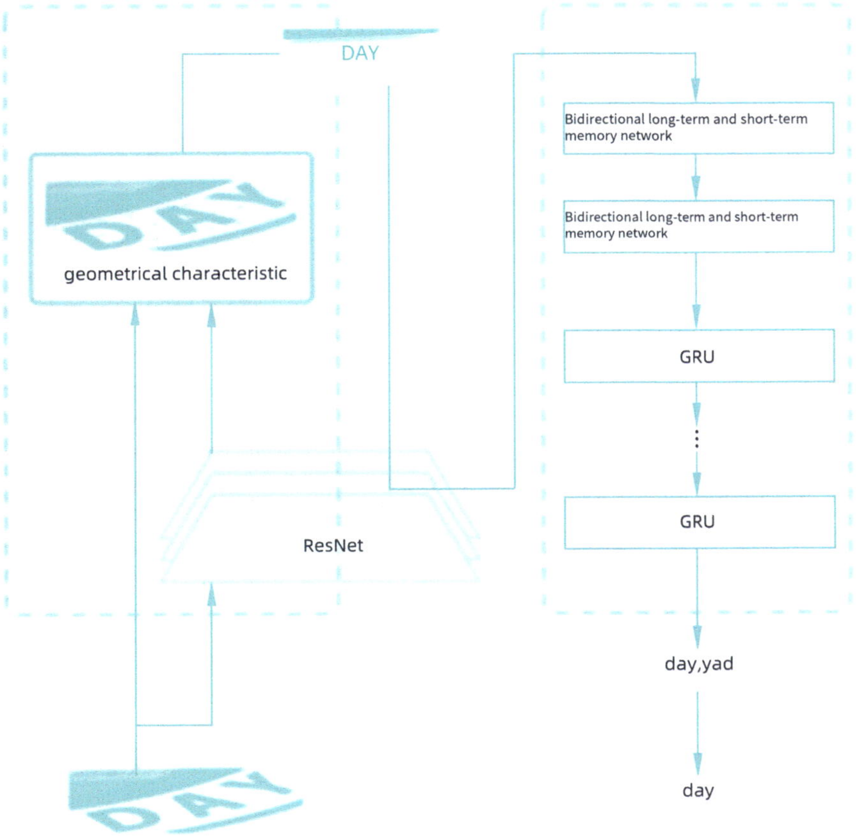

Fig. 8.7 Model structure

embedding and an attention mechanism. At each time step of decoding, the intermediate representation containing contextual information obtained from the sequence encoding network is combined with the previous hidden state of the gated recurrent unit to calculate the attention weight. The weight is used to perform a weighted sum of the current output of the gated recurrent unit, predicting the character category of the current decoding time step. The decoding process uses beam search, keeping the top K most likely results at each step. The final prediction output is the result with the highest overall probability.

8.2.3.2 Effect Display

The linearly constrained correction model has been trained and extensively evaluated on public datasets. Figure 8.8 shows an example of the evaluation results of the CUTE dataset. The left side is the input image, and the right side is the corrected image. Figure 8.9 shows four cases, with the rightmost column giving the recognition results and the correct text. Figure 8.10 shows some examples demonstrating the comparison between this method and the Aster model, where the left side is the input image, the middle is the result predicted by the Aster model, and the right side is the output of this method.

Fig. 8.8 Example of CUTE dataset evaluation results

Fig. 8.9 Four cases

Fig. 8.10 Comparison between this method and the Aster model

References

1. KULKARNIA, GIRISH S, et al. BabyTalk: Understanding and Generating Simple Image Descriptions[C]. IEEE Conference on Computer Vision and Pattern Recognition IEEE Computer Society, 2011: 1601–1608.
2. XU K, BAJ, KIROSR, et al. Show, Attend and Tell: Neural Image Caption Generation with Visual Attention [C]. International Conference on Machine Learning. PMLR, 2015: 2048–2057.
3. Yang, Zhilin, et al. "Review networks for caption generation." *Advances in neural information processing systems* 29 (2016).
4. YAO T, PAN Y, LI Y, et al. Hierarchy Parsing for Image Captioning[C]. 2019IEEE/CVF International Conference on Computer Vision (ICCV), IEEE, 2019.
5. Yang, Xu, et al. "Learning to collocate visual-linguistic neural modules for image captioning." International Journal of Computer Vision 131.1 (2023): 82–100.
6. LIG, ZHU L, LIU P, et al. Entangled Transformer for Image Captioning[C]. 2019 IEEE/CVF International Conference on Computer Vision (ICCV), IEEE, 2019.
7. Herdade, Simao, et al. "Image captioning: Transforming objects into words." Advances in neural information processing systems 32 (2019).
8. HUANG L, WANG W, CHEN J, et al. Attention on Attention for Image Captioning[C]. Proceedings of the IEEE/CVF International Conference on Computer Vision, IEEE, 2019: 4634–4643.
9. CORNIA M, STEFANINI M, BARALDI L, et al. Meshed-Memory Transformer for Image Captioning[C]. 2020 IEEE/CVF Conference on Computer Vision and Pattern Recognition (CVPR), IEEE, 2020.
10. VINYALS O, TOSHEV A, BENGIO S, et al. Show and Tell: A Neural Image Caption Generator[J]. IEEE, 2015. DOI: https://doi.org/10.1109/CVPR.2015.7298935.

11. LU J, XIONG C, PARIKH D, et al. Knowing When to Look: Adaptive Attention via A Visual Sentinel for Image Captioning[J]. IEEE, 2017.
12. CHEN X, MAL, JIANG W, et al. Regularizing RNNs for Caption Generation by Reconstructing The Past with The Present[J]. IEEE.
13. GE H, YAN Z, ZHANG K, et al. Exploring Overall Contextual Information for Image Captioning in Human-like Cognitive Style[J]. Dalian University of Technology; Mila. McGill University.
14. WANG Y, LIN Z, SHEN X, et al. Skeleton Key: Image Captioning by Skeleton-Attribute Decomposition[J]. IEEE, 2017.
15. GU J, CAI J, GANG W, et al. Stack-Captioning: Coarse-to-Fine Learning for Image Captioning[J]. 2017.
16. ANEJAJ, DESHPANDE A, SCHWING A. Convolutional Image Captioning[J]. 2017.
17. JIJ, LUO Y, SUN X, et al. Improving Image Captioning by Leveraging Intra- and Inter-layer Global Representation in Transformer Network[J]. 2020.
18. ZHOU L, PALANGI H, ZHANG L, et al. Unified Vision-Language Pre-trained for Image Captioning and VQA[J]. 2019.
19. LIX, YIN X, LIC, et al. Oscar: Object-Semantics Aligned Pre-trained for Vision-Language Tasks[J]. 2020.
20. Chen Manlong. Error analysis of machine vision thread measurement[J]. Laser Technology, 2014, 38(1): 109–113.
21. CHENG S, WANG Y, HUANG H, et al. NBNet: Noise Basis Learning for Image Denoising with Subspace Projection[J]. 2020.
22. Shi Mengjie. Overview of text clustering algorithms[J]. Modern Computers, 2014(2):3–6.
23. ZHIT, HUANG W, TONG H, et al. Detecting Text in Natural Image with Connectionist Text Proposal Network[M]. Cham: Springer, 2016
24. GRAVES A. Connectionist Temporal Classification[J]. Springer Berlin Heidelberg, 2012.

Chapter 9
Chinese Word Segmentation and Part-of-Speech Tagging

This chapter explores the fundamentals of Chinese word segmentation and part-of-speech tagging, emphasizing challenges and key algorithms in Chinese information processing. The discussion covers segmentation difficulties, including vocabulary selection, ambiguity resolution, and out-of-vocabulary word recognition. Advanced models, such as hierarchical hidden Markov models (HHMM) and bidirectional recurrent neural networks (Bi-RNN) combined with conditional random fields (CRF), are analyzed for their effectiveness in improving segmentation accuracy. This chapter concludes with an application demonstration of NLPIR-ICTCLAS, a leading system in Chinese lexical analysis, showing its high accuracy and adaptability in large-scale natural language processing tasks.

9.1 Overview of Chinese Word Segmentation

The main characteristic of the contemporary technological revolution is the use of computers for information processing. With the widespread application of computers, they have evolved from past data processing and information processing to current knowledge processing and language text information processing. Text language is the most important communication tool for humans and is the primary information carrier. The depth and breadth of people's requirements for language text processing are increasing. In our country, the application of computers to transaction processing, office automation, typesetting, information retrieval, machine translation, human–computer dialogue, etc., all involve Chinese. The information in these areas is carried in Chinese, so language text information processing has become a bottleneck in our country's information construction.

Chinese information processing technology is a crucial computer application technology. It has penetrated various fields of computer applications, such as computer networks, database technology, software engineering, etc. The "National

H. Zhang, J. Shang, *Natural Language Processing and Applications*,
https://doi.org/10.1007/978-981-97-9739-4_9

Medium and Long-term Science and Technology Development Plan" formulated by the State Council clearly indicates that Chinese information processing technology is critical in high-tech development. The focus of our country's software industry development is Chinese information processing software, and the development of Chinese information processing has received national attention. Therefore, solving the technical problems of Chinese information processing has become a "must-win battle" in our country's informationization process.

According to statistics, more than 80% of the information in the information field is carried in language text. The automatic input and output of this information, the collation and classification of texts, the extraction and retrieval of information, and language translation and other language engineering are all important foundations for constructing national economic and national defense information. Chinese information processing covers information processing tasks at multiple levels, such as words, phrases, sentences, and articles. Chinese automatic word segmentation is an important basic work in Chinese information processing. Many Chinese information processing projects involve word segmentation issues, such as machine translation, Chinese literature, automatic abstracts, automatic classification, full-text search of Chinese literature databases, etc.

Because Chinese text is written in sentences without gaps between words, word segmentation is the first problem encountered in Chinese text processing. Correct word segmentation is necessary for Chinese text processing when converting sentence writing into word writing. In the mid-1980s, automatic word segmentation technology was already valued, and various word segmentation models and software appeared one after another. In recent years, with the continuous development of national economic information technology and the widespread application of the Internet in the broad application of Chinese information processing, there has been an urgent need to realize the sharing and reuse of Chinese information, such as Chinese dictionaries and corpora, and the requirements for automatic word segmentation are also increasing. The demand for word segmentation technology is getting higher and higher. Under the strong driving force of information industry demand, automatic word segmentation has attracted attention from various aspects and has become a frontier topic in Chinese information processing.

Up to now, Chinese word segmentation technology can generally be divided into the following categories: dictionary-based word segmentation methods (mechanical word segmentation methods), statistical-based word segmentation methods, and word segmentation methods combining dictionary and statistics. The advantage of the dictionary-based word segmentation method is that the word segmentation speed is relatively fast, and the efficiency is relatively high. The word segmentation process can be transformed into matching with the words in the dictionary. The dictionary-based word segmentation method is relatively easy to implement, but its disadvantages are also obvious. First of all, with this method, the quality of the dictionary, especially its capacity and breadth, directly affects the results of word segmentation. If a word does not appear in the dictionary used by the segmentation system, then this word will not be segmented or mis-segmented as other words. For example, if the dictionary does not contain "percussion music," it is very likely to be segmented as "percussion/music," losing the meaning it should express. Second, this method cannot effectively handle ambiguity issues. The statistical-based segmentation method has good ambiguity resolution ability; its disadvantage is that it requires a large amount of corpus as input to train related models, and it may segment out high-frequency but non-word Chinese character strings in the corpus, for example, "ate an apple" and "thought for a while" in "for a while" appear at a high frequency. The statistical-based segmentation method will likely recognize it, mistakenly thinking it is a word. The method combining dictionary and statistics can effectively combine the two, using the high efficiency of the dictionary-based segmentation method and the strong ambiguity resolution ability of the statistical-based segmentation method, complementing each other and achieving good segmentation results.

9.2 Difficulty of Chinese Word Segmentation

With the deepening of Chinese information processing research, Chinese word segmentation is receiving more and more attention. In natural language processing, words, as the most minor language components that can be used independently, have no noticeable distinguishing marks between each other, making text analysis difficult. Therefore, segmenting the strings in Chinese text into reasonable word sequences for analysis is expected when processing Chinese in natural language. At present, Chinese word segmentation has been widely used in many aspects as the first link in Chinese information processing, such as Chinese text processing, information extraction, and text mining. Chinese word segmentation involves many problems, mainly including core vocabulary problems, word deformation problems, affix problems, ambiguous field problems, and out-of-vocabulary problems.

9.2.1 Core Vocabulary Problem

Many segmentation algorithms require a core vocabulary, and the words in the core vocabulary need to be segmented. However, there is currently no unified standard for which words should be included in the core vocabulary.

9.2.2 Word Deformation Problem

Many verbs and adjectives in Chinese can form deformed structures. For example, some reduplicated words, like "lingering" and "unrestrained," can be deformed into "lingering and lingering" and "unrestrained and unrestrained"; there are also some separable words, such as "eating" and "playing games," that can be deformed into "eating a meal" and "playing a few rounds of games." There is a lack of operable and reasonable norms for the segmentation of these deformed structures.

9.2.3 *Affix Problem*

In modern Chinese, some words are meaningless when used alone, such as "zhe," which is usually used to refer to people or things but is generally not used alone, such as "reporter," "the one who started to make figurines," "the third party," etc. It is impossible to segment sentences using these words. When words similar to "zhe" are used as the suffix of a sentence, their complex structure contradicts the definition of a word, such as "the developer of China's first operating system software." The standard for automatic Chinese word segmentation must support applications with various objectives, as different applications have different requirements for words. For example, in retrieval systems, retrieval rules tend to favor smaller segmentation units. For the phrase "the developer of China's first operating system software," it is required that both "software" and "development" be retrieved. In keyboard input systems, whether the pinyin "zhege" or "zege" is entered, the word "this" can be obtained, with the aim of improving user input speed and user experience.

The main problems faced by Chinese word segmentation technology can be divided into the following three aspects: the definition of the concept of "word" in Chinese, the recognition of out-of-vocabulary words, and the handling of ambiguous segmentation fields.

9.2.4 *Ambiguous Field Problem*

During the Chinese word segmentation process, fields with multiple segmentation possibilities often appear, referred to as ambiguous fields. Ambiguous fields can be divided into intersection-type ambiguous fields and covering-type ambiguous fields, among which covering-type ambiguous fields are also known as inclusive-type ambiguous fields, polysemous ambiguous fields, and combination-type ambiguous fields. For humans, it is generally understood through context, but it is difficult for computers to accurately judge how to segment. For example, for "combine into a machine," both "combine" and "into" are words, and whether to segment into "combine/into" or "combine/into" is a problem.

9.2.4.1 Intersection-Type Ambiguous Fields

Assume A, B, and C each represent a string composed of one or more characters. If in the ABC field, A, AB, BC, and C are words in the word segmentation table, this field can be segmented into A/BC or AB/C, then ABC is called an intersection-type

Table 9.1 Analysis results of intersection-type ambiguous fields

Chain length	Number of words in intersection ambiguous fields	Proportion (%)	Number of fields for intersection-type ambiguous fields	Proportion (%)
1	47,402	60.27	12,686	53.05
2	28,790	36.89	10,131	42.36
3	1217	1.56	743	3.11
4	608	0.78	324	1.35
5	29	0.04	22	0.09
6	19	0.02	5	0.02
7	2	0.00	2	0.01
8	1	0.00	1	0.01
Total	78,068	100	23,914	100

ambiguous field. For example, in the string "combine into," it can produce two segmentation results "combine/into" and "combine/into", so it belongs to the intersection-type ambiguous field. According to statistics, intersection-type ambiguous fields account for 85%–90% of all ambiguous fields, as shown in Table 9.1. Therefore, intersection-type ambiguous fields are problems that Chinese word segmentation systems need to focus on solving.

9.2.4.2 Covering-Type Ambiguous Fields

Assume A and B each represent a string composed of one or more characters. For the string AB, if AB is a word and the substring A or B in AB is also a word, then AB is called a covering-type ambiguous field. For example, in the string "immediately," "immediately" is a word, and "horse" and "on" can also form words in certain contexts (such as "he/from/horse/on/comes down"). Although covering-type ambiguous fields do not appear often in texts, potential covering-type ambiguous fields should not be ignored. The elimination of ambiguity generally requires more grammatical and semantic information and sometimes needs to be understood in the context of the surrounding text. For example, for "the student union organizes charity performance activities," should it be segmented into "student/union/organize/charity performance/activities" or "student union/organize/charity performance/activities"? It is impossible to judge with just one sentence; it must be combined with more context to understand.

9.2.5 Out-of-Vocabulary Word Problem

Out-of-vocabulary words refer to words not included in the dictionary, including various named entities (such as numerals, personal names, place names, organization names, translated names, time, and currency), Internet-new words, etc. In addition, some abbreviations (such as science associations) and terminologies (such as osteonecrosis) also fall within the scope of out-of-vocabulary words. Some named entities (such as numerals, time, etc.) have strong regularity and are relatively easy to recognize; some personal names, place names, organization names, etc., also have some common words and rules of their own and can be recognized using some rules and statistical information. However, reduplicative words such as "neat and tidy" and "shiny" that are numerous, have regular composition, and can be expressed with simple regular expressions are not considered out-of-vocabulary words. Because the composition of numeric compound words shows regularity, they are easy to recognize, and many scholars do not include them in the scope of out-of-vocabulary words. With the development of society, various new words emerge in an endless stream, many new words are formed without any rules in grammar and semantics, and many out-of-vocabulary words mixed with common words can also cause ambiguity, such as "[Wang Hui [Guo] Jia] has a bit of an emergency" and "Fever [is] not typical] one of the symptoms," etc.

If the out-of-vocabulary words in a sentence cannot be recognized, they can only be divided into individual Chinese characters, which will lose the information conveyed by the out-of-vocabulary words, causing segmentation problems. The number of Chinese words can be countless and inexhaustible, and it is almost impossible to have a dictionary covering all Chinese words.

The classification of out-of-vocabulary words is diverse, and the academic community generally accepts the classification of out-of-vocabulary words into the following five categories:

1. Abbreviations, such as "science association" and "SARS."
2. Proper names, mainly including personal names, place names, and organization names, such as "Zhang San," "Beijing," and "Microsoft"
3. Derived words are formed by adding affixes to familiar words, such as "mechanization" and "high efficiency."
4. Compounds, composed of two or more core words, are formed by combinations of verbs or nouns, such as "campus culture," "data mining," etc.
5. Numeric-type compounds, that is, the components include numbers, including time, date, phone number, address, number, etc., such as "2014" or "one million."

From the perspective of word class, out-of-vocabulary words are divided into 14 subcategories. Among them, nouns account for the largest proportion, 65.54%, verbs are second, 32.9%, and adjectives are the least, accounting for only 1.55%. Recognizing out-of-vocabulary words is an essential task in Chinese word segmentation recognition and is one of the bottlenecks restricting the improvement of segmentation performance.

The two main reasons for the generation of out-of-vocabulary words are as follows:

1. The selection and number of entries in the machine-readable dictionary.
2. The matching relationship between the machine-readable dictionary and the vocabulary in the text to be processed, including the coverage of the machine-

readable dictionary for the vocabulary in the text to be processed, Coverage refers to the proportion of the vocabulary in the text to be processed in the machine-readable dictionary. If the text to be processed contains m nonrepeated words and n of them appear in the machine-readable dictionary, then the coverage of the machine-readable dictionary for the vocabulary in the text to be processed is n/m.

Currently, there are mainly methods such as Chinese out-of-vocabulary word recognition based on decomposition and dynamic programming strategies and out-of-vocabulary word detection based on corpus learning.

From the birth of the first practical word segmentation system, CDWS (The Modern Printed Chinese Distinguishing Word System), in 1983 to the present, researchers at home and abroad have conducted extensive research on Chinese word segmentation and proposed many effective algorithms. The following introduces the Chinese word segmentation algorithm based on mechanical matching.

9.3 Chinese Word Segmentation Algorithm Based on Mechanical Matching

Mechanical matching is the most basic algorithm in automatic word segmentation, and its basic idea is as follows:

1. Establish a word library in advance that contains all possible words.
2. For a given Chinese character string S to be segmented, cut a substring of S according to a certain principle. If this substring matches a word in the word library, then this substring is a word, otherwise, this substring is not a word, and a new substring of S is cut for matching.
3. Continue to divide the remaining parts until the remaining parts are empty.

Among them, according to different principles of cutting substrings, mechanical matching can be divided into three categories: One is forward matching and reverse matching according to the direction of cutting substrings; the second is maximum matching and minimum matching according to whether long words or short words are considered first when matching; and the third is increasing word method and decreasing word method according to the strategy of re-cutting when the match is unsuccessful. An optimization of the mechanical matching method is the so-called N-shortest path method. The above methods will be explained below. For ease of understanding, the algorithm is explained using "Welcome new and old teachers to school" as an example.

9.3.1 Dictionary Matching Method

9.3.1.1 Forward Maximum Matching Method

The Forward Maximum Matching (FMM) method sets the maximum word cutting length maxlen (Chinese words are mostly two-character words, three-character words, four-character words, and a few are five-character phrases, such as "sit on the mountain and watch the tigers fight"; therefore, setting maxlen to 4 or 5 is more appropriate) based on experience. Each time you scan, you look for the longest word starting from the current position and match it with the words in the dictionary; if you can't find it, you shorten the length and continue to look for it until you find it, or the remaining substring becomes a single character. The flowchart of the forward maximum matching method is shown in Fig. 9.1.

"Teacher," match these five characters with the words in the dictionary, find that there is no such word in the dictionary, and then reduce the number of characters taken. Take the first two characters "欢 迎," find that the dictionary has this word, cut this word from the sentence, and perform forward maximum matching on the remaining seven characters "新老师生来学校" again, dividing it into "新" "老师" "生" "来" "学校." The entire sentence is divided into "欢迎/新/老师/生/来/学校".

9.3.1.2 Reverse Maximum Matching Method

The reverse maximum matching (RMM) method is different from the forward maximum matching method in that when splitting Chinese characters, the RMM method does not extract substrings in order from left to right but starts from the end of the sentence. The flowchart of the RMM method is shown in Fig. 9.2.

Still taking "欢迎新老师生来学校" as an example, set maxlen = 5; take out the last 5 of "欢迎新老师生来学校"

"欢迎新老师生来学校"means "Welcome every new lectures and students to our university."

Words are split until the segmentation process is complete. The entire sentence is divided into "欢/迎新/老/师生/来/学校".

The basic principles of these two segmentation methods are the same. The forward maximum matching method and the reverse maximum matching method basically only have a difference in direction. The example above shows that the result of the forward maximum matching method is "欢迎/新/老师/生/来/学校", and the result of the reverse maximum matching method is "欢/迎新/老师生/来/学校". Generally speaking, the accuracy of the reverse maximum matching method is slightly higher than that of the forward maximum matching method. Because people are accustomed to understanding sentences in a forward manner, the error rate of the reverse maximum matching method is slightly smaller. According to online statistics, the error rate of using the forward maximum matching method alone is 1/169, and the error rate of using the reverse maximum matching method alone is

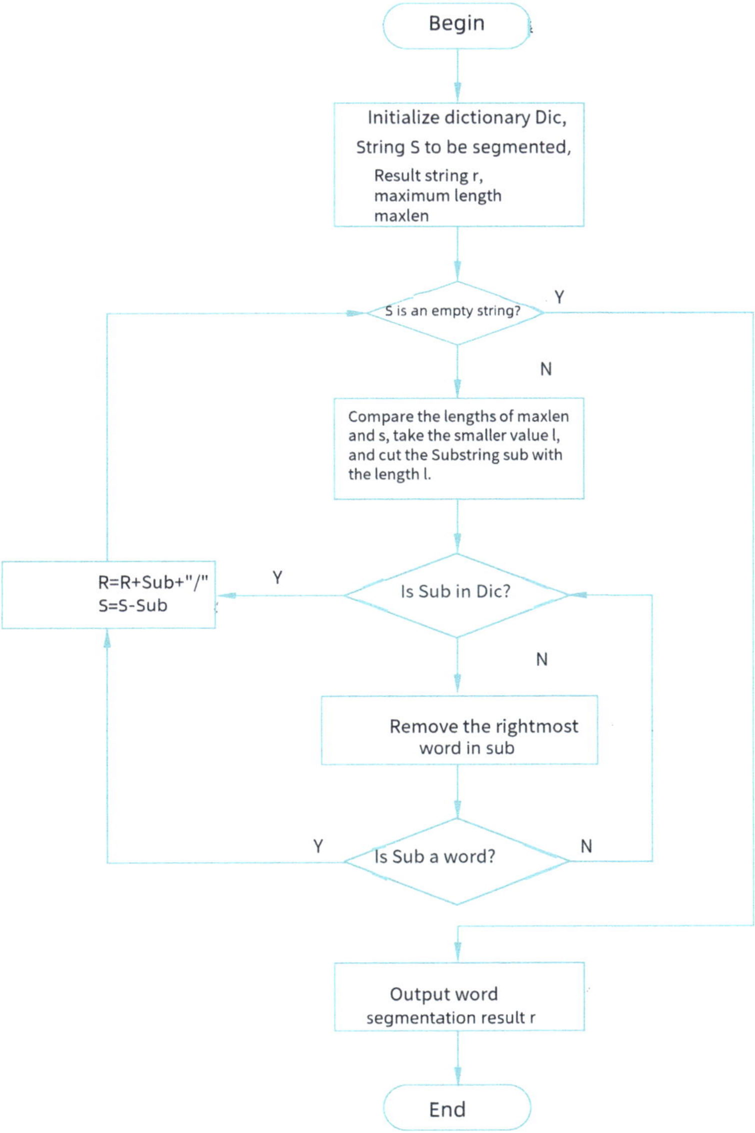

Fig. 9.1 Flowchart of the forward maximum matching method

1/245. But this precision is far from meeting actual needs. The actual segmentation system uses mechanical segmentation as a preliminary segmentation method and further improves the accuracy of segmentation by using various other language information.

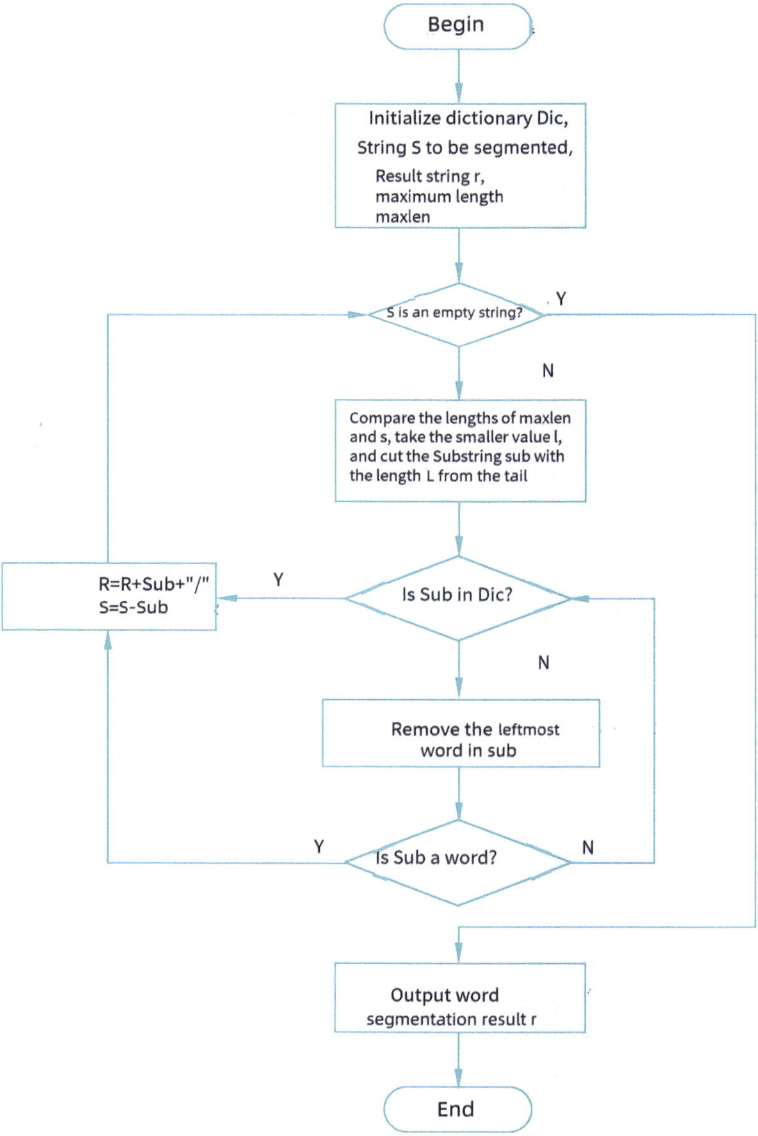

Fig. 9.2 Flowchart of the reverse maximum matching method

9.3.1.3 Bidirectional Scanning Matching Method

The bidirectional scanning matching method compares the segmentation results obtained by the forward maximum matching method and the reverse maximum matching method.

To determine the correct segmentation method. Research by M. S. Sun and K. T. Benjamin shows that about 90% of the sentences in Chinese information completely overlap and are proper when using the combination of the forward maximum matching method and the reverse maximum matching method; only about 9% of the sentences have different results from the two segmentation methods, but one of them must be correct (ambiguity detection is successful), only less than 1.0% of the sentences. The forward maximum matching method and the backward maximum matching method are wrong even though they overlap, or the forward maximum matching method and the backward maximum matching method are different. Still, both are wrong (ambiguity detection fails). This is precisely why the bidirectional scanning matching method is widely used in practical Chinese information processing systems.

The focus of this method is error detection and correction. Its basic principle is to use the forward maximum matching method and the backward maximum matching method for forward and backward scanning and preliminary segmentation and compare the results of the preliminary segmentation using the forward maximum matching method with the results of the preliminary segmentation using the backward maximum matching method. If the two results are consistent, it is judged that the segmentation is correct; if the two results are inconsistent, it is judged as a doubtful point, and a manual intervention method, a statistical frequency algorithm, or a segmentation method selected by combining the context-related information is adopted. The flowchart of the bidirectional scanning matching method is shown in Fig. 9.3.

The bidirectional scanning matching method combines the forward maximum matching method and the backward maximum matching method to improve the accuracy of word segmentation.

It has reduced the error rate, but it has also paid a specific price: one is the cost of execution efficiency. When executing the bidirectional scanning matching method, it is necessary to call both the forward and backward matching algorithms simultaneously, increasing the execution time. The second is that these two different matching algorithms may require different dictionary structures, so when executing bidirectional matching, two different dictionaries need to be loaded, increasing memory space and wasting time. In summary, while the bidirectional scanning matching method improves word segmentation accuracy, it also pays the price in terms of time and space.

The mechanical matching method is simple and easy to implement. For example, the maximum matching method embodies the principle of long words first and is most widely used in practical engineering. The mechanical matching method is relatively simple to implement, but its limitations are also obvious: Efficiency and accuracy are constrained by the capacity of the dictionary; the mechanical matching method adopts a simple, mechanical word segmentation strategy and does not involve grammatical and semantic knowledge, so it cannot effectively overcome the difficulty of ambiguous segmentation, and the segmentation accuracy is not high. Although experts have used many methods to improve the performance of the mechanical matching method, from an overall perspective, it is not easy to meet the

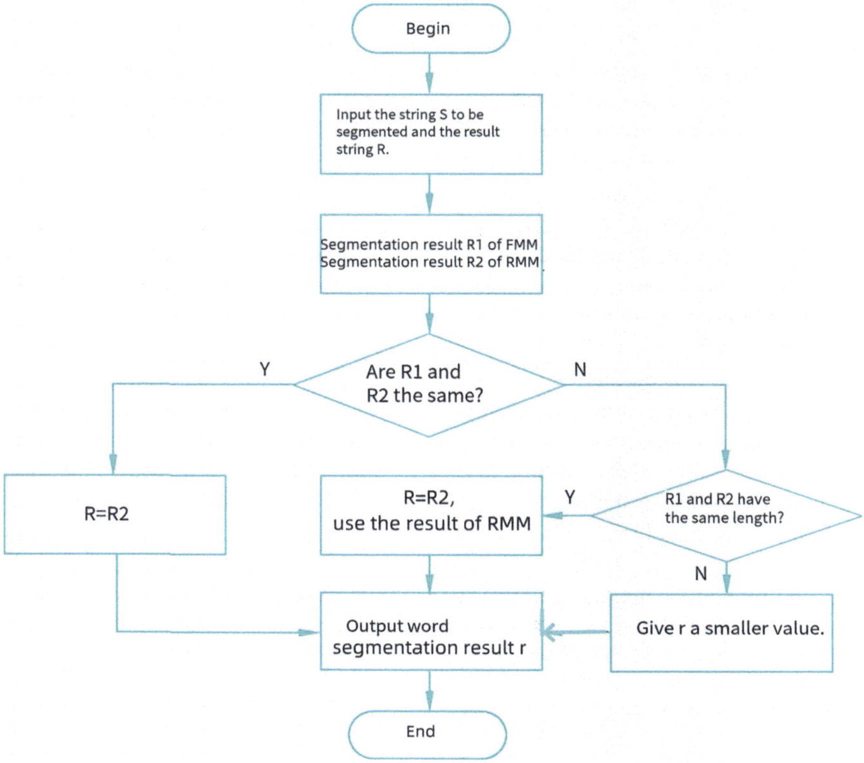

Fig. 9.3 Flowchart of the bidirectional scanning matching method

requirements for Chinese word segmentation in Chinese information processing by simply using the mechanical matching method. Based on mechanical matching word segmentation, using various language information for ambiguity correction is an essential means to break through the limitations of the mechanical matching method.

9.3.2 *N-Shortest Path Method*

The N-shortest path method is an improvement of the shortest path method. Its basic idea is that, based on the existing dictionary, each sentence is decomposed into a weighted directed acyclic graph. Each character represents a node in the graph; the edge represents possible word segmentation, the starting point of the edge is the first character of the word, and the end point is the next character at the end of the word; here, the frequency of the word is used to represent the weight of the edge, and the final result is to find N paths with the largest sum of weights in the abovementioned weighted directed acyclic graph.

The model is established as follows: Suppose the string $S = c_1, c_2, \ldots, c_n$, where $c_i, (i = 1, 2, \ldots, n)$ a single character, and the length of S is n, $n > 1$. A directed acyclic graph G is established with $n + 1$ node, and the node numbers are $V_0, V_1, V_2, \ldots, V_n \ldots$ Vn in order.

G's all possible word edges are established through the following two steps.

Step 1: Establish a directed edge $<V_k, V_{k+1}>V$, between adjacent nodes V_k and V_{k+1}; the word corresponding to the edge defaults to c_k ($k = < V_i$ step 1 $j_2 i_1$, c_j $(0 < i \leq j)$ is a word; then, establish a directed edge between nodes V_{i-1} and V_j. Assuming that words are mutually independent, the occurrence probability $P(a_i)$ of word a_i is introduced here, resulting in a unigram statistical model based on the N-shortest path method. According to the law of large numbers, when the sample data is large, the frequency of the sample is close to its probability value, so the maximum likelihood estimate of all $P(a_i)$ is equal to the word frequency, and then we have

$$P(a_i) = \frac{k_i}{\sum_{j=0}^{m} k_j} \tag{9.1}$$

According to the derivation of the formula in the literature, the final edge length formula is obtained as

$$L_a = \ln\left(\sum_{j=0}^{M} k_j + M\right) - \ln(k_i - 1) \tag{9.2}$$

where k_i is the number of times a_i appears in the training sample.

From formula (9.2), the higher the word frequency of ai, the shorter the edge length. Therefore, the string S and the words it contains correspond one-to-one with the edges of the directed acyclic graph G, as shown in Fig. 9.4.

Finally, the Dijkstra algorithm calculates the shortest path and gets the final result. The characteristic of the Dijkstra algorithm is to extend from the starting point to the endpoint to find the shortest path. Take the sentence "We are rest this afternoon" as an example, as shown in Fig. 9.5; for convenience of calculation, it is assumed that the length of all edges is 1.

In Fig. 9.5, there is a path between every two adjacent characters. "Today," "under," "afternoon," "noon rest," and "rest" can all form words, and it is assumed that they are all words existing in the dictionary. Then, each word adds a path. The arrow points to the next character at the end of the word, according to the Dijkstra algorithm, to calculate the path length. "Rest this afternoon."

All possible segmentation results and corresponding path lengths of this sentence are shown in Table 9.2. There are 11 segmentation results for this sentence. Assuming $N = 5$, the shortest path is selected.

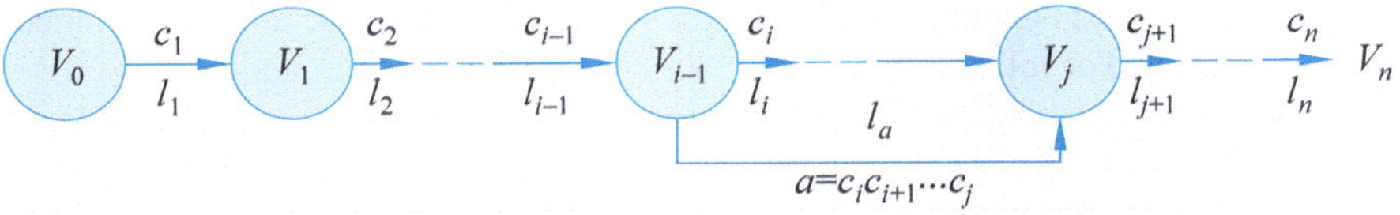

Fig. 9.4 Directed acyclic graph

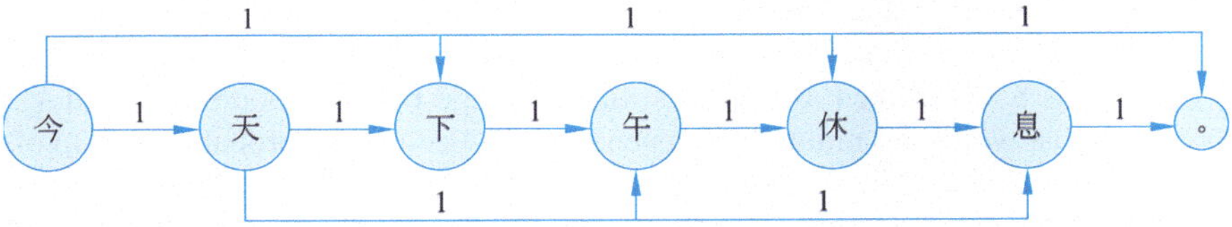

Fig. 9.5 N-shortest path method word segmentation example

9.4 Chinese Word Segmentation Algorithm Based on Statistical Language Model

The advantages of dictionary-based segmentation methods are obvious. First, it can ensure that 100% of the Chinese character strings segmented by the dictionary-based segmentation method are "words" because the segmented strings are all matched with the dictionary; second, only a Chinese dictionary is needed for Chinese word segmentation; no additional corpus is required, no additional language model needs to be established, there is no very complex calculation, and the calculation amount is small, so the efficiency of word segmentation is higher. However, the dictionary-based word segmentation method is a mechanical word segmentation method because the dictionary-based word segmentation method is just purely matching Chinese character strings in the dictionary. If a string matches in the dictionary, it is considered a word, otherwise, it is not considered a word without considering the relationship between words and grammar, which is the fundamental reason why dictionary-based word segmentation methods are prone to ambiguity.

The advantage of the statistical language model-based segmentation method introduced in this section lies in its handling of ambiguity. The most obvious difference between the statistical language model-based segmentation method and the dictionary-based segmentation method is that the statistical segmentation method abandons the dictionary; it does not need a dictionary as input when segmenting, but what it needs as input are various language models, and the training of language models requires a Chinese corpus. The so-called Chinese corpus generally contains large-scale (at least 100,000) Chinese sentences. Chinese word segmentation based on statistical language models is usually divided into three steps: First, load a large-scale Chinese corpus, then train the corresponding language model, and finally, perform Chinese word segmentation.

Table 9.2 Segmentation results and the corresponding path lengths

Serial number	Word segmentaion result	Path length
1	今/天/下/午/休/息/	6
2	今天/下/午/休/息/	5
3	今/天下/午/休/息/	5
4	今天/下午/休/息/	4
5	今天/下午/休息/	3
6	今/天下/午休/息/	4
7	今/天/下午/休/息/	5
8	今/天/下/午休/息/	5
9	今/天/下/午/休息/	5
10	今/天/下午/休息/	4
11	今天/下/午/休息/	4

9.4.1 N-gram Language Model

In the N-gram language model, for a Chinese string S, it can be regarded as a continuous string sequence, which can be a single-character sequence or a word sequence. For the string S, there is a way to segment it into w_1, w_2, …, w_n. The N-gram language model calculates how likely the string is to be segmented into w_1, w_2, …, w_n. At this time, its probability of occurrence is denoted as $P(S)$.

$$P(s) = P(w_{1,}w_{2,}\dots w_{n,}) \tag{9.3}$$

For each word w_i in the sentence, the N-gram language model assumes that the probability of w_i appearing is related to the previous $i-1$ words, and the probability value is calculated using the conditional probability formula, so we have

$$p(w_1) = \frac{p(w_{1,}w_{2,}\dots w_{i-1,})}{p(w_{1,}w_{2,}\dots w_{i-1,})} = P(w_i / w_{2,}\dots w_{i-1}) \tag{9.4}$$

Furthermore, Eq. (9.3) can be transformed into the following form:

$$p(s) = p(w_{1,}w_{2,}\dots w_{i-1,}) = p(w_1)P(w_2|w_1)P(w_3|w_1w_2)\dots P(w_n|w_1\dots w_{n-1,}) = p_{(w_i)\prod_{i-2}^{n} p(w_{1,}w_{2,}\dots w_{i,})} \tag{9.5}$$

In Eq. (9.5), the probability of sentence S is the probability of generating the first word w_1 multiplied by the following probabilities: the probability of generating the second word w_2 given the first word w_1, the probability of generating the third word w_3 given the first two words w_1w_2, and so on, up to the probability of generating the nth word w_n given the first $n-1$ words. The probability of each word w_i appearing is determined by the $i-1$ words before it. This is the core idea of the N-gram language model, which predicts unknown results based on known conditions. However, there are two drawbacks to this in practical applications: First, the sentences to be calculated may be very long and contain many words. If it is assumed that the appearance of n words is related to the previous $n-1$ words, the calculation of the probability of words in later positions will be large; second, in general, the appearance of the nth word is only more related to the adjacent few words. If all the previous $n-1$ words are included, it will lead to the generation of a lot of interference data, which will cause the calculation results to deviate from the actual situation. Therefore, in actual use, the parameter N is selected according to conditions such as resource limitations, and in general, N is 1, 2, 3, corresponding to the unigram, bigram, and trigram language models.

9.4.2 Mutual Information Model

Given a large-scale corpus containing hundreds of thousands of Chinese sentences, the first task to complete Chinese word segmentation is to find words in the corpus. The mutual information model provides a way of thinking to judge whether the combination is a word based on the closeness between the characters in the corpus. If two characters always appear next to each other, then these two characters can primarily be judged as a Chinese word. If two characters have never appeared adjacent or only appeared adjacent once or a few times but have appeared separately many times, then it can be concluded that the closeness between these two characters is not so high, and thus these two characters cannot form a word.

The formula for self-information in information theory is

$$I(x) = -\mathrm{Log}_2 p(x) \tag{9.6}$$

Probability itself is a measure of the certainty of an event, and information is a measure of the uncertainty of an event. Therefore, the smaller the probability of an event, the greater the information is. When two events are independent, the information of their joint event is equal to the sum of their individual information, expressed as

$$I(x,y) = I(x) + I(y) \tag{9.7}$$

In information theory, the formula for mutual information is

$$\text{Mutual information} = \log 2 \frac{\text{Posterior probability}}{\text{priori probability}} \tag{9.8}$$

When combined with a word segmentation system, mutual information is used to represent the strength of the combination between two characters, and its formula is

$$M(x,y) = \log_2 \frac{p(x,y)}{p(x)p(y)} = \log 2 \frac{p(x|y)}{p(x)} \tag{9.9}$$

where x and y represent the objects being studied, $p(x, y)$ represents the probability of x and y appearing at the same time, $p(x)$ is the probability of the Chinese character x appearing, and $p(y)$ is the probability of the Chinese character y appearing. Suppose their occurrence times in the corpus are $n(x)$, $n(y)$, $n(x, y)$, and the total word frequency in the corpus is n, here, we use sample frequency to replace probability, that is

$$p(x,y) = \frac{n(x,y)}{n} \tag{9.10}$$

$$p(x) = \frac{n(x)}{n} \tag{9.11}$$

$$p(x) = \frac{n(y)}{n} \tag{9.12}$$

Mutual information describes the closeness between two characters. When the threshold is set to 1, if $MI(x,y) > 1$, according to formula (9.9), we have

$$\log 2 \frac{p(\mathrm{x}|\mathrm{y})}{p(x)} > 1 \tag{9.13}$$

From this, we can deduce that $p(x,y) > p(x)p(y)$. There is a positive correlation between $MI(x,y)$ and $p(x,y)$. The larger the MI value, the closer the two characters are, such as the words "awkward" and "crawl." Conversely, the greater is the possibility of the two characters breaking apart. When the value of MI is greater than a certain threshold, it is considered that x and y are basically forming words; if $MI(x,y) \approx 0$, it is considered that the relationship between x and y is very weak and basically irrelevant, and if $MI(x,y) < 0$, it is considered that x and y are unrelated, that is, the two characters will not form words.

The following is an application example based on the mutual information model. For a text, first, take each character at the beginning as the object of word segmentation and judge whether m forms a word by calculating the closeness between adjacent fields in the string m. Here, m is taken as the number of words in a Chinese sentence. The judgment of closeness is obtained by counting the word frequency in the corpus. For m, if the mutual information between any two adjacent characters is greater than the threshold 1, it is considered that m forms a word and is added to the corpus; otherwise, discard the tail character and judge whether m minus 1 character forms a word. When m decreases to 2, judge whether the mutual information between the remaining two characters is greater than the threshold of 2. When MI is greater than threshold 2, add this two-character word to the corpus; otherwise, add these two characters as one-character words to the corpus. Repeat this until the entire string is processed. In the entire implementation process, the selection of two thresholds is the key to the effectiveness of the entire model.

9.4.3 Maximum Entropy Model

Entropy is a measure of the average uncertainty of a random event. Section 9.4.2 mentioned self-information; in fact, entropy is the expectation of self-information. The calculation formula for entropy is

$$H(X) = -\Sigma P(x)\log_2 p(x) \tag{9.14}$$

The role of the maximum entropy model is to choose a suitable distribution to predict possible events under known conditions. Its main idea is that when only partial knowledge about the unknown distribution is mastered, the probability distribution with the maximum entropy that conforms to this knowledge should be selected. $H(x)$ is a measure of the uncertainty of the experimental results before the experiment ends and the amount of information obtained from the experiment after the experiment ends. The larger $H(x)$ is, the greater the uncertainty, and the more information is obtained after the experiment ends.

The popular explanation of the principle of maximum entropy is not to put all eggs in the same basket, that is to say, the more evenly the eggs are distributed, the smaller the risk. The principle of maximum entropy is to accommodate all uncertainties, reduce the risk of an event to the lowest, that is, under the condition of having limited knowledge, and not make biased assumptions when predicting unknown events.

From the perspective of probability theory, the goal of the maximum entropy model is to take known events as constraints and obtain the probability distribution that can maximize entropy.

The probability distribution of the chemical refers to constructing a statistical model capable of generating the training sample distribution $p(x,y)$ and establishing a feature equation. This feature must be able to express the characteristics of the data in the training samples more completely. For example, in Chinese word segmentation task, the following feature functions can be introduced:

$$f(x,y) = \begin{cases} 1, & \text{When } x \text{ and } y \text{ satisfy certain conditions} \\ 0, & \text{when the conditions are not met} \end{cases} \tag{9.15}$$

Specified is the expectation of the empirical distribution (x,y) and the feature function $f(x,y)$, called the empirical expectation, and its formula is as follows:

$$p(f)=\sum_{x,y}p(x,y)f(x,y) \tag{9.16}$$

p(f) is the mathematical expectation of the probability distribution *p(x,y)* established by the model, called the model expectation, and its formula is as follows:

$$p(f)=\sum_{x,y}p(x,y)f(x,y)=\sum_{x,y}p\sim(x,y)f(x,y) \tag{9.17}$$

where *p*(*x*) is the empirical distribution of the random variable *x* in the training sample, that is, the probability of appearing in the sample. The constraint is the mathematical expectation of the feature function obtained by the model, which is equal to the empirical mathematical expectation of the feature function obtained from the training sample, and it is required.

$$p\sim(f)=p(f) \tag{9.18}$$

According to Eqs. (9.16) to (9.18), we can get Eq. (9.19):

$$\sum_{x,y}p\sim(x)p(y|x)f(x,y)=\sum_{x,y}p\sim(x,y)f(x,y) \tag{9.19}$$

Equation (9.19) is the constraint condition in the model. There may be multiple constraint conditions in actual application. When solving the model, we need to introduce the Lagrange function $\Lambda(p,\lambda)$:

$$\Lambda(p,\lambda)=H(p)+\sum_{i=1}^{m}\lambda i\left(p(fi)-p\sim(fi)\right) \tag{9.20}$$

In Eq. (9.20), *H*(*p*) is the maximum entropy of the model to be solved.

Assuming that the variable λ in the Lagrange function $\Lambda(p,\lambda)$ is fixed, we can find the maximum value of the unconstrained Lagrange function $\Lambda(p,\lambda)$. The maximum value of the Lagrange function $\Lambda(p,\lambda)$ when λ is fixed is denoted as $\psi(\lambda)$, and *p* when $\Lambda(p,\lambda)$ is the maximum value is denoted as $p\lambda$, that is,

$$p\lambda=\text{argmax}\ \Lambda(p,\lambda) \tag{9.21}$$

$$p\in P$$

$$\psi(\lambda)\equiv\Lambda(p,\lambda) \tag{9.22}$$

$\psi(\lambda)$ is a dual function, i.e., $\lambda*$ = argmax $i\in\{1,2,\ldots,m\}\psi(\lambda i)$ are dual functions. $p*$ = argmax $p\in$ C H(*P*) is called the primal problem, $p*$ = argmax $p\in$ C H(*P*) is the dual problem. The solution to the initial problem can be obtained by solving the dual problem, so the solution can be obtained by solving $\lambda * = \text{argmax } i \in \{1,2,\ldots,m\}\psi(\lambda i)$.

9.5 NLPIR-ICTCLAS: Chinese Word Segmentation Algorithm Based on Hierarchical Hidden Markov Model

This section presents a shallow lexical analysis based on the hierarchical hidden Markov model, aiming to integrate analysis tasks such as word segmentation, disambiguation, and recognition of out-of-vocabulary words into a relatively unified theoretical model. In terms of disambiguation, the model uses the N-shortest path strategy for ambiguity elimination; in terms of out-of-vocabulary word recognition, the method proposed in this section adopts a two-layer hidden Markov model (Hidden Markov Model, HMM) to recognize common names, place names, and complex place names and organization names that contain names and place names, calculates the probability of out-of-vocabulary words, and adds the calculation results of out-of-vocabulary words to the binary word segmentation graph.

The class-based hidden Markov model segmentation method realizes the unified competition and selection of out-of-vocabulary words and common words. The Chinese lexical analysis system NLPIR-ICTCLAS has been developed based on this theoretical framework. This section also provides other detection and processing methods in Chinese word segmentation for ambiguity and out-of-vocabulary words.

The specific practices of shallow lexical analysis based on the hierarchical hidden Markov model (Hierarchical Hidden Markov Model, HHMM) are as follows:

1. In the preprocessing stage, the N-shortest path coarse segmentation method is adopted to quickly obtain N best coarse segmentation results that can cover ambiguity.
2. On the coarse segmentation results, the lower-level hidden Markov model is used to recognize common non-nested names and place names, and the higher-level hidden Markov model is used to recognize complex place names and organization names that contain names and place names.
3. The probability of the recognized out-of-vocabulary words calculated by scientific methods is added to the class-based segmentation hidden Markov model. Both out-of-vocabulary words and ambiguities are not treated as special cases and participate in the competition of various candidate results together with common words.
4. Perform hidden Markov tagging of word properties on the globally optimal word segmentation results.

This method has been applied to the Chinese lexical analysis system NLPIR-ICTCLAS of the Institute of Computing Technology of the Chinese Academy of Sciences and has achieved good word segmentation and tagging effects.

NLPIR-ICTCLAS has achieved remarkable results in the second stage of machine translation evaluation by the 973 expert group and the first Chinese word segmentation competition held by SIGHAN and is currently one of the best Chinese lexical analysis systems.

9.5.1 Hierarchical Hidden Markov Model

The hidden Markov model is a classic statistical method for describing random processes and has been widely used in natural language processing. Compared with complex natural language phenomena, the traditional HMM is still slightly simple. Therefore, this section extends and generalizes HMM and proposes a hierarchical hidden Markov model. Here, HHMM is formalized as a six-tuple:

$$M = \langle \Omega X, \Omega O, A, B, \Pi, D \rangle \tag{9.23}$$

where ΩX is a finite state set, ΩO is a finite set of observation results, A is a state transition matrix, B is a probability matrix from the state to observation value, Π is the initial state distribution, and D is the depth of M.

The main differences between HHMM and HMM are as follows:

1. The representation of the state in ΩX is $qd = (d\in\{1, 2,\ldots, D-1\})$, where i represents the state in the current layer.
2. Each internal state $qd = (d\in \{1, 2,\ldots, D-1\})$ has substates, and the number of substates is denoted as $|qd|$.

The transition matrix of the state sequence of HMM is $A(qd) = (a_{ij}\ (q_d))$, where $a_{ij}\ (q_d) = P(q + 1\ |q + 1, qd)$.

3. Similarly, in each layer of HMM, only the Dth layer has observable terminators, that is, the output from the state to the observation.

The probability matrix is $B(qD) = (bk\ (qD))$, where $bk\ (qD) = P(ok\ |qD)$. ok is the observed value, belonging to a finite end.

Therefore, the parameter set of HHMM can be represented as

$$\begin{aligned}\lambda &= \left\{\lambda\left(q^d\right)\right\} d \in \{1,2,\ldots,D\} \\ &= \left\{\left\{A\left(q^d\right)\right\} d \in \{1,\ldots,D-1\}, \left\{\Pi\left(q^d\right)\right\} d \in \{1,\ldots,D-1\}, \left\{B\left(q^D\right)\right\}\right\}\end{aligned} \tag{9.24}$$

The point here introduces the unified model of HHMM. It provides a framework for Chinese lexical analysis based on HHMM. This model includes atomic segmentation, simple out-of-vocabulary word recognition, nested out-of-vocabulary word recognition, class-based hidden Markov segmentation, and class-based hidden

Markov tagging, which are five levels of hidden Markov models, as shown in Fig. 9.6.

The main functions of its parts are as follows:

1. N-shortest path rough segmentation. It can quickly generate N best rough segmentation results, and the rough segmentation result set can cover as many ambiguities as possible. In the entire lexical analysis framework, the binary segmentation word graph is a key intermediate data structure that integrates the recognition of out-of-vocabulary words, disambiguation, and segmentation. The recognition of out-of-vocabulary words, disambiguation, and segmentation are organically integrated.
2. Atomic segmentation. This is a preprocessing process of lexical analysis, the task of which is to segment the original string into a sequence of word atoms. Word atoms are the smallest processing units of word segmentation, including individual Chinese characters, punctuation, and non-Chinese strings composed of single-byte characters and numbers. For example, the word atom sequence corresponding to "2002.9, NLPIR-ICTCLAS's open source code began to be released" is "2002.9/,/NLPIR/-/ICTCLAS/的/自/由/源/码/开/始/发/布/". In this layer of HMM, the terminators are all characters in written language; the state set is word atoms, and the training and solving of the model are relatively simple, so they are not repeated.

9.5.2 Class-Based Hidden Markov Segmentation Algorithm

The class-based Hidden Markov Model (HHMM) word segmentation algorithm is at the second level, that is, it is performed after all unregistered words are recognized. First, all words can be divided into nine classes. Each word in the core dictionary corresponds to the class of the word itself. Assuming the number of words included in the core dictionary Dict is |Dict|, then the total number of word classes defined is

‡ Dic# +6.

Given a word segmentation atomic sequence S, some possible word segmentation result of S is denoting W=(W1,W2,... ,Wn),W corresponding to the class sequence is denoted by C=(C1,C2,... At the same time, take the word segmentation result W# with the highest probability as the final word segmentation result, then

$$W^{\#} = \arg_{w} \max P(w) \tag{9.25}$$

Expand using Bayes' theorem:

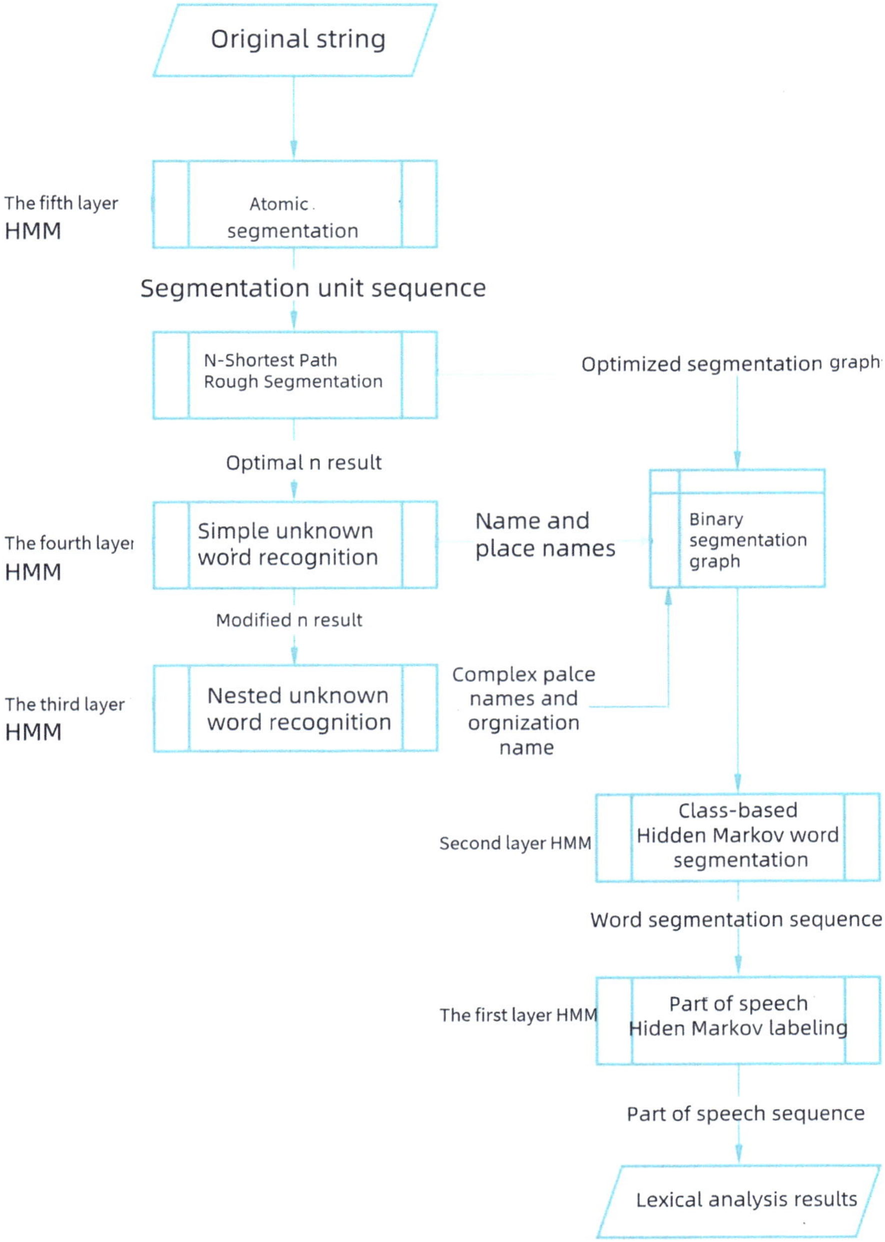

Fig. 9.6 Framework of Chinese lexical analysis based on HHMM

$$W^{\#} = \arg_w \max P(W|C)P(C) \tag{9.26}$$

Considering the word classes in Fig. 9.7 as states and words as observations, using a first-order HMM to expand, we get

$$W^{\#} = \arg_w \max \prod_{i=1}^{n} p(\omega_i|c_i)P(c_i|c_{i-1}) \tag{9.27}$$

where c_0 is the start marker BEG of the sentence.

$$W^{\#} = \mathrm{argmax}_w \sum_{i=1}^{n} \left[-\ln P(\omega_i|c_i) - \ln P(c_i|c_i - 1)\right] \tag{9.28}$$

For convenience of calculation, negative logarithms are often used for operation; then, we have simplified. Therefore, in the process of word segmentation, only the $p(\omega_i|c_i)$ of unregistered words needs to be considered. When applying the class-based word segmentation HMM in actual problems, whether segmentation ambiguity can be integrated and eliminated in this model is a difficult question; another key question is how to determine the unregistered word ω_i, identify its category c_i, and calculate a reliable $p'(\omega_i|c_i)$.

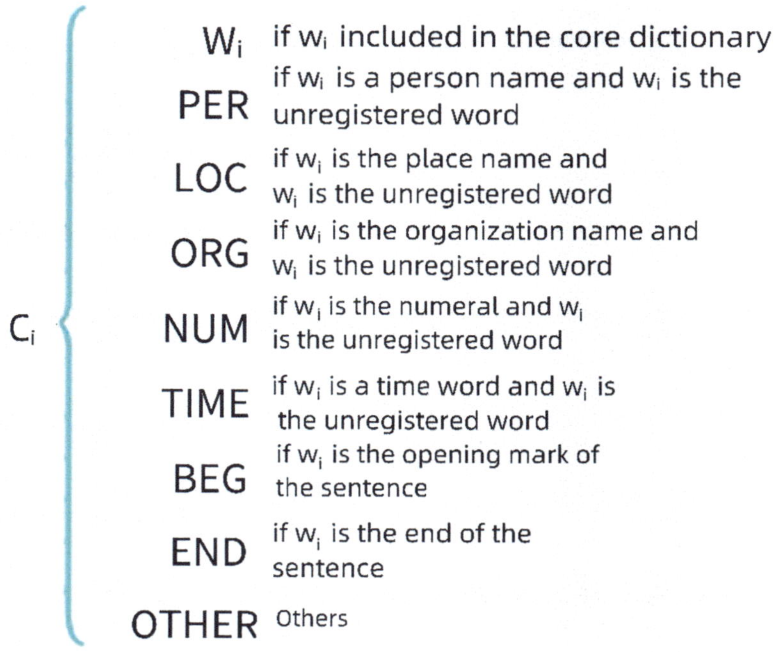

Fig. 9.7 Classification of words

9.5.3 N-Shortest Path Segmentation Disambiguation Strategy

From a morphological perspective, ambiguity is divided into cross-ambiguity and combination ambiguity. "Combine/into/molecule/when" is a typical cross-ambiguous field, and "this/person/hand/has/mole" forms a combination ambiguous field. From the perspective of disambiguation, ambiguity can be divided into global and local ambiguity. Global ambiguity refers to ambiguity that can only be accurately eliminated by combining the context of the current sentence. For example, "The ping pong paddle is sold out," in the absence of context, can be reasonably segmented into "ping pong/bid/finished" and "ping pong paddle/sold/finished." In contrast, local ambiguity can be eliminated within the sentence. According to statistics from large-scale corpora, local ambiguity accounts for the vast majority, and global ambiguity can almost be ignored.

The N-shortest path segmentation disambiguation strategy is adopted here. The basic idea is as follows: In the initial stage, keep the N results with the highest segmentation probability $P(W)$ as the candidate set of segmentation results, and after unrecognized word recognition, part-of-speech tagging, and other lexical analysis, calculate the accurate optimal result through the final evaluation function. The N-shortest path method is a generalization and integration of the least segmentation method and the complete segmentation method, avoiding the problem of the least segmentation method discarding a large number of correct results on the one hand and solving the problem of the whole segmentation method having a large search space and low operation efficiency on the other hand. This method can maximize the retention of ambiguous fields and unrecognized words by retaining a few high-probability rough segmentation results. Common segmentation methods are often too arbitrary, making whether to segment judgments too early in the initial stage, only retaining one result that the algorithm thinks is optimal. This result often makes mistakes due to the existence of ambiguity or unrecognized words, and the subsequent remedial measures are usually time-consuming and laborious, and the effect is not very good.

Table 9.3 provides a comparison of the eight shortest path methods and four common segmentation methods in terms of ambiguity inclusion in segmentation results.

At the same time, an open ambiguity test was conducted on the only segmented annotation result finally selected. The test set is 120 pairs of common combination ambiguities and 99 pairs of common cross-ambiguities collected by the Computational Linguistics Institute of Peking University. The final success rates of eliminating combination ambiguity and cross ambiguity are 80.00% and 92.93%, respectively.

Table 9.3 Comparison of eight shortest path methods and four common segmentation methods

Method	Maximum number of splits	Average number of cuts	Correct segmentation coverage %
Eight shortest path	8	5.82	99.92
Maximum match	1	1	85.46
Minimum split	1	1	91.80
Maximum probability	1	1	93.50
Full segmentation	>3424507	>391.79	100.00

Note:

1. The maximum number of segmentations refers to the maximum number of possible segmentation results for a single sentence

2. The average number of segmentations refers to the average number of segmentation results for a single sentence

Number of sentences/total number of sentences

9.6 Lexical Analysis Based on Bidirectional Recurrent Neural Network and Conditional Random Fields

9.6.1 Overview

Word segmentation and part-of-speech tagging are essential techniques in Chinese language processing, widely used in lexical analysis, semantic understanding, machine translation, information retrieval, and other fields. In lexical analysis, it is necessary to determine the position of each character in its corresponding word in a sentence. Accordingly, the word segmentation problem is usually transformed into a sequence tagging problem for each character. The tags are $\{B, M, E, \text{and } S\}$; these four letters, respectively, represent Begin, Middle, End, and Single. Traditional lexical analysis techniques include dictionary-based maximum matching word segmentation (its disadvantage is that it heavily relies on the dictionary and cannot handle word segmentation ambiguity and out-of-vocabulary words well), full segmentation path selection (its idea is to represent all possible segmentations as a directed acyclic graph, with each possible segmented word as a node in the graph), character sequence tagging method (tagging each character in the sentence), methods based on Hidden Markov Networks, and methods based on Conditional Random Fields (CRF), among others. Traditional methods may have some limitations, for example, algorithms like maximum forward matching rely entirely on the dictionary for word segmentation and do not consider the semantic information of the text, and their performance could be better in tasks such as semantic disambiguation and new word recognition. Some researchers have proposed a word segmentation model based on hidden Markov networks, which effectively solves problems such as new word recognition. With the widespread application of the Conditional Random Field (CRF) algorithm, there has been some progress in its application in lexical analysis. With the development of computer hardware such as GPUs, computational techniques based on large-scale neural networks have been developed and applied.

Deep neural network methods generally do not require manual feature construction and have achieved more significant performance improvements than traditional machine learning methods in many applications. In addition, some variants of the classic RNN algorithm (such as LSTM and GRU) can effectively capture information in a more extended time domain than the traditional RNN algorithm. Since a unidirectional RNN can only capture unidirectional dependencies in the sequence, this section proposes a lexical analysis algorithm based on bidirectional recurrent neural network learning of the forward and backward dependencies in the entire text sequence. Since the GRU merges the gates in the LSTM, it trains faster and has fewer parameters to process, so it uses GRU for sequence feature learning, replaces the manually constructed features with the learned features, and uses CRF for the final sequence tagging.

9.6.2 Sequence Tagging Based on Bidirectional Recurrent Neural Networks

First, each character in the training set is converted into a corresponding number through the dictionary V, and then the character-embedding matrix $E \in R|V| \times DE$ is initialized, where $|V|$ represents the total number of characters in the dictionary and DE represents the length of the character vector. Each row of E represents the vector of its corresponding Chinese character. Using this method, each character in the dataset and its neighboring characters are first converted into the corresponding character vectors. For convenience of description, the range of a few characters adjacent to a certain character will be referred to as a local window. Then, all the character vectors in the local window are concatenated, and the concatenated vector is sent into the RNN to obtain the vectorized representation of the current character. Finally, the vectorized representation of the character is used, combined with the CRF algorithm, to determine the final tag of the current character sequence.

For the convenience of matrix operations, the entire article in the training set is segmented by sentence. Considering the actual situation of processing Chinese, the maximum length of a sentence can be set to a certain empirical threshold. Sentences in the training set that exceed this threshold length are discarded, and for sentences shorter than this threshold, placeholders can be used to pad them. In the experiments in this section, this threshold can be set to 30. In this way, a sentence is converted into a real-number matrix, with one dimension of the matrix being the width of the local window.

First, a sentence is converted into a vector sequence through embedding, and then it is used as the input of the recurrent neural network. At the same time, in order to better capture the dependencies between the characters in the sequence, a bidirectional stacked recurrent neural network is used for sequence feature learning. The recurrent neural network model outputs a feature vector at each step, and these feature vectors will be fed into a fully connected network for further feature extraction. Here, GRU is used as the recurrent unit of the recurrent neural network. Then, the Softmax probability vectors for each annotation are calculated and output. In sequence annotation, the annotation result at the current position is likely to be related to the information at the previous and next positions. Adding local window data information at the current position will help improve the level of sequence annotation.

Suppose a sentence s has n characters, $|s| = n$. uses $t\ (1 \leq t \leq n)$ to represent the t-th position in sentence s.

$$s^{(t:t+w)} = \left\{s^t, s^{t+1}, \ldots, s^{t+w-1}, s^{t+w}\right\} \tag{9.29}$$

Take $s^{(t:t+w)}$ as the input of the bidirectional recurrent neural network at the t-th moment, use two-directional recurrent neural networks to process this character vector sequence separately, and then concatenate the outputs of the recurrent neural networks in these two directions. The formalized representation of the related operations is as follows:

$$\rightarrow ht = \rightarrow \mathrm{GRU}\left(s^{(t:t+w)}, \rightarrow ht-1\right) \tag{9.30}$$

$$\leftarrow ht = \leftarrow \mathrm{GRU}\left(s^{(t:t+w)}, \leftarrow ht-1\right) \tag{9.31}$$

$$\leftrightarrow ht = \left[\rightarrow ht, \leftarrow ht\right] \tag{9.32}$$

As can be seen, merging the outputs of the recurrent neural networks in the left and right directions of the current character for annotating the character at the current position takes into account the influence of the surrounding characters on the character at the current position. Its formalized representation is as follows:

$$v^t = \mathrm{Softmax}\left(W \cdot \leftrightarrow ht + b\right) \tag{9.33}$$

where W and b are the parameters of the fully connected layer, and the output is the probability vector of each annotation category.

Finally, the annotation corresponding to the position with the highest probability is taken as the annotation of the current sequence information.

In terms of the loss function of the model, the cross-entropy is calculated for the difference between the predicted output value and the real annotation, and then the Adam (adaptive moment) method is used to minimize the cross-entropy loss function. Adam is a method that uses different parameters to adaptively adjust different learning rates. The difference between Adam and Adadelta and RMSprop lies in the way they calculate the decay of historical gradients. It does not use historical square decay, and its method is similar to momentum, which makes the model less likely to fall into saddle points during optimization and can make the model converge faster.

9.6.3 Deep Neural Network Model Integrating Conditional Random Fields

When seeking the final annotation result on the probability vector, if only a greedy strategy is used, it is difficult to consider the constraints on one output to the next, which are obvious drawbacks. In related work, HMM or CRF algorithms are also used to calculate and utilize this inherent constraint relationship between vocabularies. The output of the Bi-GRU network replaces the traditionally handcrafted features based on CRF segmentation, and CRF is used for the final annotation output after Bi-LGRU. This method considers the constraint relationship between characters and can reduce some unreasonable outputs. Its formal representation is as follows:

$$\text{tag} = \text{CRF}\left(vt\,i\right), i = 0,1,2,3 \tag{9.34}$$

Due to the use of the CRF layer, it is no longer just taking the difference between the single moments.

Output and the real mark as the basis for calculating loss, but scoring the entire sentences predicted annotation jointly. For this problem, the overall optimization goal is to make the possibility of outputting the correct annotation sequence *Y* when the input *X* is the greatest. This possibility is represented as

$$P\left(Y|X\right) = \frac{\Omega\left(Y|X\right)}{\sum_{yEQ} \Omega\left(Y'|X\right)} \tag{9.35}$$

where $\Omega(Y|X)$ represents the possibility (probability) of outputting the annotation sequence *Y* when the input is *X*, *n* is the sequence length, *Qn* represents all possible prediction sequences, and *Y*′ represents one of the possible prediction sequences.

The scoring function Ω considers the influence between two adjacent characters, and the definition of Ω is shown in formulas (9.36) and (9.37):

$$\Omega\left(Y|X\right) = \sum_{t=2}^{n} \omega\left(X,t,y_{t-1},y_t\right) \tag{9.36}$$

$$\omega\left(X,t,y_{t-1},y_t\right) = \exp(s\left(X,t\right)y_t + Ay_{t-1},y_t \tag{9.37}$$

where Ay_{t-1}, y_t represents the transition probability matrix from the y_{t-1} annotation to the *yt* annotation, ω represents the probability of making the current decision considering the output of the previous moment and the original input, *X* represents the input character sequence, y_{t-1} and y_t represent the predicted output of the previous moment and the current moment, $s(X,t)$ is a $|Q|$ long probability vector, representing the probability of outputting various labels at time *t*, and its definition is as follows:

$$s(X,t) = W_s h_t + b_s \tag{9.38}$$

where W_s is the weight matrix, h_t is the hidden state output by the bidirectional RNN network, b_s is the bias term, $s(X,t)$ y_t represents the output probability of y_t, i.e., $s(X,t)$ $y_t \in [0,1]$.

9.7 Application and Analysis

9.7.1 NLPIR-ICTCLAS Application Demonstration

The big data semantic intelligence analysis platform NLPIR [1] is a free Chinese word segmentation and tagging software based on the HMM model developed by the Zhang Huaping team, with the advantages of high accuracy, fast speed, and strong adaptability. NLPIR, catering to the needs of big data content processing, integrates 13 functions such as precise network collection, natural language understanding, text mining, and network search technology and provides client tools, cloud services, and secondary development interfaces using NLPIR-ICTCLAS.

9.7.2 LTP

LTP (Language Technology Platform) is a complete set of open Chinese natural language processing systems developed by the Harbin Institute of Technology's Social Computing and Information Retrieval Research Center over 10 years. LTP has established XML-based language processing result representation rules and provides a complete set of rich, efficient, and high-precision Chinese natural language processing modules, application programming interfaces, visualization tools, and language technology clouds that can be used in the form of network services. The LTP interface [2] is shown in Fig. 9.8.

9.7.3 Jieba Segmentation

Jieba (jieba) segmentation is a very popular open-source Chinese word segmentation package, with high performance, high accuracy, and high scalability. It supports three segmentation modes: precise mode, full mode, and search engine mode, and currently mainly supports Python.

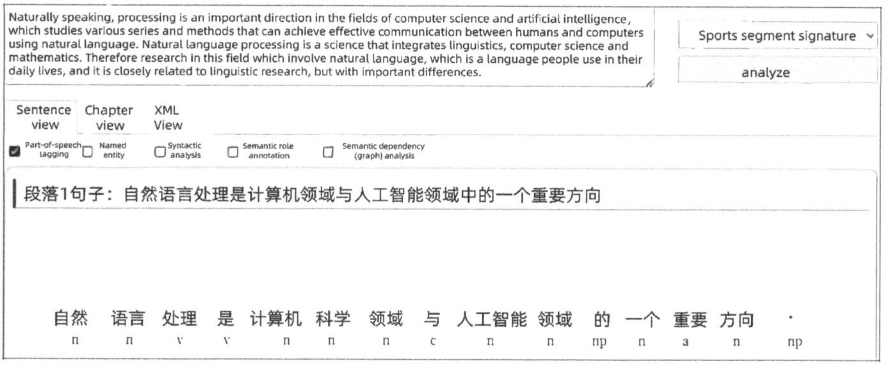

Fig. 9.8 LTP interface

9.7.4 *PKUSeg*

PKUSeg is a Chinese word segmentation toolkit developed by the Language Computing and Machine Learning Research Group of Peking University. It is simple to use, supports fine-grained domain segmentation, effectively improves the accuracy of word segmentation, and has features such as multidomain word segmentation, high word segmentation accuracy, support for user self-training models, and part-of-speech tagging.

References

1. http://www.nlpir.org
2. http://ltp.ai/demo.html

Part III
Semantic Analysis

Chapter 10
Sentiment Analysis

With the rapid development of information technology and the mobile Internet, quickly and accurately mining users' emotional tendencies toward events and products is essential to grasping social public opinion, adjusting product strategies, and understanding user preferences. Sentiment analysis plays a vital role in this. Especially in recent years, with the development of multimodal sentiment analysis and emotional dialogue, the practical application prospects of sentiment analysis have been further broken through. This chapter systematically introduces sentiment analysis and its methods, mainly including sentiment analysis methods based on sentiment dictionaries, machine learning, and deep learning, and finally gives examples of sentiment analysis applications.

10.1 Overview of Sentiment Analysis

With the rapid development of social media, many user comment information has appeared on platforms such as Weibo, Zhihu, and Facebook. Most of this information contains users' subjective emotional colors or emotional tendencies, such as joy, anger, sorrow, affirmation, denial, etc. With the rapid development of e-commerce, the amount of user comment information on the Internet has increased in recent years. There is a lot of helpful information behind many user comment information. The sentiment information contained in the comment information can be extracted by analyzing this user comment information, thereby understanding the user's views on a specific event or product and providing a reference for the corresponding analysis task. For example, before making a purchase, people can search for relevant comments on the product online to learn more about the product; manufacturers can also understand consumers' views on the product by searching for relevant comments and then adjusting the quality or quantity of the product. As the number of online users increases, user comment information rapidly expands. It is

H. Zhang, J. Shang, *Natural Language Processing and Applications*,
https://doi.org/10.1007/978-981-97-9739-4_10

impossible to deal with this massive amount of user comment information by relying solely on manual analysis and processing. There is an urgent need for an automated analysis method to analyze the massive user comment information, so sentiment analysis was born and developed rapidly.

Text sentiment analysis, also known as opinion mining, refers to the mining and analysis of the text's subjectivity, viewpoint, emotion, and emotional polarity through computing technology and making classification judgments on the emotional tendency of the text. With the development of computer and network technology, people have begun to study how to make computers understand and use human social and natural language. This research has achieved fruitful results, which have laid the foundation for text sentiment analysis. Text sentiment analysis is a crucial research branch in the field of natural language understanding, involving theories and methods in statistics, linguistics, psychology, and artificial intelligence.

Sentiment analysis can help intelligent systems that need to understand user emotions, and its applications are extensive. For example, sentiment analysis-based technology can be used to conduct public sentiment surveys and public opinion analysis, which have higher accuracy and efficiency than traditional questionnaire surveys, and the cost is significantly reduced. In corporate decision-making, analyzing platform feedback and related market information can provide practical decision-making support, helping enterprises make better-informed decisions. In the consumer field, sentiment analysis serves a dual purpose: it helps consumers form comprehensive evaluations of products, while also enabling enterprises to understand market perceptions of their products and make targeted optimizations. In application-oriented research, sentiment analysis can leverage user-generated content on social networks to predict public sentiment and trends.

10.1.1 Research Task

The core task of text sentiment analysis is to mine user opinions from the text, analyze them, and judge their emotional tendencies. Sentiment analysis can be seen as a classification task, usually with two processing strategies: One strategy is to directly classify the text into neutral, positive, and negative, that is, a three-classification at once; another approach is to first classify the text into subjective and objective, that is, to determine whether the text contains emotional color, and then to classify the subjective text containing emotional color into positive or negative. This strategy is generally referred to as a two-step binary classification strategy.

According to the granularity of the text, text sentiment analysis can be divided into recognition and analysis at three levels: words, sentences, and articles.

Word sentiment analysis, as the cornerstone of text sentiment analysis, plays a pivotal role in determining text sentiment and setting the stage for sentence and article sentiment analysis. The main research on word-based sentiment analysis includes emotion word extraction, emotion word judgment, corpus, emotion dictionary, etc., underscoring its significance in the process.

Sentence sentiment analysis, as the heart of text sentiment analysis, is where the analysis results of emotional words are integrated to give a comprehensive result of sentence sentiment analysis. Moreover, a sentence can be seen as a short article, and the result of sentence sentiment analysis largely determines the result of article sentiment analysis, further highlighting its central role.

Article sentiment analysis is the most uncertain because it needs to integrate the sentiment analysis results at all levels of the text, combined with the context and domain knowledge base, to make judgments.

10.1.2 Research Hotspots

Table 10.1 is the model performance ranking on the SST-2 dataset (2021). The SST-2 dataset is a binary sentiment analysis benchmark dataset, where the data are divided into positive and negative categories.

From Table 10.1, we can see that, except for the model ranked third, the other six models are all pre-trained Transformer models. The 6:1 ratio of the listing fully illustrates the performance of the pre-trained Transformer model. However, we can also see some problems from here: From the timeline, the model proposed in 2019 can still top the list 2 years later, in 2021, which shows that the performance of traditional sentence-level sentiment analysis is temporarily impaired. The progress has

Table 10.1 Model performance ranking on the SST-2 dataset (2021)

Ranking	Model	Accuracy %	Paper proposing the model	Years
1	SMART-RoBERTa Large	97.5	SMART: Robust and Efficient Fine-Tuning for Pre-trained Natural Language Models through Principled Regularized Optimization	2019
2	T5-3B	97.4	Exploring the Limits of Transfer Learning with a Unified Text-to-Text Transformer	2019
3	MUPPET Roberta Large	97.4	Muppet: Massive Multitask Representations with Pre-Fine-Tuning	2021
4	ALBERT	97.1	ALBERT: A LiteBERT for Self-Supervised Learning of Language Representations	2019
5	T5-11B	97.1	Exploring the Limits of Transfer Learning with a Unified Text-to-Text Transformer	2019
6	StructBERTRoBERTa ensemble	97.1	StructBERT: Incorporating Language Structures into Pre-training for Deep Language Understanding	2019
7	XLNet	97.0	XLNet: Generalized Autoregressive Pretraining for Language Understanding	2019

entered a bottleneck period, and it has become quite difficult to continue to improve performance in a short period of time. Therefore, people have shifted their focus from sentence-level sentiment analysis and started to study multimodal sentiment analysis, attribute-level sentiment analysis, cross-domain sentiment analysis, dialogue emotion analysis, etc.

10.2 Classic Methods

Sentiment analysis is a significant research hotspot in natural language processing at home and abroad, and its results have a promoting effect on fields such as data mining, information retrieval, and public opinion analysis. Existing sentiment analysis technologies are mainly divided into methods based on sentiment dictionaries, machine learning, and methods based on deep learning. At present, the results obtained by the process based on deep learning are quite high in terms of accuracy and precision, but correspondingly, this method requires a large amount of annotated data and hardware computing resources.

10.2.1 Sentiment Analysis Method Based on Sentiment Dictionary

The sentiment analysis method based on the sentiment dictionary first performs word segmentation on the text to be classified, then refers to the existing text dictionary, checks the sentiment polarity corresponding to each word in the word segmentation result, obtains the classification result of each word, and finally calculates the overall sentiment classification result. From its processing process, it is easy to understand that the sentiment analysis method based on the sentiment dictionary mainly includes two aspects of work: the construction of the sentiment dictionary and the text sentiment calculation based on the sentiment dictionary.

The earliest sentiment analysis method is based on the sentiment dictionary. The early sentiment dictionaries were obtained through manual construction, and the sentiment dictionaries were relatively simple, with fewer sentiment words and only some common words, such as "great," "happy," "beautiful," etc. Hong Wei and others [1] pointed out that when using rule-based methods for analysis, most researchers focus on constructing the sentiment dictionary because the calculation methods are basically the same. The core of the machine learning method is to obtain text features. Liu Kaiyuan [2] mentioned that machine learning classifiers mainly include KNN, maximum entropy, SVM, etc. Cao Haitao [3] used PAD emotional semantic features on SVM based on these several machine learning methods and found that the experimental effect is better than other features. With the rise of deep learning, many researchers have begun to use neural networks for sentiment classification. Because LSTM can save sequence information, Li and some researchers [4] proposed a method of using LSTM to extract text sentiment features in sentiment analysis. Hassan and others [5] used LSTM in deep learning methods to replace the pooling layer, reducing the loss of local detail information. In general, these methods are analyzed on a sentence-by-sentence basis, extracting features of sentences, classifying them with neural networks, or scoring sentences according to syntax and sentence-type rules. The features they focus on are relatively single, and they do not put the elements of sentiment analysis into the same system, which affects the accuracy of sentiment analysis.

Traditional sentiment dictionary construction methods are mainly divided into three types: manual annotation, using public dictionary resources, and using statistics to calculate datasets. This book divides the construction methods of sentiment dictionaries into four types: methods based on the semantics of candidate sentiment words, methods using statistical knowledge, graph-based methods, and deep learning models.

Semantic-based methods first select some seed sentiment words and then use synonyms and antonyms to expand the sentiment words. For example, by using other people's public sentiment dictionaries, choose some words as seed words and then judge the related sentiment words through the type of associated words and the seed words before and after the associated words.

10.2.2 Sentiment Analysis Methods Based on Machine Learning

Although the sentiment analysis method based on the sentiment dictionary has achieved good results in some fields, due to the rapid expansion of information today, many new words constantly appear, and the dictionary needs to be frequently expanded and updated. In addition, the construction of the sentiment dictionary is usually for a specific field, time period, and language environment. The same vocabulary may express emotions in different fields, time periods, and language environments. For example, "hehe" originally expressed joy, but now it is often used to express speechlessness. Therefore, the sentiment analysis method based on the sentiment dictionary sometimes cannot achieve the ideal effect for expanding and updating these new words and variant words, and it often cannot achieve good results for cross-domain sentiment analysis. At the same time, the sentiment words in the sentiment dictionary are limited, so it works well for short text analysis, but more is needed for long text. In view of the various shortcomings of the sentiment analysis method based on the sentiment dictionary, the sentiment analysis method based on machine learning is essential.

Convolutional neural network (CNN) has become a commonly used machine learning model by many scholars. Yang Rui and others [6] studied the text classification method based on convolutional neural networks. Zhang [7] and others' research shows that the deep convolutional neural network based on characters performs well for text classification. The convolutional neural network model is applied to many tasks. Convolutional neural networks can extract local N-gram features in text, but they may not be able to capture long-distance dependencies, while long short-term memory (LSTM) networks can solve this problem by sequentially modeling text. Convolutional neural networks and recurrent neural networks (RNNs) are often combined with models based on sequences or tree structures. Experiments show that the convolutional neural network is a substitute method that can avoid the problem of high computation of neural networks, but compared with other methods, it requires more training time.

Due to the ability of recurrent neural networks (RNNs) to capture information in relatively flexible computations, they have been widely used in supply chain management. Compared to convolutional neural networks (CNNs), RNNs have two important features. First, CNNs have different parameters at each layer, but RNNs have the same parameters at each layer. In RNNs, one stage's output depends on the previous stage's output, requiring much memory space. Therefore, RNNs have an advantage over CNNs in processing sequential information. In addition, there are some other extensions of RNNs, such as bidirectional recurrent neural networks (BRNNs) [8]. RNNs contain a forward layer and a backward layer to learn information from tokens from the front and back. When RNNs handle document-level sentiment classification problems, they establish sentence representations and aggregate them into document representations, thus obtaining hierarchical representations. Furthermore, combining long short-term memory (LSTM) networks and RNNs has resulted in Bidirectional Long-Term Memory (BLSTM) networks, which can access all input directions for context and more information. Miao et al. [9] proposed a voice conversion method based on BLSTM and WaveNet to improve voice quality. Therefore, BLSTM can also consider the connections between and within sentences.

Recursive neural networks (RNNs) are a generalization of RNNs, applying the same set of weights recursively on directed acyclic networks, but the input segment is a tree structure. CNN models are language-driven as they explore tree structures and try to learn complex combination semantics. The tree structure of recursive neural networks includes constituency trees and dependency trees. In constituency trees, leaf nodes represent words, internal nodes represent phrases, and the root node represents the entire sentence. In dependency trees, each node can define a word connected to other nodes with dependency connections. In recursive neural networks, the vector representation of each node is calculated from all its child nodes using a weight matrix. Ren et al. [10] proposed an algorithm for a new hybrid parameter recursive neural network composed of two virtual unidirectional recursive neural networks. Support vector machines (SVMs), as supervised learning models capable of effectively analyzing data, are a new type of machine learning method based on statistical learning theory. They are used for regression analysis and classification applications related to machine learning algorithms. In recent years, they have gradually become a research hotspot in machine learning due to their excellent performance. Support vector machines can handle categorizing commonly used emotional expressions, which are evaluated based on the measurements' accuracy, precision, and recall. An improved sentiment analysis method with advanced preprocessing has been proven to provide better results. Cai et al. [11] proposed a three-layer emotional dictionary, which can link emotional words with the corresponding entities and aspects, reducing the multiple meanings of emotional words. This model more comprehensively predicts the process features of emotional evolution by describing and calculating emotional dynamic features. In the future, classification techniques that blend other models can be used to improve accuracy.

The Bidirectional Encoder Representation from Transformer (BERT) is a neural network-based natural language preprocessing technique. The BERT model can be fine-tuned through the input and output layers to create models for various text analysis tasks. The core of BERT is the Transformer technology, which is very suitable for natural language processing tasks based on encoding–decoding models and attention mechanisms. Compared to the support vector machine model, BERT can perform better when the data volume is large, and the processing performance will significantly improve. For example, the COVID-19 pandemic evolved into a global pandemic from 2020 to 2022. Public health issues are related to public infection prevention and the psychological state of the public experiencing the pandemic. Therefore, analyzing social media generates negative emotions.

Analyzing social media data can help understand the public's experience during the COVID-19 pandemic and provide references for preventing other diseases. Wang et al. [12] analyzed the evolution of public sentiment over time during the COVID-19 pandemic and the topics related to negative emotions on Weibo. The experiment shows that BERT has superior feature extraction capabilities, which can improve the performance and stability of sentiment classification and speed up the convergence speed. BERT can be used to analyze the sentiment of the exact text in three languages. Li Yanhui et al. [13] provided an effective solution for the sentiment analysis problem of multilingual text.

10.2.3 Deep Learning-Based Sentiment Analysis Method

The deep learning method was first applied to natural language processing by Collobert et al. [14] in 2011, to solve problems such as part-of-speech tagging, Subsequently, Grefenstette et al. [15] proposed a wide convolutional model and chose to use k-max pooling instead of traditional CNN's max pooling to retain more features. CNNs are often used to capture local features. RNNs, due to their feedback loop structure, can maintain memory information and have been well applied in time series models. However, RNNs have certain drawbacks. They perform well on short texts, but when the text length increases, RNNs may experience gradient vanishing and gradient explosion. Long short-term memory networks and gated recurrent units (GRU) introduced a gate mechanism based on traditional RNNs, effectively solving the problems of RNNs. Socher et al. [16] obtained more text features by constructing Tree-LSTM. In order to enhance the model's ability to handle historical information, an external memory unit was additionally introduced based on LSTM, but due to the increase in a large number of parameters, the model's accuracy could have improved significantly. Chen et al. [17] achieved good classification results using a bidirectional LSTM with an attention mechanism. Wang et al. [18] established AE-LSTM and ATAE-LSTM neural network models by modeling LSTM using hidden layers and attention mechanisms and finally obtained aspect-level sentiment classification. However, LSTM also has drawbacks; although it can get the semantic information of the text context, it needs to acquire local information about the text.

In order to take advantage of both RNN and CNN, researchers began to merge these two types of models for text sentiment analysis. Wang et al. [19] constructed a fusion model by fusing a single-layer RNN and a single-layer CNN and applied it to short-text data.

Experiments were conducted on the dataset, proving that the fusion model performs better than a single model. Yoon et al. proposed a multichannel CNN_BiLSTM model combined with a dictionary, where the text and dictionary are input into the multichannel CNN simultaneously. The text features are obtained and then input into the BiLSTM for classification after combining them. Vo and others proposed a parallel multichannel CNN and LSTM model, which inputs the text into multiple channels of CNN and LSTM for word embedding, then combines the text features output by these two models, and finally inputs them into a fully connected neural network for classification. Experiments show that this model performs better in Vietnamese sentiment classification than a single model.

10.2.4 Advanced Models

Table 10.2 shows the SOTA methods for sentiment analysis tasks on 10 datasets. It can be seen that the SOTA methods on these 10 datasets basically use the Transformer model and its improved version, which can highlight the excellent performance of large-scale or super-large-scale pre-trained language models trained with a large amount of data on specific sentiment analysis tasks.

10.3 Applications and Analysis

NLPIR sentiment analysis has a rich sentiment classification, including positive and negative and specific emotional attributes such as good, joy, surprise, anger, evil, sorrow, and fear. NLPIR also provides sentiment analysis about specific characters and can calculate the specific positive and negative scores. Figure 10.1 is an example of NLPIR sentiment analysis results. NLPIR sentiment analysis provides two modes: sentiment judgment of the whole text and sentiment judgment of specified objects. NLPIR sentiment analysis mainly uses two technologies:

Table 10.2 SOTA methods for sentiment analysis tasks on 10 datasets

S/N	Dataset	Optimal model
1	SST-2 Binary Classification	SMART-RoBERTa Large
2	IMDb	XLNet
3	SST-5 Fine-Grained Classification	RoBERTa-large + Self-Explaining
4	Yelp Binary Classification	XLNet
5	MR	VLAWE
6	Yelp Fine-Grained Classification	XLNet
7	User and Product Information	MA-BERT
8	Amazon Review Polarity	BERT large
9	Amazon Review Full	BERT large
10	CR	RoBERTa + DualCL

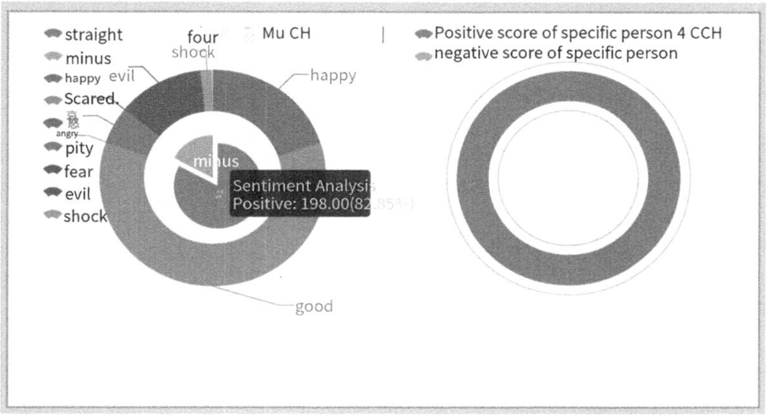

Fig. 10.1 Example of NLPIR sentiment analysis results

1. Automatic recognition and weight calculation of sentiment words. Using co-occurrence relationships, adopting the Bootstrapping strategy, iterating repeatedly, generating new sentiment words and weights.
2. Deep neural network for sentiment judgment. Based on the deep neural network, the sentiment words are expanded and calculated, and integrated into the final result.

In addition, Zhang Baohua and others [20] have constructed a hierarchical system for sentiment analysis based on the structural attributes of natural language itself and the commonly used features in sentiment analysis work, as shown in Fig. 10.2. This sentiment analysis hierarchy can incorporate the features used in sentiment analysis into the same system, realize the conversion between various features, and provide more features for sentiment analysis work.

The weight of the sentiment unit can be obtained from both the construction of the sentiment unit and the context of the sentiment unit. However, both methods

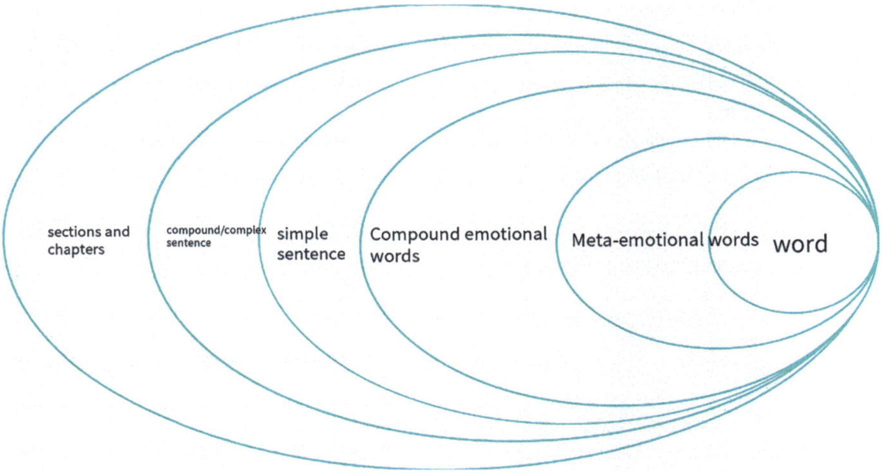

Fig. 10.2 Hierarchy of sentiment analysis

only emphasize one aspect. The real emotional weight of the sentiment unit is related to both the construction of the sentiment semantic unit and the context of the sentiment semantic unit. Only by combining these two methods can we get an accurate emotional weight. For a sentiment unit W_S,

Suppose W_S let $W_S = L_0......L_N$, the upper layer of W_S is C_s, then its emotional weight calculation formula is as follows:

$$\text{weight}(W_S) = \lambda F(\text{score}(C_s)) + (1-\lambda)(P(W_s^* | L_0 ......L_n)) \tag{10.1}$$

where $0 < \lambda < 1$ F is the calculation function score, and C_s is the weight of C_s of the compound sentiment word composed of W_S, which is derived from the sentiment value of its sentence. $P(W_s^* | L_0......L_n)$ represents the method of calculating the weight of the sentiment unit based on the known sentiment dictionary and the construction of the current sentiment unit.

The quality and quantity of the dictionary obtained by either the character-based method or the context-based method depend on the selected sentiment dictionary, and both methods will generate new sentiment words. If these new sentiment words are added to the original sentiment dictionary, additional sentiment words will be generated, so this method is a continuous iterative process.

The specific implementation steps are as follows:

1. Initialize the sentiment dictionary Ds; set the weight to 1 and −1 according to the praise and criticism of the sentiment words.
2. Discover new words from the dataset C, then segment the words, select nouns, verbs, adjectives, and adverbs as candidate sentiment words # $W_s^{\#}$ according to the part of speech.

3. For the adjectives in the candidate words, calculate the sentiment tendency probability $P(L_i| W_s^*)$ of their constituent characters separately.
4. Calculate the tendency and weight of the adjectives in the candidate words to get a new sentiment dictionary DL.
5. Split the sentences of C into simple sentences S_s according to the conjunctions and punctuation marks and then calculate the sentiment value of a single sentence based on the compound sentiment words C_s in the simple sentence.
6. For the sentiment single sentences with a sentiment score of 0, make the following deductions: If the sentence is the first sentence, guess the sentiment value of the sentence based on the relationship between the sentence and the next sentence; if the sentence is a middle sentence, first determine whether the simple sentence is the end of the previous sentence or the beginning of the next sentence, then guess the sentiment value of the sentence based on the relationship; if the sentence is the last sentence, guess its sentiment value based on the relationship between sentences.
7. For the candidate sentiment words, find all the sentiment single sentences that contain them, calculate their sentiment weights, and get a new sentiment dictionary DC.
8. Set the confidence level and frequency threshold for DL and select sentiment words based on the threshold.
9. Merge DL and DC to get DN. If there is a contradiction between the two, DC's will prevail.
10. Use DN as the new sentiment dictionary, repeat (4)–(10) until the sentiment polarity of the words in the dictionary no longer changes and the weight changes within a very small range.

The baselines chosen by the author mainly include the following:

1. Character-based method
2. Context-based method
3. Point mutual information algorithm
4. Word2vec-based construction method
5. The sentiment semantic unit representation method combining character construction and context, referred to as the hierarchical system

These five methods were used to conduct experiments on the JD comment dataset and build corresponding sentiment dictionaries. Since different sentiment dictionary construction methods result in different numbers of sentiment words, when comparing the results, the TOP10, TOP100, and TOP200 positive and negative sentiment words in the sentiment dictionary constructed by each method are selected for accuracy comparison. The experimental results are shown in Tables 10.3, 10.4, 10.5, and 10.6.

The point mutual information method has the worst results from the experimental results. This is because the point of mutual information only considers the information between words when calculating the weight of sentiment words and does not consider the larger context and the emotional attributes brought by the semantics of

Table 10.3 Accuracy of TOP10, TOP100, and TOP200 positive and negative sentiment words in the fruit sentiment dictionary

	TOP 10		TOP 100		TOP 200	
Method	Positive	Negative	Positive	Negative	Positive	Negative
Word formation	1	1	0.98	0.94	0.97	0.93
Context	0.9	1	0.98	1	0.98	0.95
Point mutual information	0.8	0.7	0.85	0.83	0.82	0.82
Word embedding	1	1	0.97	0.96	0.96	0.92
Hierarchy	1	1	1	1	0.98	0.97

Table 10.4 Accuracy of TOP10, TOP100, and TOP200 positive and negative sentiment words in the iPad sentiment dictionary

	TOP 10		TOP 100		TOP 200	
Method	Positive	Negative	Positive	Negative	Positive	Negative
Word formation	1	1	0.99	1	0.95	0.94
Context	1	1	0.99	0.96	0.98	0.94
Point mutual information	0.9	0.9	0.88	0.86	0.83	0.83
Word embedding	0.9	0.9	0.97	0.95	0.93	0.91
Hierarchy	1	1	1	1	0.99	0.95

Table 10.5 Accuracy of TOP10, TOP100, and TOP200 positive and negative sentiment words in the clothes sentiment dictionary

	TOP 10		TOP 100		TOP 200	
Method	Positive	Negative	Positive	Negative	Positive	Negative
Word formation	1	1	1	0.96	0.95	0.91
Context	1	1	1	0.97	0.96	0.95
Point mutual information	0.8	0.8	0.82	0.87	0.8	0.87
Word embedding	0.9	1	0.95	0.96	0.93	0.93
Hierarchy	1	1	1	1	0.98	0.96

Table 10.6 Average accuracy

	Average	
Method	Positive	Negative
Word formation	0.96	0.92
Context	0.97	0.95
Point mutual information	0.82	0.84
Word embedding	0.94	0.92
Hierarchy	0.98	0.96

the word itself. The context-based method is better than the construction-based method, which shows that the context significantly impacts the weight of sentiment words. Compared with the previous construction method, the hierarchical system method has an increase of about 3% in accuracy, and the obtained sentiment dictionary is more accurate.

References

1. Hong Wei, Li Min. Review of Text Sentiment Analysis Methods [J]. Computer Engineering and Science, 2019, 41(4): 750–757.
2. Liu Kaiyuan. Chinese Weibo Sentiment Analysis Based on Dictionary and Machine Learning [J]. Electronic Technology and Software Engineering, 2016(22): 73
3. Cao Haitao. Research on Chinese Weibo Sentiment Analysis Based on PAD Model [D]. Dalian: Dalian University of Technology, 2013.
4. Li D, Qian J. Text Sentiment Analysis based on Long Short-Term Memory [C]. First IEEE International Conference on Computer Communication and the Internet (ICCCI), IEEE, 2016:471–475.
5. Hassan A, Mahmood A. Deep Learning Approach for Sentiment Analysis of Short Texts [C]. 3rd International Conference on Control, Automation and Robotics (ICCAR), IEEE, 2017:705–710.
6. Yang Rui, Chen Wei, He Tao, et al. Research on convolutional neural network text classification method integrating topic information (J). Modern Intelligence, 2020, 40(4): 42–49.
7. Zhang X, Zhao J, Yang L. Character-level Convolutional Networks for Text Classification [J]. Advances in Neural Information Processing Systems, 2015, 1(9):649–657.
8. Fan Hao, Li Pengfei. Research on Short Text Sentiment Analysis Based on Fast Text Word Vector and Bidirectional GRU Recurrent Neural Network – Taking Weibo Comment Text as an Example [J]. Information Science, 2021, 39(4): 15–22.
9. Miao X, Zhang X, Sun M, et al. A BLSTM and Wave Net-based Voice Conversion Method with Waveform Collapse Suppression by Post-processing [J]. IEEE Access, 2019, 7: 54321–54329.
10. Ren H Q, Wang W Q, Qu X W, et al. A New Hybrid Parameter Recurrent Neural Network for Online Handwritten Chinese Character Recognition [J]. Pattern Recognition Letters, 2019, 128(6): 400–406.
11. Cai Y, Yang K, Huang D P, et al. A Hybrid Model for Opinion Mining based on Domain Sentiment Dictionary [J]. International Journal of Machine Learning and Cybernetics, 2019, 10(8): 2131–2142.
12. Wang T, Lu K, Chow KP, et al. COVID-19 Sensing: Negative Sentiment Analysis on Social Media in China via BERT Model [J]. IEEE Access, 2020, 8: 138162–138169.
13. Li Yanhui, Zheng Chaomei, Wang Weili, et al. A multi-dimensional multi-sentiment analysis method for mixed-language text (J). Computer Engineering, 2020, 46(12): 113–119.
14. Collobert R, Weston J, Bottou L, et al. Natural Language Processing (almost) from Scratch [J]. Journal of Machine Learning Research, 2011, 12: 2493–2537.
15. Grefenstette E, Blunsom F. A Convolutional Neural Network for Modelling Sentences. In ACL, 2014.
16. Socher R, Perelygin A, Wu J, et al. Recursive Deep Models for Semantic Compositionality over a Sentiment Treebank [C]. Proceedings of the 2013 Conference on Empirical Methods in Natural Language Processing. 2013: 1631–1642.
17. Chen P, Sun Z, Bing L, et al. Recurrent Attention Network on Memory for Aspect Sentiment Analysis [C]. Proceedings of the 2017 Conference on Empirical Methods in Natural Language Processing. 2017: 452–461.
18. Wang Y, Huang M, Zhu X, et al. Attention-based LSTM for Aspect-level Sentiment Classification [C]. Proceedings of the 2016 Conference on Empirical Methods in Natural Language Processing. 2016: 606–615.
19. Wang X, Jiang W, Luo Z. Combination of Convolutional and Recurrent Neural Network for Sentiment Analysis of Short Texts [C]. Proceedings of COLING 2016, the 26th International Conference on Computational Linguistics: Technical papers. 2016: 2428–2437.
20. Zhang Baohua, Li Nanlin, Zhang Huaping, et al. Representation method of sentiment unit based on hierarchical structure [J]. Chemical Industry Progress, 2021.

Chapter 11
New Word Discovery

New word discovery is one of the basic tasks in natural language processing. Its main work is to mine existing corpora to identify new words. This chapter defines new word discovery and then details the rule-based, statistical model-based, and deep learning-based methods. It also gives the latest review on multilingual new word discovery and finally demonstrates two new word discovery experiments.

11.1 Overview of New Word Discovery

Language, a living entity, evolves in response to the changing societal landscape. A key indicator of this evolution is the introduction of new words into the vocabulary. Throughout history, Chinese vocabulary has been a reflection of the era it represents, offering a unique perspective on the societal, political, economic, and cultural dynamics, and the values of its people.

Changes in lifestyle, etc. For example, in the 1970s and 1980s, "work points," "grain tickets," and "cloth tickets" were familiar terms to people. With the deepening of reform and opening up, new words continue to emerge in everyday life, such as "scientific development," "people-oriented," "government transparency," "notebook computers," "virtual reality," etc. These words truly reflect the rapid development of society and the economy and the increasing frequency of foreign exchanges.

Currently, two concepts of new words and out-of-vocabulary words have emerged in the field of natural language processing. Out-of-vocabulary words are usually defined as words that do not appear in the dictionary. Although new words are also words that do not appear in the dictionary and belong to out-of-vocabulary words, they are different from general out-of-vocabulary words. New words are a dynamic concept with time added, and out-of-vocabulary words are concepts relative to the dictionary. Here, the definition of new words is grasped from two aspects.

H. Zhang, J. Shang, *Natural Language Processing and Applications*,
https://doi.org/10.1007/978-981-97-9739-4_11

1. From the perspective of the dictionary, new words refer to those produced by various means, with new forms, new meanings, or new usages that basic vocabulary does not have.
2. From the perspective of time, new words are usually those that first appeared within a certain period of time or after a certain point in time, with new word forms, new meanings, or new usages.

New words from internet slang have the following typical features:

1. Novelty. The most important feature of new words is novelty, which conforms to the current trend of the times. Whether evolved from existing words or creatively proposed by users, these new words have new meanings and express new ideas.
2. Periodicity. The generation of new words generally relies on current discussions of hot topics. Usually, some new words gradually disappear as the heat of the event decreases, but some new words are retained. Therefore, different new words have different existence cycles.
3. Fast spread. New words are generated based on the network platform and can spread quickly with the help of the network platform. At the same time, the meaning of new words is generally simple, direct, and easy to understand. Therefore, new words can be accepted by people in a very short time and used on various occasions.
4. Irregularity. The formation of new words is relatively free and casual, without fixed format requirements, and does not fully comply with word formation rules, resulting in some novel word formation methods, and there are no restrictions on length and word formation symbols.

At present, the main methods of new word recognition include rule-based methods, statistical-based methods, and methods combining rules and statistics. The rule-based method uses morphological principles, constructs templates with semantic information or part-of-speech information, and then discovers new words through matching. The statistical method identifies new words by statistically analyzing the composition or feature information of words in the corpus. The advantage of the rule-based method is high accuracy and strong pertinence, but it is challenging to write and maintain rules manually. The rules are generally domain-related, so the adaptability and portability could be better. The advantages of the statistical method are flexibility, strong adaptability, and good portability, but this method requires a large-scale corpus for model training due to the use of language. Knowledge is limited; generally, there are problems of data sparsity and low accuracy. Most researchers currently use a combination of rules and statistics, hoping to leverage the advantages of these two methods.

11.2 Overview of the Frontier of Multilingual New Word Discovery

There are roughly two types of existing word discovery algorithms.

One is the new word discovery algorithm based on word formation, also known as the rule-based algorithm. This algorithm builds a rule library based on language features. The process of rule construction is often complex, and the model's transferability is relatively poor.

The other is the new word discovery algorithm based on statistics, which is currently mainly divided into the following two categories:

1. Based on the Discovery of Frequent Patterns in the Corpus. Huang and others [1] proposed an algorithm for new word discovery using adjacent entropy and mutual information as features. This type of algorithm requires the iterative discovery of frequent items and the acquisition of context information, and its time complexity and space complexity are high, making it unsuitable for processing large-scale corpora.
2. Use Annotation Models for New Word Discovery. Peng and others [2] used the CRF model to calculate the confidence of Chinese fragments, and extract new words while segmenting. This type of algorithm is based on the local features of a word's context, and its accuracy is not high. In 2017, Zhang Huaping and others [3] combined the above two methods, used the CRF model to extract candidate words, a bigram model to rescan the corpus, and extracted features such as the left and right entropy and mutual information of the candidate word set. This method can effectively avoid the dependence on the global state in traditional algorithms and achieve rapid discovery of new words in large-scale corpora.

In terms of feature selection for new word discovery, Luo and others [4] compared nine common methods of calculating the internal features of a word. Experiments show that the use of mutual information is the most effective. Huang and others [5] proposed a pattern-based framework that integrates these statistical features together to detect new words.

In recent years, there has been a significant breakthrough in pretrained language models. These models have been proven to effectively complete a variety of tasks.

Tasks. For the problem of new word discovery, in 2019, McCrae [6] proposed an "adjective + noun" new word phrase recognition classifier based on pretrained language models. The experimental results showed that the combination of deep learning models and frequency features yielded the best results, but this method did not take into account the context of the phrase.

In terms of noise word filtering in new word discovery, in 2017, Liang et al. [7] defined the overlapping score from the perspective of the stability of the external environment of the new word, which is used to filter noise words. In 2018, Zhang Jing et al. [8] used word vectors to construct a set of weakly formed word strings to filter out candidate words with weak word formation ability. Its performance surpassed the effect of the overlapping score and showed that the filtering effect of character vectors containing word position information is optimal. In 2019, Qian et al. [9] proposed the WEBM model, which calculates the cosine similarity of word fragments based on word vectors, sets a similarity threshold, and filters noise words. The experimental results showed that WEBM has a great advantage in detecting new words from large-scale Chinese corpora. In 2022, Zhang Le et al. [10] proposed the MWEC model based on WEBM, introduced an external knowledge base to train multi-semantic word vectors, and applied it to candidate word set pruning, solving the problem of one word with multiple meanings in Chinese.

Current research on new word discovery mainly focuses on modern language corpora, with little involvement in ancient Chinese corpora. In 2017, Xie et al. [11] proposed the AP-LSTM algorithm, a supervised new word discovery algorithm specifically for ancient Chinese corpora. In 2019, Liu Yutong et al. [12] proposed the new word discovery algorithm for ancient Chinese, AP-LSTM-CRF, which uses the association rule algorithm of data mining and deep learning methods to effectively mine new words in ancient Chinese corpora and verified the effectiveness of the model on the Song Ci and Song History datasets.

In practical applications, especially for multiple languages, it is very difficult to obtain a large amount of annotated corpora, so many scholars are committed to exploring unsupervised mining methods to achieve new word discovery.

In 2008, Humbley [13] implemented NEOROM, a new word detection system for Latin. In 2017, Cartier [14] implemented a system that can automatically recognize new words in seven different scripts (Chinese, Czech, French, Greek, Russian, Polish, Portuguese, and Slavic), tracking the life cycle of new words through newspaper corpora.

For some Asian texts, such as Chinese, Japanese, Thai, etc., there are no clear boundaries between words. In 2015, Uchiumi et al. [15] proposed a nonparametric Bayesian model that directly builds a class N-gram language model from strings, integrating character and word-level information. The experimental results on standard datasets in Chinese, Japanese, and Thai show that the accuracy of this algorithm is better than previous results. In 2014, Falk et al. [16] used a statistical method to discover new words in French corpora, transformed the task into a supervised classification problem, and discussed the impact of three groups of features: form-related features, morpho-lexical features, and topic features. In 2018, Klosa et al. [17] proposed a semi-automatic new word detection method for German and discussed the impact of this method on the compilation of professional dictionaries.

11.3 Rule-Based Research Methods

The main idea of the rule-based new word discovery method is to establish a rule library, a professional library, or a pattern library according to the word formation features and external characteristics of new words and then discover new words through rule matching. Rule-based new word discovery methods can be divided into two categories: one is to build a new word rule library by observing the word formation rules of new words and extracting new words through rule matching, the other is to create a new word filtering rule library, specifically formulate non-new word formation rules, and filter non-new words. The key to both methods is to implement rule matching through regular expressions, and the main research results are based on the part-of-speech rules of Chinese characters.

11.3.1 Rule Extraction Method

The rules of word formation for new words mainly include regular regulations and special rules. Regular rules are based on basic word formation methods, including noun + noun, adjective + adjective, verb + verb, adjective + noun, verb + noun, adjective + verb, noun + verb, noun + adjective, noun + quantifier, and verb + adjective. For the sake of convenience, let A, B, C, and D represent four Chinese characters. Then, AB will represent a pair, ABC will represent a triplet, and ABCD will represent a quadruplet. The following are examples of regular word formation rules:

1. Noun Word Formation Rules
   - If A is a noun, B is a noun, verb, or adjective, then AB is a new binary tuple word.
   - If A is a noun, B is a quantifier, then AB is a new binary tuple word.
   - If A is an adjective, B is a verb, then AB is a new binary tuple word.
2. Adjective Word Formation Rules
   - If A is an adjective, B is a noun, then AB is a new binary tuple word.
   - If A is an adjective, BC is a noun, then ABC is a new ternary tuple word.
3. Verb Word Formation Rules
   - If A is a verb, BC is a noun, then ABC is a new ternary tuple word.
   - If A is a verb, B is a verb, and A and B are the same, then AB is a new binary tuple overlapping the word.

Special word formation rules refer to rules that are specifically designed and extracted for the characteristics of new word formation and do not fully comply with conventional word formation rules.

The key to discovering new words by formulating word formation rules is to establish an accurate and effective rule library and represent the word formation rules through regular expressions. The main research results are rule libraries established for the formation of Chinese words.

11.3.2 *Rule Filtering Method*

In recent years, the use of rule methods in research has mainly focused on rule filtering methods. The advantage of this method is that it does not need to summarize the word formation rules of new words but to formulate some rules that do not comply with word formation and use these rules to filter some meaningless phrases from the list of new word candidates. This method is universal. The more commonly used filtering rules are as follows:

1. If A is an adverb, B is from other parts of speech, and A is at the beginning of the sentence, then AB is filtered.
2. If A is part of other parts of speech, B is an adverb, and B is at the end of the sentence, then AB is filtered.
3. If A is from other parts of speech, B is a particle, and B is at the end of the sentence, then AB is filtered.
4. If the phrase AB, ABC, or ABCD contains a conjunction, it is filtered.
5. If A is a quantifier and B is not, then filter AB.
6. If A is a preposition and B is not a noun, then filter AB.
7. If the phrase AB, ABC, or ABCD contains a proper noun, then filter it.
8. If the phrase AB, ABC, or ABCD contains names of people, places, and organizational entities, etc., then filter it.
9. If the phrase AB, ABC, or ABCD contains onomatopoeic words, then filter it.
10. If the phrase AB, ABC, or ABCD contains non-morphemic words, then filter it.

The rule-based filtering method is a way to exclude garbage strings that do not conform to Chinese word formation. Given the increasing complexity and diversity of new word formation methods, traditional new word discovery methods may filter out new words with new types of word formation, reducing the accuracy of new word recognition. Therefore, the rule-based filtering method has high requirements for the specified rules.

11.4 Research Methods Based on Statistical Models

If a group of adjacent words appears together multiple times, then this group of adjacent words is likely a new word. This is a general idea based on statistics. Generally, the new word discovery method based on statistical models first uses statistical strategies to extract candidate strings, then uses linguistic knowledge to exclude garbage strings that are not new words or calculates relevance and looks for the combination of characters with the highest relevance. This method is simple to implement, applicable to any field, and does not require establishing a rule base. It requires a large amount of training corpus, and the quality of new words could be higher. The new word discovery method based on statistical models is generally limited to finding shorter new words because it treats all words equally, it is not convenient to describe the internal and external structural features of words, thus ignoring the influence of word formation patterns and word formation ability on new words.

The new word discovery method based on statistical models mainly uses the relevant probability knowledge in statistics to judge new words. It involves the three indicators of cohesion, information entropy, and the new word inverse document frequency (IDF).

11.4.1 Cohesion

When extracting words from a piece of text, the first thing to consider is what kind of text fragment is considered a word. All text fragments with a frequency higher than a certain threshold can be extracted and output as the vocabulary of this corpus. High frequency is just one indicator; a frequently appearing text fragment may not be a word but a phrase composed of multiple words. For example, "movie" and "theater" are two words, but in daily life, we are more inclined to regard "movie theater" as a word, which brings out the concept of word cohesion. Understanding this concept is crucial for practical applications in natural language processing and text analysis.

Cohesion refers to the internal cohesion of a new word. For this, the concept of pointwise mutual information (PMI) is introduced to measure the collocation and association of words.

$$p_{\mathrm{mi}}(x,y) = \ln \frac{P(x,y)}{P(x)P(y)} \tag{11.1}$$

Let (x, y) be a pair of words, where the probabilities of word x and word y appearing separately are $P(x)$ and $P(y)$, respectively, and the probability of them appearing together is $P(x, y)$. When the probability of $P(x, y)$ is greater than $P(x)$ $P(y)$, i.e., $p_{\mathrm{mi}}(x, y) > 0$, it indicates that the probability of these two words appearing together is much greater than the probability of them appearing separately. In this case, the pair of words will be treated as a single word.

For example, words frequently used in daily life, such as "dragonfly," "linger," "camel," "carrot," etc., are words with very high cohesion in text fragments. This is because each character of these words almost always appears with its corresponding character, and they are never used in other contexts.

11.4.2 Information Entropy

Information is a relatively abstract concept, generally referring to all content communicated in human society, including news, messages, and objects transmitted and processed by communication systems. In some studies, it is necessary to quantify information. Information entropy can be understood as the amount of information needed to eliminate uncertainty, i.e., the amount of information that an unknown event may contain. For example, "Which teams can make it to the top eight in the World Table Tennis Championships?" This random variable has high uncertainty. The amount of information introduced to eliminate uncertainty is represented by information entropy. The more information is introduced to eliminate uncertainty, the higher the information entropy, and vice versa. For example, "The Chinese team enters the top eight of the World Table Tennis Championships" has high certainty, requires very little information to be introduced, and therefore has low information entropy.

Shannon's formula for calculating information entropy is as follows:

$$H(X) = -\sum_{x \in X} P(x) \ln P(x) \tag{11.2}$$

where $P(x)$ is the probability of a certain event occurring. Using the Lagrange multiplier method, it can be proven that the more equal the probabilities of various random events, the greater the information entropy, otherwise, the smaller.

Applying information entropy in natural language processing can reflect the uncertainty of the context of a word.

11.4.3 New Word IDF

Generally speaking, rare words contain more information than common words. Consider querying a word, if it is very rare in the entire text set, a document containing this word is likely to be related to the query target. Therefore, higher weights are assigned to rare words.

For common words, such as "very good," "of," "can" and other frequent words, positive weights are assigned, but these weights are less than the weights of rare words.

Next, we use inverse document frequency to evaluate new words:

$$idf_t = \log_{10} \frac{N}{df_t} \tag{11.3}$$

where df represents document frequency, which is the number of documents containing word t, N is the number of documents in the document set.

idf_t is an indicator reflecting the amount of information of word t. The larger the idf_t value, the higher the probability that word t is a new word.

11.5 Research Methods Based on Deep Learning

This section introduces the application of the BiLSTM + CRF model in new word discovery.

The characteristic of the BiLSTM model is that it can extract the relationship rules between words and then add various features to the model, resulting in a better prediction effect. In traditional methods, only artificial features are used to extract new words. Compared with this, the BiLSTM model can comprehensively use artificial features and features extracted by the model itself, thereby making the model get better prediction results.

1. Feature Introduction

The features used in new word discovery are as follows:

(1) Part of Speech. Because the problem to be studied is the combination of multiple old words that gives rise to a new word, the part of speech of each old word that can be combined to give a new word can be considered a factor.
(2) Word Length. The length of the old word combined to give a new word is also a factor to consider. If the sum of the lengths of each old word is too large, then its possibility of forming a new word is relatively low.
(3) Contextual Entropy. This feature is essentially the degree of freedom between words. The greater the contextual entropy of a word, the less chance it has to appear with other words. If a word often appears independently in the text, the possibility of it merging with other old words into a new word is smaller; otherwise, the possibility is more significant.
(4) Word Cohesion. This feature is the opposite of contextual entropy, referring to the degree of connection between words. If two words often appear together, it means that the cohesion between these two words is greater, and the combination of these two words is more likely to form a new word.

2. Application of the BiLSTM + CRF Model with Integrated Features in New Word Discovery

The BiLSTM + CRF model used in new word discovery is divided into the following four types:

(1) Model Based on Word Feature Combination. The selected features (part of speech, word length, contextual entropy, and word cohesion) are added to the model separately, which have different degrees of impact on the model. And there are connections between these features. For example, contextual entropy and word cohesion describe the degree of connection between words. Adding these two features to the model simultaneously improves the prediction effect.

(2) Model Based on Feature Vector Length. Before integrating different combinations of features into the model, we have the power to control the weight of each feature. This control is exerted through the length of the feature vector. If the vector length of a feature is larger, then the weight of this feature is also larger, and its impact on the model is also larger. Therefore, under different feature combinations, the vector length of each feature is also one of the factors that we can use to optimize the model, giving us a sense of empowerment in the model optimization process.
(3) Model Based on the Length of the Input Text. Deep learning models can extract the relationships between words from the input text, thereby finding words that often appear together and discovering new words. This underscores the potential of the model to uncover intricate word relationships. Therefore, the longer the text input into the model, the more fully the model can extract the word relationship information in the text, instilling a sense of optimism about the model's capabilities.
(4) Model Based on Semantic Features. Semantics is the smallest unit of meaning information of a word. In natural language, the meaning of a word can often be described by other words, and different words can also explain the meaning of various words, and so on, until the meaning is indivisible, that is, different words cannot describe it, such words are semantics.

11.6 Application and Analysis

11.6.1 Open-Domain New Word Discovery for Social Media

With the development and popularity of new internet social applications, such as Weibo and WeChat, social media has gradually become an important carrier of information transmission and has integrated into people's daily lives. However, social media has characteristics such as wide domain distribution and a high degree of colloquialism, which challenge the analysis of such texts. Social media texts often come with a large number of out-of-vocabulary words. If these out-of-vocabulary words cannot be effectively and timely recognized, it will directly affect the results of upper-level analysis tasks based on word segmentation (such as sentiment computation and dependency syntax analysis).

The new word discovery algorithm for written language materials such as news can handle small amounts of data, and the morphology and syntax are relatively regular and formal. Researchers mostly use time-consuming frequent item discovery algorithms or manually marked garbage string templates to filter incorrect new word results. Considering that social media has the characteristics of colloquialism, wide sources, and large data volumes, the above algorithm has certain limitations. First, memory usage will increase linearly or even exponentially with the scale of the text; therefore, it is limited by computer hardware resources and needs help to handle large corpora. Second, a large scale of colloquial language materials will make the construction of garbage string templates more complicated, requiring more manual annotation and significantly affecting the template's precision and recall. From another perspective, social media texts cover a wide range of fields and are not limited to certain specific fields. Especially for words like "Shenma," which do not belong to a specific field, it is not suitable to use new word discovery algorithms related to the field.

This section proposes a method of using a word-tagging segmentation algorithm based on the CRF model for candidate word extraction. After filtering named entities, such as people's names, a candidate word set is formed using the maximum entropy model. Then, the candidate word set is combined with the bigram segmentation model to re-segment the text corpus, thereby obtaining the global features of the candidate words in the corpus and using statistical methods for garbage string filtering and new word discovery. This method can combine the efficient out-of-vocabulary word recognition method of the CRF model with the garbage string filtering method based on global features, and all steps are of linear time complexity.

11.6.1.1 New Word Discovery

The algorithm proposed here combines the two types of statistical new word discovery methods mentioned above. First, the CRF word segmentation model is used for candidate word extraction. This approach speeds up the extraction process and reduces memory usage compared to frequent pattern extraction. Then, a bigram language model is used to rescan the corpus, extracting two features left and right entropy, and mutual information to address the limitations of the second method, which relies solely on local features.

The new word discovery algorithm has four steps: candidate word extraction, named entity filtering, new word feature selection, feature calculation, and candidate word sorting.

Candidate Word Extraction

Use the CRF-based character tagging model (the CRF model from now on) to segment the corpus and extract words with a frequency exceeding a certain threshold as candidate words.

The CRF model views the process of Chinese word segmentation as a sequence tagging problem of Chinese character boundaries, usually using the BMES tagging set, that is, the first character of a word is tagged as B, the last character is tagged as E, the middle characters are tagged as M, and a single character word is tagged as S. When using the CRF model for word segmentation, the word itself and the context become factors determining whether a segment constitutes a word. Therefore, it has a high recall rate for out-of-vocabulary words that do not exist in the dictionary. However, limited by the window size of feature selection, the probability of a segment becoming a word is only determined by the context of this segment's feature window size.

The correct segmentation of out-of-vocabulary words by the CRF model depends on the composition of the out-of-vocabulary words themselves and the context features. In some contexts, segmentation errors may occur, but switching to different contexts may result in correct segmentation, and the erroneous segmentation results are highly correlated with the context. That is, in the case of a large enough corpus, most out-of-vocabulary words can be correctly segmented. Experiments show that more than 92% of the wrongly segmented words produced in the CRF model have a word frequency of less than three. Therefore, using the CRF model for new word extraction is feasible. The segmentation results of the CRF model are saved in the form of (word, word frequency) as a word list. In the word list, out-of-vocabulary words with a frequency more significant than a certain threshold can be selected as candidate words. The time complexity of this algorithm is linear, and only the word list needs to be stored in memory.

The threshold for discovering new words varies according to the corpus scale, and the two are positively correlated. Here, the arctan function selects the corresponding threshold according to the corpus size.

$$\text{ThresFreq} = \beta \arctan\left(\alpha \| D \|\right) \tag{11.4}$$

where, $|D|$ is the total number of words in the corpus, and the arctan function is used to prevent the threshold from increasing linearly with the increase in the scale of the corpus. According to experimental results and experience, β is 50, and α is 10^{-7}.

Named Entity Filtering

From the experimental results, it was found that about one-fourth of the new words in the candidate word set formed by the CRF word segmentation results are named entities, and this ratio is even as high as one-half in news corpora. At present, named entity recognition has achieved good results, and there is no need to build a word list specifically for named entities. At the same time, in order to reduce memory usage in subsequent processing, it is necessary to filter out named entities in the candidate word set.

In terms of filtering organization names, use finely segmented corpora when training the CRF model. Such named entities can be segmented into fine-grained words, thereby filtering out organization names such as "Beijing Institute of Technology" and "Institute of Computing Technology, Chinese Academy of Sciences."

In terms of filtering Chinese names, because the regularity of Chinese names is very strong, whether it is the surname as the first character or Chinese characters such as "Ling" and "Wen" that often appear as names, the recognition of names is simpler than the recognition of other words or named entities.

The maximum entropy model is suitable for this type of feature-based classification task. The feature selection is as follows: B is the first character, E is the last character, and M is the middle character. The samples use the 63,704 names with the highest frequency in the Chinese name database as positive examples, and the 85,144 words (names have been filtered out) in the "People's Daily" corpus word list as negative examples. From these 148,848 samples, 90% are randomly selected as the training set and the remaining 10% as the test set. The experimental results show an accuracy rate of 94.7%.

In this step, because it is only necessary to classify the candidate word set, the calculation time can be ignored compared to other steps (the size of the word list is much smaller than the size of the corpus).

New Word Feature Selection

Compared to traditional news corpora, social network texts are primarily colloquial and often contain incorrect words, dialects, or other non-standard language characters and symbols.

If you filter out garbage strings through manual annotation templates, it requires a huge amount of human resources, and the error rate is high. Therefore, statistical methods are employed to filter candidate words, ranking them from high to low based on their likelihood of forming valid words.

The CRF model uses characters as the processing unit for word segmentation and annotates the boundaries of the characters to construct words. Therefore, this type of word segmentation model mainly produces two types of word segmentation errors that affect the discovery of new words. The first type is separation type word segmentation errors, such as the wrong segmentation result "Si Luan Xiang" (corresponding to "Hu Si Luan Xiang"), and the second type is combination type word segmentation errors, that is, two consecutive words cannot be correctly segmented, such as "Eat hot pot" (corresponding to "Eat/Hot pot").

The overall idea of feature selection is to find two types of features that can filter the above two main types of word segmentation errors. For the first type of error, use adjacent entropy features for filtering; for the second type of error, select the language model to calculate the mutual information features for filtering.

Adjacent Entropy

Adjacent entropy is a feature that calculates the richness of the context of candidate words. The richer the context of the candidate word, the higher the probability of it forming a word, and the higher its adjacent entropy. The calculation formula for adjacent entropy is as follows:

$$H_{\mathrm{L}} = \sum_{w \in W_{\mathrm{L}}} -\frac{n_{\mathrm{w}}^{\mathrm{L}}}{c} \times \log \frac{n_{\mathrm{w}}^{\mathrm{L}}}{c} \tag{11.5}$$

$$H_{\mathrm{R}} = \sum_{w \in W_{\mathrm{R}}} -\frac{n_{\mathrm{w}}^{\mathrm{R}}}{c} \times \log \frac{n_{\mathrm{w}}^{\mathrm{R}}}{c} \tag{11.6}$$

$$H_{\mathrm{ADJ}} = \min\left(H_{\mathrm{L}}, H_{\mathrm{R}}\right) \tag{11.7}$$

where, H_L represents the left information entropy of the candidate word, W_L represents the set of words that appear to the left of the candidate word, $n_{\mathrm{w}}^{\mathrm{L}}$ represents the number of times the word w appears to the left of the candidate word. Similarly, H_R, W_R, and $n_{\mathrm{w}}^{\mathrm{R}}$ represent the corresponding features on the right side of the candidate word, and C represents the frequency of the candidate word.

Choosing the minimum of the left and right information entropy as the adjacent entropy of the candidate word can effectively filter out the internal segmentation errors in CRF segmentation, such as the incorrect segmentation result "思乱想 (corresponding to '胡思乱想')," in the corpus it can only appear "胡" on the left, so the left information entropy is 0, thus achieving filtering.

Mutual Information

Mutual information reflects the internal features of the candidate word; the larger the value, the higher the cohesion of the candidate word, and the greater the probability of forming a word. Using mutual information can effectively filter out CRF model combination type segmentation errors like "吃火锅"(corresponding to "吃/火锅").

At this time, the calculation of mutual information can be introduced into the bigram model, thereby making its ability to filter out combination type error candidate words stronger. The mutual information calculation formula proposed by Chen Fei and others [18] is as follows:

$$M(w) = MI(w_1, w_2) = \frac{P(w_1, w_2)}{P(w_1)P(w_1)} \tag{11.8}$$

where w_1 and w_2 are parts of w, and $P(w)$ represents the probability of word w appearing in the corpus. This calculation method has certain defects. First, idioms like "排山倒海" are not composed of just two parts. Second, for some sequences that often appear together but do not form words, such as "了一", the filtering effect of the above mutual information calculation method is not good.

This section combines the bigram model with the above formula to propose an improved mutual information calculation method. The formula is as follows:

$$M(w_1, \ldots w_n) = \log \frac{P(w_1, \ldots w_n)}{P(w_1)P(w_2|w_1)\ldots P(w_n|w_{n-1})} \tag{11.9}$$

$$M(\mathbf{w}) = \min(Mw_1, w_2, \ldots w_n) \tag{11.10}$$

where w_1, w_2, … , w_n represent the various components of the candidate word w, and the value of M is the minimum value obtained by Mw for different sequences w_1, w_2, … , w_n. Here, $P(w_n|w_{n-1})$ is calculated using the Jelinek–Mercer smoothing method.

$$P(w_n|w_{n-1}) = \lambda P^{'(w_n|w_{n-1})} + (1-\lambda)P(w_n) \tag{11.11}$$

$P^{'(w_n|w_{n-1})}$ is learned from the annotated corpus, and $P(w_n)$ is learned from the corpus for new word discovery. For words like "了一", because the probability P(一了) of the bigram model is significantly larger than the unigram model's P(一), the improved mutual information calculation formula can effectively increase the distinction for these types of words.

Feature Calculation and Candidate Word Sorting

After obtaining the candidate word set, a second scan of the corpus is needed to calculate the adjacency entropy and mutual information values of each word in the candidate word set. Here, the bigram model is chosen, and each word in the candidate word set is added to the user dictionary of the bigram model segmentation program in the form of (candidate word, word frequency), and the corpus is re-segmented.

The N-gram model segmentation is to find the word sequence $w_1w_2 \ldots w_n$ for the text $T = C1C2C3C4 \ldots Cn$, composed of character sequence C.

$$w_1w_2\ldots w_n = \underset{w_n \in V,\ w_1w_2\ldots w_n = T}{\operatorname{argmax}} P(w_1w_2\ldots w_n|C1C2\ldots Cn) \tag{11.12}$$

$$P(w_1w_2\ldots w_n|C1C2\ldots Cn) \propto P(w_1w_2\ldots w_n) \tag{11.13}$$

$$P(w_1w_2\ldots w_n) = P(w_1)P(w_2|w_1)\ldots P(w_n|w_{n-1}\cdots w_1) \tag{11.14}$$

The bigram model assumes that the probability of a word appearing is only related to the previous word, so $P(w_1w_2\ldots w_n)$ can be simplified to

$$P(w_1w_2\ldots w_n) = P(w_1)P(w_2|w_1)\ldots P(w_n|w_{n-1}) \tag{11.15}$$

As above, $P(w_n|w_{n-1})$ is calculated using the Jelinek–Mercer smoothing method.

The sequence $w_1w_2\ldots w_n$ can be obtained with a time complexity of $O(n)$ using dynamic programming methods such as Beam Search.

This section chooses to use the bigram model for re-scanning instead of calculating features in the segmentation results of the CRF model. First, using the CRF model will cause additional disk space consumption. Second, the segmentation of unregistered words in the CRF segmentation model mainly depends on the boundary information of the context, so the candidate words in the corpus are not necessarily correctly segmented at all positions, and it is necessary to use a segmentation algorithm based on the word list and language model to re-segment and calculate features. Finally, the speed of bigram segmentation is fast enough.

In the process of bigram segmentation, record the words appearing on both sides of each candidate word and the frequency of the candidate word itself; the former is used for the calculation of adjacency entropy, and the latter is used for the calculation of mutual information.

Once the adjacency entropy and mutual information values of each candidate word are obtained, remove the 10% of candidate words with the lowest adjacency entropy and mutual information, and use linear interpolation to obtain the weight of each candidate word, that is,

$$\text{Weight}_w = \alpha \frac{H_{\text{ADJ}}(w)}{H_{\text{ADJ}}^{\max}} + (1-\alpha)\frac{M(w)}{M^{\max}} \tag{11.16}$$

where H is the maximum value of the adjacency entropy of all candidate words, and $M^{\max}$ represents the maximum value of the mutual information of the candidate words. After using Eq. (11.16) to calculate the words in the candidate word set, the candidate words are sorted in descending order based on their weight. The larger the weight, the higher the ranking, and the higher the probability of the candidate word becoming a word. Finally, the top 30–40% of the sorted results can be extracted as new words.

11.6.1.2 Experiment

Experimental Data

Using a web crawler to crawl NetEase's news, sports, technology, and education column corpora, a total of about 3.2 GB of pure text was obtained. Additionally, by crawling around 40 million Chinese Weibo posts from Twitter, and after removing duplicate content and converting traditional characters into simplified characters, we gathered 3.4 GB of pure text (both encoded in UTF-8). This resulted in a combined total of 6.6 GB of pure text, forming the test corpus.

In addition to test the accuracy of new word discovery, the 82,902 candidate words obtained from candidate word extraction and three examples of each candidate word were put on the web for annotation in a crowdsourcing manner, producing the annotated set R. The annotation content is whether the candidate word is a word. In order to ensure the consistency of the annotation, the standard for words is defined as follows:

1. Those that do not change their meaning after splitting are not words, such as "专用飞机" (special aircraft).
2. Obvious common phrases composed of two words are not words, such as " 的是"(is).
3. Numbers and names are not words, place names and organization names are words.

At the time of the writing this book, a total of 12,764 valid annotations had been obtained, of which 8365 are positive examples and 4399 are negative examples (some words with low frequency were not used in later experiments).

Experimental Results

In the experiment, the accuracy of each 10% area of the sorting result was tested separately, as shown in Table 11.1. The accuracy is the ratio of the number of words marked as positive examples in the corresponding area of the test word set to the number of candidate words, that is,

$$P_{\text{ra}} = \frac{\left|\{w|, w \in T_{\text{ra}} \cap R|, L_{\text{w}} = \text{True}\}\right|}{T_{\text{ra}} \cap R} \tag{11.17}$$

where T_{ra} is the new word discovery result to be tested, ra represents the selected ratio, and $T30\%$ refers to testing the top 30% of the ranking results. R is the annotation set, L_{w} represents the result of word w in the annotation set R, True means w is a word, and false means w is not a word. False means w is not a word. Similarly, PI represents the accuracy when only mutual information is used as a feature, and PDJ represents the accuracy when only left and right entropy is used as a feature.

In addition, for the sorting results divided into areas, the accuracy of each area is tested, that is,

Table 11.1 The impact of different features on the accuracy of each region, $\alpha = 0.2$

	10%	20%	30%	40%	50%	60%	70%	80%	90%	100%
P_{ra}	0.872	0.849	0.830	0.812	0.800	0.782	0.767	0.754	0.733	0.712
$P_{\text{ra}}^{\text{MI}}$	0.842	0.838	0.822	0.803	0.790	0.776	0.761	0.741	0.725	0.712
$P_{\text{ra}}^{\text{ADJ}}$	0.755	0.75	0.762	0.766	0.766	0.761	0.756	0.744	0.731	0712
$P_{\text{ra}}^{'}$	0.872	0.828	0.789	0.755	0.744	0.688	0.670	0.652	0.554	0.453
$P_{\text{ra}}^{'\text{MI}}$	0.842	0.834	0.792	0.742	0.731	0.700	0.671	0.596	0.594	0.561
$P_{\text{ra}}^{'\text{ADJ}}$	0.755	0.744	0.786	0.778	0.768	0.733	0.725	0.650	0.611	0.493
R_{NUM}	4.828	9.657	14.486	19.314	24.143	28.972	33.800	38.629	43.458	48.287

$$P'_{\text{ra}} = \frac{\left|\left\{w|, w \in T'_{\text{ra}} \cap R|, L_{\text{w}} = \text{True}\right\}\right|}{T'_{\text{ra}} \cap R} \tag{11.18}$$

where T'_{ra} is the area of the new word discovery result to be tested, e.g., testing the top 30% of the sorting results ($T'30\%$). Test new words in the 20–30% range. As can be seen from Tables 11.1 and 11.2, using mutual information as a feature to extract new words is better than using adjacent entropy, and the effect of combining the two features is significantly better than a single feature.

For the selection of the linear interpolation coefficient α when the two features are combined, the accuracy P30% of the top 30% of the new word results and the accuracy P50% of the top 50% of the new word results are used to test the impact of different values of α on the experimental results. The experimental results (see Table 11.2) show that the accuracy is relatively high at the [0.1, 0.2) interval.

As can be seen from Table 11.3, when mutual information is used as a feature, long words composed of characters that do not form words alone are ranked first, such as "phthalate" and "Stakhovsky." Adjacent entropy tends to filter out common words, such as "appear" and "too much," among others. These two features reflect the possibility of a candidate word forming a word from different angles and are complementary.

The new word discovery method used in the comparative experiment is based on the CRF model proposed by Peng et al. [19]. Was implemented. Different from the algorithm proposed by Gu Sen [20] in this method, identifying frequent items combines segmentation results instead of suffix arrays. Considering the size limitations of different algorithms for the corpus, the first 100 MB of the 6.6 GB corpus is selected as the test corpus. The comparison results are shown in Table 11.4. The number of new words recalled by different algorithms varies greatly, especially since only 220 words are the same among the 2463 and 2779 new words, respectively, recalled by this method and the algorithm proposed by Gu Sen. Therefore, the number of recalled new words is used instead of the recall rate. In terms of accuracy, 100 words are randomly selected from the first 1000 new words recalled by the algorithms of Peng et al. and Gu Sen for judgment, and the accuracy values POP1000 and POP1000 are statistically calculated.

The highest-weighted new words recalled by different algorithms are shown in Table 11.5. The algorithm of Peng et al. is similar to this method in terms of the type characteristics of the results because both are based on CRF model segmentation and tend to find fine-grained words. The difference between the two is that this method uses global features to calculate weights, while the algorithm of Peng et al. uses local context features. The latter's local noise has a significant impact on

Table 11.2 The impact of different "values on accuracy"

	0.0	0.1	0.2	0.3	0.4	0.5	0.6	0.7	0.8	0.9	1.0
*P*30%	0.822	0.837	0.830	0.822	0.809	0.795	0.778	0.776	0.768	0.765	0.762
*P*50%	0.790	0.799	0.800	0.793	0.787	0.784	0.777	0.775	0.771	0.770	0.766

Table 11.3 New words with the highest weight when using mutual information, adjacency entropy, and their combination as features

Mutual information	Adjacency entropy	Mutual information + adjacency entropy
Paracetamol for children	Too much	Petitioners
Beetle beetle	Surprise	Comparable
The son wants to support his parents but they are no longer there	Will it?	Transfer to
Alzheimer's disease	Sign	Animation
Stakhovsky	Run	Medical insurance

Table 11.4 The number of new words recalled by different algorithms and their accuracy

Algorithm	Number of new words recalled	P^3_{TOP1000}	P^2_{TOP1000}
Algorithm by Peng et al	5062	0.63	0.60
Gu Sen's algorithm	2779	0.72	0.76
Our method	2463	0.81	0.80

Table 11.5 New words with the highest recall weight from different algorithms

Our method	Algorithm by Peng et al	Gu Sen's algorithm
Albef	Meat industry	State Food and Drug Administration
San Lipu	Extinguish	Per capita disposable income of urban residents
Chaya mountain	Medical insurance	Pistorius
Su ning	Leading type	Beijing Gongmei Group Co., LTD
Bloomberg	Bad	State Administration of Press Publication, Radio, Film and Television

accuracy. Gu Sen's algorithm is exactly the opposite of this method, tending to recall coarse-grained words related to its candidate set mining algorithm based on frequent patterns. This algorithm is more conducive to finding named entities or some common phrases. To some extent, this method and Gu Sen's algorithm can complement each other.

We first tested the relationship between running time and corpus size for the algorithm efficiency experiment. We extracted different sizes from the 6.6 GB test corpus, namely 0.5, 1, 1.5, 2, 4, and 6.3 GB, and used this method for new word discovery for each corpus, calculating the running time and processing speed. The final results are shown in Fig. 11.1. The running time of the new word discovery is proportional to the size of the corpus, and the processing speed does not change with the size of the corpus, which is consistently stable at around 2.6 MB/s. The experimental results verify that this method has a time complexity of $O(n)$.

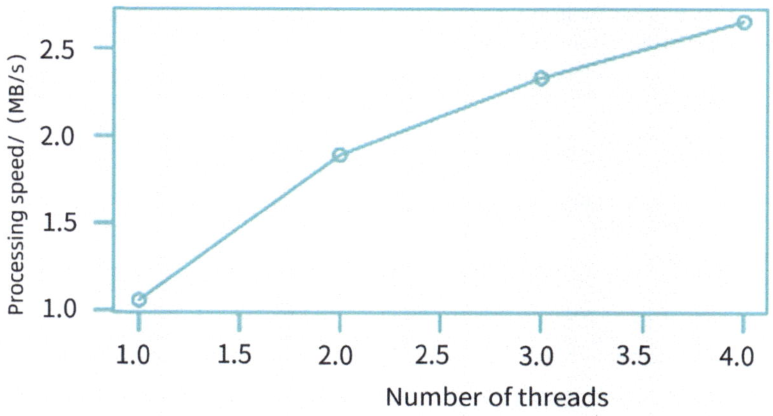

Fig. 11.1 Relationship between thread count and processing speed

11.6.2 Examples of New Word Discovery in Multiple Languages

In practical applications, obtaining a large annotated corpus is challenging, especially for minor languages. Therefore, exploring unsupervised new word discovery algorithms is more valuable for minor languages. This section proposes a universal multi-language new word discovery algorithm. Compared with the single-language situation, this algorithm only differs in the morphological segmentation part for multi-language processing.

This section conducts new word discovery experiments on Chinese, Japanese, and English, three languages used in different countries, and Tibetan and Uighur, two minority languages in China. The datasets used in this section for Chinese, Japanese, and English are the news commentary corpus provided by WMT2019. The size of the Tibetan dataset is about 34.93 MB of pure text corpus crawled from the China Tibetan Network and China Tibetan News website using a crawler; the Uighur dataset is the Uighur-Chinese bilingual corpus XJIPC-corpus-CWMT2017 provided by the Xinjiang Institute of Physics and Chemistry, Chinese Academy of Sciences.

Based on the characteristics of different languages, the cleaned and pure text corpus is morphologically segmented.

The segmentation points of the three languages used in different countries are as follows:

1. There is no clear separation mark between Chinese words, and any adjacent characters may combine to form a word, so the Chinese corpus is segmented by character. In each round of iteration, the newly discovered words are added to the dictionary of the word segmentation program NLPIR-ICTCLAS [21] for pre-segmentation.

2. The Japanese corpus is very similar to the Chinese one, so the same segmentation method is adopted.
3. English words are separated by spaces, so the English corpus is segmented by space.

The key points of segmentation for two Chinese minority languages are as follows:

1. Tibetan is composed of syllables, each of which is composed of several characters. A syllable in Tibetan can be simply understood as a character in Chinese, and Tibetan text is segmented using "·" as the syllable separator.
2. Uighur is similar to English, with words separated by spaces. Uighur text is segmented using space as the separator.

Based on the results of morphological segmentation, a word list is established. After removing the stop words, an inverted index is established for the word list. The weights of each word in the word list are calculated based on frequency, left and right information entropy, and mutual information. The results are then linked and filtered based on context and weight calculations.

In the Chinese dataset, the top 20 results for new word discovery are as follows: Saudi Arabia, Morocco, Internet, satellite TV, Mohammed, twentieth century, Lebanon, caution, Sunni, Nayef, brewing, blasphemy, left-wing party, sacrifice, barrier, brother, invitation, burial, Kemal, Turkey. In the Japanese dataset, the top 20 results for new word discovery are as follows: multicultural society, crown prince, Europe, framework, worker, arrogance, pregnancy, supervision, black tea, soft, blasphemy, embryonic cell, committee, luxury, caricature, savings, arrest, encouragement, European, exposure. In the English dataset, the top 10 results for new word discovery are as follows: Prime Minister, climate change, central banks, advanced economies, trade intensity, sustainable development, developing countries, billion people, financial crisis, GDP growth.

In the Uighur dataset, the top five results for new word discovery are as follows: "promotion," "strengthening," "million," "done," and "finance."

The results obtained by the method in this section are basically the words of the corresponding corpus, and hence prove this method's applicability.

References

1. Huang J H, Powers D. Chinese Word Segmentation Based on Contextual Entropy [C]. Proceedings of the 17th Pacific Asia Conference on Language, Information and Computation. 2003: 152–158.
2. Peng F, Feng F, McCallum A. Chinese Segmentation and New Word Detection Using Conditional Random Fields [C]. Proceedings of the 20th International Conference on Computational Linguistics. 2004: 562–568.
3. Zhang Huaping, Shang Jianyun. Open Domain New Word Discovery for Social Media [J]. Journal of Chinese Information Processing, 2017, 31(3): 55–61.

4. Luo S, Sun M. Two-character Chinese Word Extraction based on Hybrid of Internal and Contextual Measures [C]. Proceedings of the Second SIGHAN Workshop on Chinese Language Processing. 2003: 24–30.
5. Huang M, Ye B, Wang Y, et al. New Word Detection for Sentiment Analysis [C]. Proceedings of the 52nd Annual Meeting of the Association for Computational Linguistics (Volume 1: Long Papers). 2014: 531–541.
6. McCrae JP. Identification of Adjective-noun Neologisms Using Pretrained Language Models [C]. Proceedings of the Joint Workshop on Multiword Expressions and WordNet (MWE-WN2019). 2019: 135–141.
7. Liang Y, Yin P, Yiu S M. New Word Detection and Tagging on Chinese Twitter Stream [C]. International Conference on Big Data Analytics and Knowledge Discovery. Cham: Springer, 2015: 310–321.
8. Zhang Jing, Huang Kaiyu, Liang Chen, et al. Unsupervised New Word Extraction from Chinese Social Media Data [J]. Journal of Chinese Information Processing, 2018, 32(3): 17.
9. Qian Y, Du Y, Deng X, et al. Detecting New Chinese Words from Massive Domain Texts with Word Embedding [J]. Journal of Information Science, 2019, 45(2): 196–211.
10. Zhang Le, Leng Jidong, Lv Xueqiang, et al. MWEC: A Chinese new word discovery method based on multi-semantic word vectors [J]. Data Analysis and Knowledge Discovery, 2022, 6(1): 113–121.
11. Xie T, Wu B, Wang B. New Word Detection in Ancient Chinese Literature [C]. Asia-Pacific Web (APWeb) and Web-age Information Management (WAIM) Joint Conference on Web and Big Data. Cham: Springer, 2017: 260–275.
12. Liu Yutong, Wu Bin, Xie Tao, et al. A new word discovery method based on ancient Chinese corpus [J]. Journal of Chinese Information, 2019, 33(1): 46–55.
13. Humbley J. Les Dictionnaires de NéOlogismes, LeurÉVolution Depuis 1945: une Perspective Européenne. 2008.
14. Cartier E. Neoveille, A Web Platform for Neologism Tracking [C]. Proceedings of the Software Demonstrations of the 15th Conference of the European Chapter of the Association for Computational Linguistics. 2017.
15. Uchiumi K, Tsukahara H, Mochihashi D. Inducing Word and Part-of-Speech with Pitman-Yor Hidden Semi-Markov Models [C]. Proceedings of the 53rd Annual Meeting of the Association for Computational Linguistics and the 7th International Joint Conference on Natural Language Processing (Volume 1: Long Papers). 2015: 1774–1782.
16. Falk I, Bernhard D, Gérard C. From Non Word to New Word: Automatically Identifying Neologisms in French Newspapers [C]. The 9th edition of the Language Resources and Evaluation Conference. 2014.
17. Klosa A, Lüngen H. New German Words: Detection and Description [J]. 2018.
18. Chen Fei, Liu Yiqun, Wei Chao, et al. Open domain new word discovery based on conditional random field method [J]. Journal of Software, 2013, 24(5): 10–14.
19. Peng F, Feng F, McCallum A. Chinese Segmentation and New Word Detection Using Conditional Random Fields [C]. Proceedings of the 20th International Conference on Computational Linguistics. 2004: 562–568.
20. Gu Sen. A new word discovery algorithm based on large-scale corpus [J]. Programmer, 2012(7): 54–57.
21. Hang HP, Yu H K, Xiong D, et al. HHMM-based Chinese Lexical Analyzer ICTCLAS [C]. Proceedings of the Second SIGHAN Workshop on Chinese Language Processing. 2003: 184–187.

Chapter 12
Named Entity and Keyword Extraction

This chapter provides an in-depth exploration of named entity recognition (NER) and keyword extraction technologies. It begins with definitions and challenges of NER and keyword extraction, progressing to an analysis of key techniques such as rule-based, machine learning, and deep learning methods. The chapter highlights advancements in models like Bidirectional Long Short-Term Memory - Conditional Random Field (BiLSTM-CRF) and Iterated Dilated Convolutional Neural Network - Conditional Random Field (IDCNN-CRF), emphasizing their efficiency in sequence labeling tasks. Additionally, keyword extraction methods are introduced, including statistical, graph-based, and embedding approaches such as term frequency-inverse document frequency (TF-IDF) and TextRank. Finally, practical examples are presented, including the NLPIR-ICTCLAS-DocExtractor, showing its performance in real-world entity extraction tasks.

12.1 Overview of Named Entity and Keyword Extraction

12.1.1 Named Entity Recognition

12.1.1.1 Definition of Named Entity Recognition

With the explosive growth of data, it is undoubtedly a time-consuming and laborious task for humans to find useful information from massive texts, hence the emergence of information extraction research. Named entity recognition (NER), as one of its key technologies, has received widespread attention from academia and industry over the years. Named entity recognition is also the foundation of many natural language processing applications, such as entity relationship extraction, knowledge graph construction, intelligent question answering, etc.

H. Zhang, J. Shang, *Natural Language Processing and Applications*,
https://doi.org/10.1007/978-981-97-9739-4_12

Named entities were first proposed [1] at the sixth Message Understanding Conference in 1996, which only defined some general entity categories, such as personnel, location, organization, etc. In the 1990s, the early work of NER systems was mainly to extract information from news articles. Subsequently, the focus of NER shifted to processing military documents and reports. The later stages of the automatic content extraction (ACE) evaluation also included several informal text styles, such as blogs and text transcripts from conversational telephone voice dialogues. Since 1998, in the fields of molecular biology, bioinformatics, and medical natural language processing, there has been great interest in entity recognition. The most common entities of interest in this field are the names of genes and gene products. Against the backdrop of the CHEMDNER competition, the recognition of chemical entities and drugs has aroused widespread interest in academia and industry, and the resulting corpora and systems have helped to more effectively access the information about compounds and drugs (chemical entities) described in text repositories [2].

Named entity recognition, also known as proper noun recognition, is a basic task in natural language processing and has a very wide range of applications. Named entities generally refer to entities in the text that have specific meanings or strong preferentiality, usually including personal names, place names, organization names, dates and times, proper nouns, etc. The NER task usually consists of two parts:

1. Recognition of entity boundaries.
2. Determination of entity type (personal name, place name, organization name, or other).

According to the number of entity types, NER tasks can be divided into two categories:

1. Coarse-grained NER. There are few types of entities, and each named entity corresponds to one entity type.
2. Fine-grained NER. There are many types of entities, and each named entity may have multiple corresponding entity types.

The NER system extracts the above entities from unstructured input text and can identify more categories of entities according to business needs, such as product names, models, prices, etc. Therefore, the extension of the entity concept can be very wide; as long as it is a special text fragment required by the business, it can be called an entity.

From an academic perspective, the named entities involved in NER generally include three major categories (entity class, time class, and number class) and seven subcategories (person names, place names, organization names, time, date, currency, and percentage).

In practical applications, the NER model usually only needs to identify person names, place names, organization names, and dates and times. Some systems will also give out proper noun results (such as abbreviations, conference names, product names, etc.). Currency, percentage, and other numerical entities can be identified by regular expressions. In addition, in some application scenarios, entities within a specific domain will be given, such as book names, song names, journal names, etc. [3].

The tuple list of **s** = 1, w_2, …, w_N>, the purpose of the NER task is to output a list of tuples likes, I_e, t> each of which corresponds to a named entity in s. Here, $I_s \in [1,n]$, $I_e \in [1,n]$ are the start and end indices of the named entity, and t is the entity type from the predefined set of entity types. Figure 12.1 shows an example of three named entities recognized by the NER system from a given sentence. Figure 12.1 gives an example of the NER system identifying 3 named entities from a given sentence.

Research Difficulties in Named Entity Recognition

The main difficulties in Named Entity Recognition research are reflected in the following three aspects:

1. The Limitations and Effectiveness of Domain-Specific Named Entity Recognition. Named Entity Recognition has achieved good results in a limited number of domains and entity types, such as the recognition of person names, place names, and organization names in news corpora. However, these technologies cannot be well transferred to other specific domains, such as military, medical, biological, and small language languages. On the one hand, data from different domains often have unique domain characteristics, such as diseases, symptoms, drugs, and other special named entities in the medical field, which makes it impossible for models in the news field to complete the recognition task; on the other hand, due to the lack of domain resources, the lack of annotated datasets makes it difficult to carry out model training directly. The Chinese Named Entity Recognition process must often combine Chinese word segmentation and shallow language analysis.

The reliability of word segmentation and syntactic analysis systems directly determines the effectiveness of Named Entity Recognition, making Chinese Named Entity Recognition more difficult.

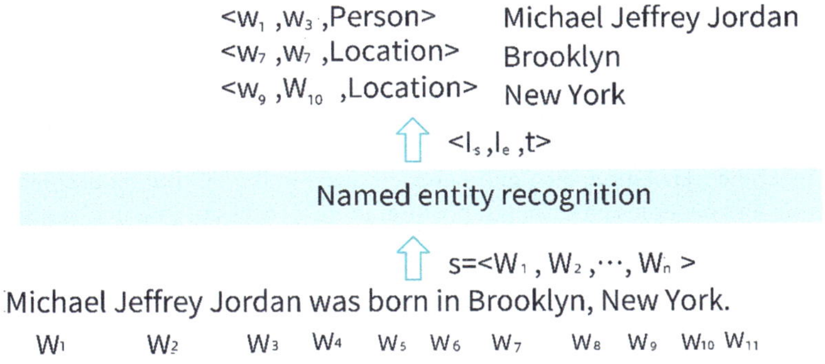

Fig. 12.1 Named entity recognition example

2. Diversity and Ambiguity of Named Entity Expressions. The diversity and ambiguity of natural language bring significant challenges to natural language understanding. The extension of named entities is different in different cultures, domains, and backgrounds, which is a fundamental problem that Named Entity Recognition technology must solve. Currently, there is no commonly followed strict naming convention for the definition and type determination of named entities. After obtaining a large amount of text data, due to issues such as different knowledge representation granularity, varying confidence levels, and lack of normative constraints, phenomena such as diverse naming entity expressions and unclear references often occur.
3. The Complexity and Openness of Named Entities. Traditional entity types only focus on a few types, such as names, place names, and organization names. The complexity of named entities is reflected in the complex and diverse types of entities in actual data, requiring the recognition of fine-grained entity types and assigning named entities to more specific entities. There is no strict naming convention in the industry to follow, and there are also relatively long names of ethnic minorities or translated foreign names in the names. The openness of named entities refers to the fact that the content and type of named entities are permanently the same and will evolve, eventually becoming invalid, making it difficult to establish a large and comprehensive database. The complexity and openness of named entities pose a huge challenge to named entity analysis and are vital issues that must be resolved urgently.

Although NER is a primary task in the field of natural language processing, in recent years, only some researchers have devoted themselves to the pure research of NER problems. Through the analysis of the papers on NER included in ACL2021, it is found that most of the research is aimed at the exploration of the subdivision direction of NER, and the focus includes:

1. Nested and discontinuous problems in NER.
2. Few-shot learning or weakly supervised learning.
3. Practice in certain specific fields (such as the biomedical field).

Application Fields of Named Entity Recognition

Named Entity Recognition, as a basic task in natural language processing, is the foundation of many upstream tasks such as information extraction, knowledge graph, machine translation, etc., and is widely used in the field of natural language processing and occupies an essential position in the process of practical application of natural language processing technology.

The application fields of Named Entity Recognition mainly include the following five:

1. Event Detection. Named entities, such as characters, time, and place, are the fundamental components of an event. When constructing an event summary,

these relevant entities can be highlighted. In the event search system, these entities can serve as index keywords, enhancing the semantic description of the event.
2. Information Retrieval. Named entities are not just additional elements, they also significantly improve and enhance the effects of retrieval systems. For instance, when a user inputs "major," what they may want to retrieve is "Chongqing University," not the meaning of its corresponding adjective. The inclusion of named entities in the system is crucial for its efficiency. Moreover, search engines are moving toward semantic understanding and computing answers, where named entities play a vital role.
3. Semantic Network. The semantic network generally includes concepts, instances, and their corresponding relationships. For example, "country" is a concept, "China" is an instance, and "China is a country" expresses the relationship between the entity and the concept. A large part of the instances in the semantic network are named entities.
4. Machine Translation. Named entities, especially names, place names, organization names, etc., are not just translated as ordinary words. They often have special translation rules, and accurately identifying them in the text is crucial for improving machine translation. The translation effect of named entities has direct significance in the accuracy of the translated text.
5. Question and Answer System. Accurately identifying the various components of the question, related fields, and related concepts is the focus and difficulty of the question-and-answer system. Currently, most question-and-answer systems can only search for answers, not calculate answers. The search for answers is a keyword match, and users manually extract answers based on search results. A more friendly way is to calculate the answer and present it to the user. Some questions in the question-and-answer system need to consider the relationship between entities. For example, when searching for "the 45th President of the United States," the current search engine will return the answer "Trump" in a special format.

Problem Characteristics

Comparing Chinese and English language characteristics, named entities in English have relatively obvious morphological signs. For example, the first letter of each word in entities such as names and place names must be capitalized, and spaces naturally separate each word in English sentences. Recognizing entity boundaries in English is relatively easy compared to Chinese, so its task focuses on determining the entity type. Compared with English, Chinese characters are closely arranged, Chinese sentences are composed of multiple characters, and there is no space between characters and words. This unique language characteristic increases the difficulty of named entity recognition.

Frontier Datasets

Since the first NER task was introduced, the academic community has created many excellent datasets for NER. The mainstream named entity recognition datasets are mostly English datasets and datasets in other languages (such as Chinese, German, etc.) are few.

CoNLL2002 and CoNLL2003 were created based on news articles in four different languages (Spanish, Dutch, English, and German), focusing on four types of entities: person names (Person), place names (Location), organization names (Organization), and others (miscellaneous including all other types of entities). Rajeev Sangal and Singh created the Southeast Asian language NER dataset in 2008, which includes named entity types such as person names, titles, tense expressions, abbreviations, object numbers, brands, etc. Benikova and others proposed the German NER dataset based on German Wikipedia and online news in 2014, and its entity types are similar to CoNLL.

Chinese-named entity recognition datasets include the People's Daily (1998) dataset, the Weibo NER dataset (weibo-ner), the MSRA dataset, OntoNotes4, SIGHANNER, etc.

For a comprehensive list of NER datasets in languages other than English, please refer to the GitHub repository. Table 12.1 provides a snapshot of some English-named entity recognition datasets.

12.1.2 Keyword Extraction

12.1.2.1 Definition of Keyword Extraction

Document keywords represent the thematic and key content of a document and are the smallest units for understanding its content. Keyword extraction, also known keyword annotation, is the process of extracting words or phrases from a text that

Table 12.1 Some English named entity recognition datasets

Dataset	Field	License	Availability
CONLL 2003	News	DUA	Easy to find
NIST-IEER	News	None	
MUC-6	News	LDC	NLTK data
OntoNotes 5	Various fields	LDC	
BBN	Various fields	LDC	LDC 2003T13
GMB-1.0.0	Various fields	None	
GUM-3.1.0	Wiki	Several	LDC 2013T19
WikiGold	Wikipedia	CC-BY 4.0	
Ritter	Twitter	None	LDC 2005T33
BTC	Twitter	CC-BY 4.0	

are most relevant to the meaning expressed by the text. Automatic keyword extraction from documents is a technology for identifying or annotating representative words or phrases in a document that has this function.

Application Fields of Keyword Extraction

In the early days of literature retrieval, which did not support full-text search, keywords were used as the words for searching papers, and keywords had to be arranged in the papers. Even today, this requirement is still present in many documents. However, with the development of the Internet and the advent of the era of big data, a large amount of text information has emerged. These texts are no longer limited to papers that provide keywords, and many texts do not offer keywords, which requires manual or computer program extraction of keywords.

At the same time, in natural language processing, keywords play an important role in text clustering, classification, and summarization. For example, by extracting the keywords from all the news of a certain day, you can roughly know what happened that day. By extracting the keywords from the literature in a certain academic field within a recent period of time, you can understand the current academic research hotspots in this field.

Cutting-Edge Datasets

The more classic English keyword extraction datasets include NLM_500, FAO780, FAO30, SemEval2010, and others. Chinese keyword extraction datasets include SemEval2017Task10 and the China Academic Journal Full-text Database.

12.2 Classic Algorithms

12.2.1 Classic Named Entity Recognition Algorithms

As shown in Fig. 12.2, named entity methods can generally be divided into rule-based methods and machine learning-based methods [4]. Machine learning-based methods can be divided into supervised learning methods and unsupervised learning methods based on the degree of dependence on the corpus. Supervised learning

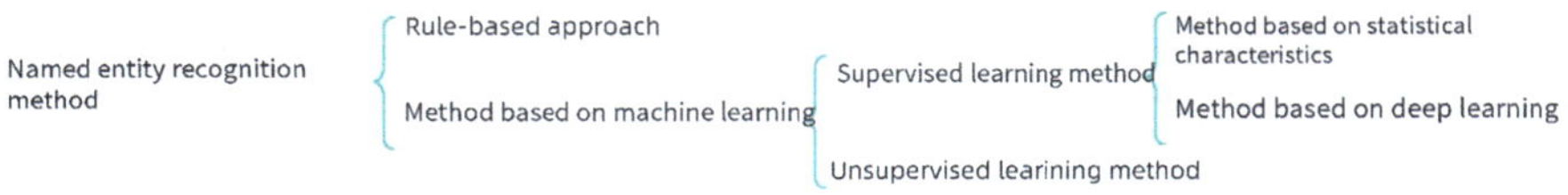

Fig. 12.2 Classification of named entity recognition methods

methods can be further divided into methods based on statistical features (traditional machine learning) and methods based on deep learning.

12.2.1.1 Rule-Based Methods

In the early research of named entity recognition, rule-based named entity recognition methods were the main focus. Rule-based named entity recognition methods are often manually constructed by linguistic experts, and the features used include statistical information, punctuation, keywords, demonstrative words, and direction words, prepositions (such as final words), central words, etc. These methods mainly use pattern and string matching (forward/reverse/bidirectional longest matching, dictionary tree, AC automaton, etc.) to process text to achieve named entity recognition. These systems mainly rely on establishing knowledge bases and dictionaries, including dictionaries composed of feature words and existing common knowledge dictionaries. It is worth noting that using these methods requires assigning weights to each rule, and when encountering rule conflicts, choose the rule with the highest weight to determine the type of named entity.

Rau [5] published a paper on extracting and recognizing company names at the 7th IEEE Artificial Intelligence Applications Conference and described the extraction of formula-type named entities from the text through manual rule writing. It can be found that when the extraction rules can accurately reflect language phenomena, the accuracy of rule-based named entity recognition methods is high, and there is no need to annotate the corpus in advance. However, the limitations of this method are also very obvious. The construction of rules and dictionaries depends on specific languages, domains, and text styles. It is not easy to expand on other entity sets, needs better portability, and this method needs to discover new words and unrecorded words. Some famous rule-based NER systems include LaSIE-II [6], NetOwl, Facile, SAR, Fastus, and LTG.

The shortcomings of the rule-based method were quickly exposed:

1. Although this method can achieve higher recognition results on specific corpora, the better the recognition effect, the more rules need to be formulated, and the feasibility of manually formulating these rules is too low.
2. It is almost impossible for this method to recognize infinitely changing named entities by formulating a limited number of rules.
3. The rules are extremely dependent on domain knowledge. When the domain difference is large, the previously formulated rules often cannot be transplanted, and rules have to be re-formulated.

These inherent shortcomings have led researchers to adopt new research ideas, and at this time, machine learning is on the rise in the field of natural language processing, and named entity recognition has naturally turned to the camp of machine learning.

12.2.1.2 Machine Learning-Based Methods

The named entity recognition method based on statistical features belongs to the traditional machine learning-based, supervised learning method. This method usually converts the named entity recognition task into a sequence labeling task, and the model uses the annotated corpus for training. Compared with classification problems, the current prediction label in sequence labeling problems is related to the current input features and the previous prediction labels; there is a mutual dependency relationship between the prediction label sequences. For example, I-LOC cannot appear before B-LOC, I tags cannot appear at the beginning of the sequence, and so on. The specific steps of the machine learning method based on statistical features are as follows:

1. Annotation of Corpus. Generally, the Inside-Outside (IO), Begin-Inside-Outside (BIO), and Begin-Inside-Outside-End-Single (BIOES) annotation systems are used for manual or automatic annotation of the text in the corpus.
2. Feature Definition. By statistically analyzing the language information contained in the training corpus, specific word features, context features, dictionary and part-of-speech features, stop word features, core word features, and semantic features are mined from the training corpus. The quality of feature selection determines the effectiveness of the named entity recognition model.
3. Model Training. Commonly used statistical models include the hidden Markov model (HMM) [7], maximum entropy (ME) [8] model, support vector machine (SVM) model, and conditional random field (CRF) model [9].

Hidden Markov Model

The hidden Markov model was developed in the 1970s and is a statistical model. There are two sequences in this model's system:

1. The observable sequence, which can be directly obtained through observation. In named entity recognition, this refers to each word itself.
2. The hidden state transition sequence, i.e., the annotation of each word.

The most likely annotation sequence for the observation sequence is sought, i.e., generating annotations based on a series of input words, thereby obtaining entities.

Maximum Entropy Model

The principle of maximum entropy is a general principle of statistical learning. Applying it to classification problems results in the maximum entropy model.

Assume the classification model is a conditional probability distribution $P(Y|X)$, where X represents the input and Y represents the output. This model represents the output Y with conditional probability $P(Y|X)$ for a given input X. In the NER model, X represents a character, and Y represents the label corresponding to the character.

Given a training dataset $T = \{(x_1, y_1), (x_2, y_2)... (x_N, y_N)\}$, the goal is to use the principle of maximum entropy to select the best classification model.

According to the principle of maximum entropy, the model should first ensure that all known constraints are met. How are these constraints obtained?

The idea is to extract several features from the training data T and then require that the expectations of these features on T with respect to the empirical distribution are equal to their mathematical expectations in the model with respect to $p(x, y)$. In this way, one feature corresponds to one constraint.

The feature function is a description of a certain fact between the input x and the output y, defined as:

$$f(x,y) = \begin{cases} 1 & x \text{ and } y \text{ satisfy a certain fact} \\ 0 & \text{otherwise} \end{cases} \tag{12.1}$$

The empirical distribution refers to the distribution obtained by statistics on the training data T. Two empirical distributions need to be examined, namely the joint empirical distribution of x, y and the distribution of x. Their definitions are as follows:

$$\tilde{p}(x, y) = \frac{\text{count}(x, y)}{N} \tag{12.2}$$

$$\tilde{p}(x) = \frac{\text{count}(x)}{N} \tag{12.3}$$

where count(x, y) represents the number of times (x, y) appears in the data T, and count(x) represents the number of times x appears in the data T.

For any feature function $f(\cdot)$, let $E_{\tilde{p}}(f)$ represent the mathematical expectation of f on the training data T with respect to $\tilde{p}(x, y)$. $E_p(f)$ represents the mathematical expectation of f on the model with respect to $p(x, y)$. According to the definition of expectation, we have:

$$E_{\tilde{p}}(f) = \sum_{x,y} \tilde{p}(x,y) f(x,y) \tag{12.4}$$

$$E_p(f) = \sum_{x,y} p(x,y) f(x,y) = \sum_{x,y} \tilde{p}(x) p(y|x) f(x,y) \tag{12.5}$$

For the probability distribution $p(y|x)$, it is hoped that the expectation of the feature f should be the same as the feature expectation obtained from the training data. Therefore, a constraint can be proposed:

$$E_{\tilde{p}}(f) = E_p(f). \tag{12.6}$$

In this way, there are as many features as there are feature functions and constraints. Assume that the set of models that satisfy all constraints is C, then the model with the maximum conditional entropy in the model set C is called the maximum entropy model.

Conditional Random Field Model

The conditional random field model is the mainstream model for named entity recognition at present. Assume X represents the observed sequence to be tagged, Y represents the hidden state sequence, $P(Y|X)$ represents the conditional probability of Y given X, the random variable Y satisfies:

$$P(Y_v|\mathbf{X}, Y_w, \ w \neq v) = P(Y_v|\mathbf{X}, Y_w, \ w \sim v). \tag{12.7}$$

This holds for all v. In other words, for a given node v, it is independent of all nodes that are not adjacent to it.

In the task of named entity recognition, the main model used is the chain-structured CRF model, that is, the linear chain conditional random field model.

In named entity recognition tasks, the main model used is the chain-structured CRF, i.e., the linear chain CRF. Suppose the observed sequence $\boldsymbol{X} = (X_1, \mathrm{X}_2, \ldots, X_n)$ and the corresponding state sequence $\boldsymbol{Y} = (Y_1, Y_2, \ldots, Y_n)$, . Given the random variable sequence X, the conditional probability distribution $P(Y|X)$ of the random variable sequence Y constitutes a conditional random field, satisfying:

$$P\left(Y_i|X, Y_1, \ldots, Y_{\{i-1\}}, Y_{\{i+1\}}, \ldots, Y_n\right) = P\left(Y_i|X, Y_{\{i-1\}}, Y_{\{i+1\}}\right) \tag{12.8}$$

12.2.1.3 Methods Based on Deep Learning

Compared with methods based on machine learning, the key advantage of methods based on deep learning lies in their ability to learn representations.

Applying deep learning to named entity recognition has three core advantages. First, thanks to nonlinear transformations, models based on deep learning produce nonlinear mappings from input to output. Compared with linear models (such as log-linear HMM models and linear chain CRF models), models based on deep learning can learn complex features from data through nonlinear activation functions. Second, methods based on deep learning save time when designing features for named entity recognition. Methods based on statistical features require many engineering skills and domain expertise, while models based on deep learning can automatically learn useful features. Third, deep neural named entity recognition models can be trained end-to-end using gradient descent algorithms. This feature allows quite complex named entity recognition systems to be designed.

In recent years, with the development of hardware computing capabilities and the introduction of word embeddings, neural networks have become a model that can effectively handle many natural language processing tasks. In the task of named entity recognition, the traditional way of processing neural networks is as follows: first, all words in the sentence are represented as word embeddings, then the embedding sequence of the sentence is input into the neural network, and features are automatically extracted, and finally, each label is predicted through a layer of Softmax. Among them, neural networks often use RNN, LSTM, BiLSTM, etc.

However, using traditional neural network methods to tag each word is an independent classification, which cannot directly use the tags predicted by the previous context, leading to the prediction of illegal tag sequences. To solve this problem, the academic community has proposed the neural network-CRF model for sequence tagging. Representative models include BiLSTM-CRF, IDCNN-CRF, etc.

BiLSTM-CRF Model

First, introduce entity labels.

Suppose the dataset has two types of entities: Person and Organization. Assuming the use of the BIO tagging system, there will be five types of entity tags, as shown in Table 12.2.

Assume x is a sentence containing five words (w_0, w_1, w_2, w_3, and w_4). In sentence x, $[w_0, w_1]$ is a person's name, $[w_3]$ is the name of an organization, and the rest is nonentity information.

First, each word in the sentence is a vector containing word and character embeddings. Word embeddings are usually pretrained, while character embeddings are randomly initialized. All embeddings will be adjusted during the training iteration process.

Second, the input of the BiLSTM-CRF model is the word embedding vector, and the output is the predicted tag corresponding to each word, as shown in Fig. 12.3.

As shown in Fig. 12.4, the input of the BiLSTM layer represents the scores corresponding to each word category. For example, w_0, the output of the BiLSTM node is 1.5 (B-Person), 0.9 (I-Person), 0.1 (B-Organization), 0.08 (I-Organization), and 0.05 (O). These scores will be used as input for the CRF layer. All scores output by

Table 12.2 Entity tags

Entity name	Entity meaning
B-Person	The beginning of a person's name
I-Person	The middle of a person's name
B-Organization	The beginning of the organization's name
I-Organization	The middle part of the organization's name
O	Nonentity information

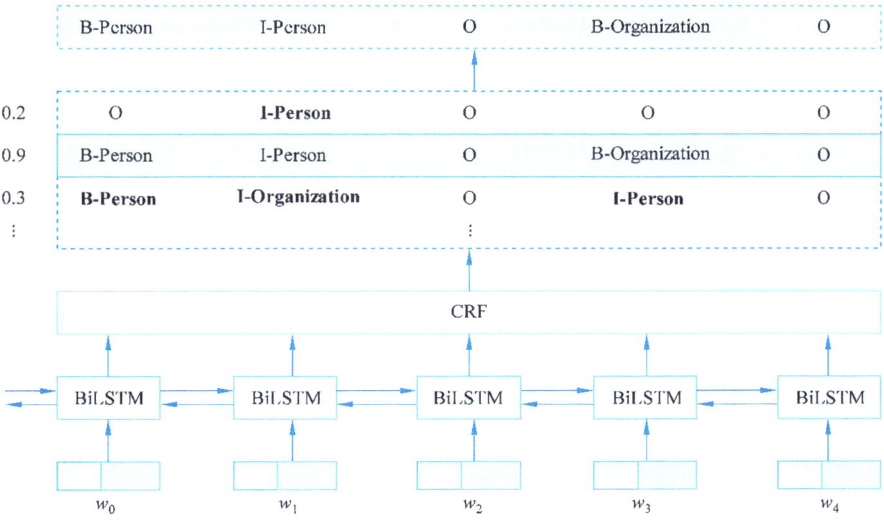

Fig. 12.3 Example of BiLSTM-CRF model input and output

the BiLSTM layer will be used as the input for the CRF layer, and the category with the highest score in the category sequence is the final prediction result.

The BiLSTM-CRF model consists of a BiLSTM layer and a CRF layer.

BiLSTM, or bidirectional LSTM, can capture both forward and backward information, unlike unidirectional LSTM models, which can only capture forward information. This makes the use of text information more comprehensive and improves the effect.

A linear layer is added after the final output layer of the BiLSTM network to project the hidden layer output results generated by the BiLSTM network to an interval with certain label feature significance, as shown in Fig. 12.5.

Even without the CRF layer, a named entity recognition model based on the BiLSTM network can still be trained. For example, if the score of w_0, B-Person is the highest (1.5), then it can be B-Person is selected as the prediction result. Similarly, w_1 is I-Person, w_2 is O, w_3 is B-Organization, w_4 is O. The ideal output without the CRF layer is shown in Fig. 12.6.

However, the actual output obtained may be as shown in Fig. 12.7.

This is because the role of the CRF layer is to add some constraints to the prediction results. The CRF layer can learn the constraints of the sentence during training. For example:

- The beginning of a sentence should be "B-" or "O", not "I-".
- In the "B-label1I-label2I-label3…" pattern, categories 1–3 should be the same entity category. For example, "B-Person I-Person" is correct, while "B-Person I-Organization" is wrong.

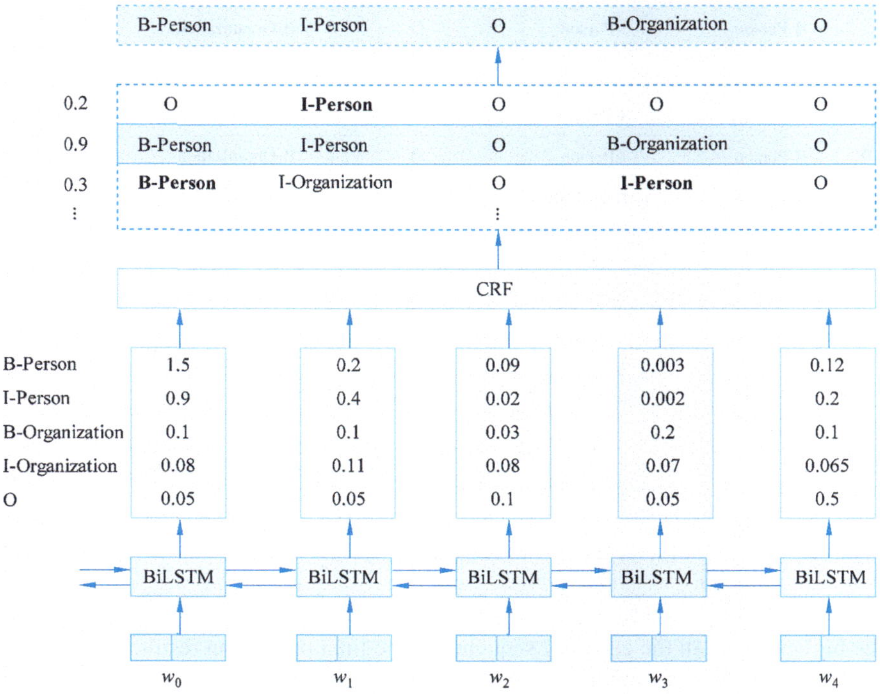

Fig. 12.4 Example of BiLSTM-CRF model input and output II

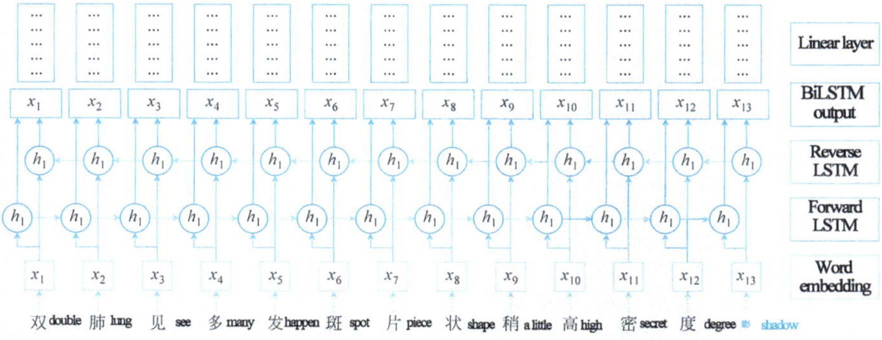

Fig. 12.5 BiLSTM layer structure

- The label 'O-/I-' is incorrect; the beginning of a named entity should be marked with 'B-' instead of 'I-'.

For sequence labeling, a common CNN has a shortcoming; after convolution, the neurons in the last layer may only get a small piece of information from the original input data. Every word in the entire input sentence may affect the current position's

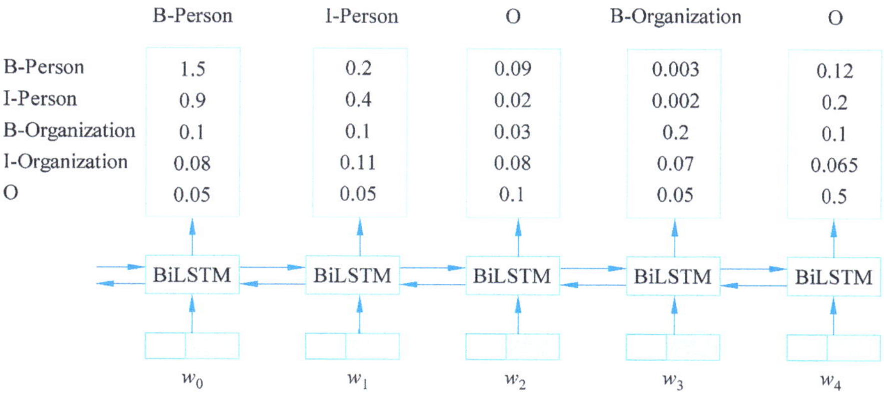

Fig. 12.6 Ideal output without CRF layer

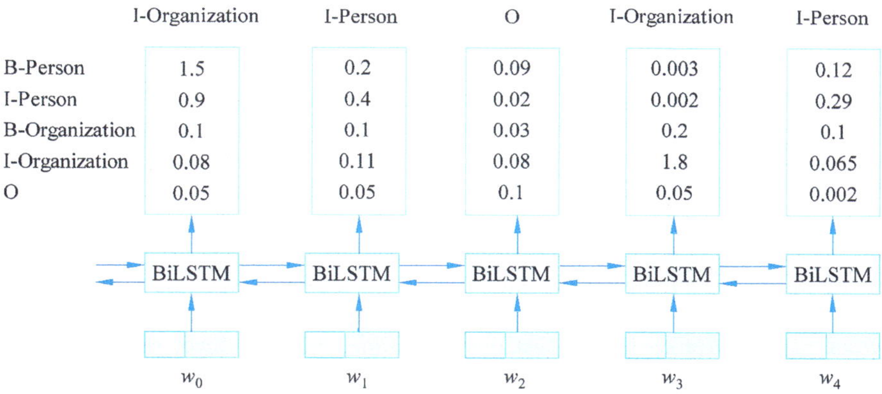

Fig. 12.7 Possible actual output without CRF layer

annotation, which is the so-called long-distance dependency problem for named entity recognition. More convolution layers need to be added to cover all input information, leading to more layers and more and more parameters. More regularization, such as dropout, needs to be added to prevent overfitting, which brings more hyperparameters and makes the entire model large and difficult to train. Because of these disadvantages of CNN, for most sequence labeling problems, people still choose network structures like BiLSTM, trying to use the network's memory to annotate the current word through the information of the whole sentence.

But this brings another problem: BiLSTM is essentially a sequence model that is less powerful than CNN in utilizing GPU parallel computing. How can we make full use of a GPU firepower like CNN and remember as much input information as possible with a simple structure like LSTM?

Fisher Yu and Vladlen Koltun proposed the dilated CNN model. The idea is simple: the normal CNN filter works on a continuous area of the input matrix, and the convolution operation is continuously performed. The dilated CNN adds a dilation width to this filter. When it works on the input matrix, it will skip all input data in the middle of the dilation width, and the size of the filter itself remains unchanged. In this way, the filter can obtain data from a larger area on the input matrix, which looks like it has been dilated.

In specific use, the dilation width expands exponentially with increased layers. As the number of layers increases, the number of parameters increases linearly while the receptive field expands exponentially, quickly covering all input data.

The trend of the receptive field is shown in Fig. 12.8.

As can be seen from Fig. 12.8, the receptive field expands exponentially. The original receptive field is a 1 × 1 area located at the center point.

1. The original receptive field expands outward with a stride of 1, resulting in 8 1 × 1 areas, forming a new receptive field with a size of 3 × 3.
2. Upon further expanding outward with a stride of 2, the receptive field expands to 7 × 7.
3. Upon further expanding outward with a stride of 4, the receptive field expands to 15 × 15.

An IDCNN block with a maximum dilation stride of 4 is shown in Fig. 12.9.

Corresponding to the text, the input is a one-dimensional vector, each element is a word embedding.

IDCNN generates logits for each word of the input sentence, which are exactly the same as the logits output by the BiLSTM model. It adds a CRF layer and decodes the annotation results using the Viterbi algorithm.

In sequence labeling, it is common to connect a CRF layer at the end of a network model like BiLSTM or IDCNN. BiLSTM or IDCNN calculates the probability of each tag for each word. At the same time, the CRF layer introduces the sequence's transition probability and calculates the loss feedback to the network.

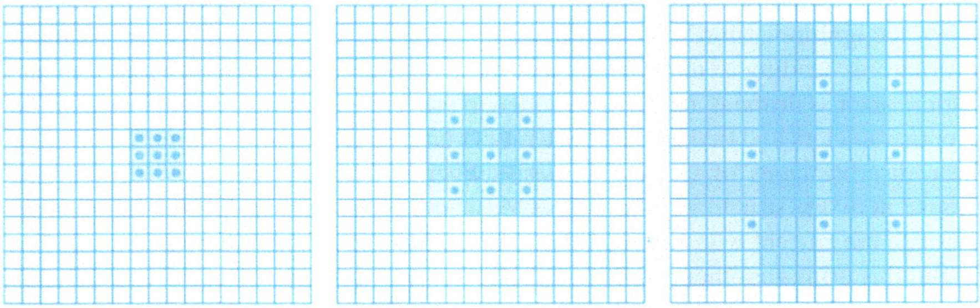

Fig. 12.8 Trend of the receptive field

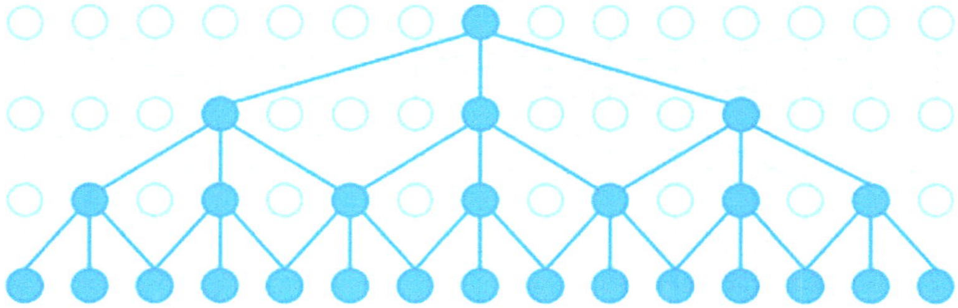

Fig. 12.9 An IDCNN block with a maximum dilation stride of 4

12.2.2 Classic Keyword Extraction Algorithms

12.2.2.1 TF-IDF Algorithm

Term frequency-inverse document frequency (TF-IDF) is a common weighting technique in information retrieval and text mining.

TF-IDF is a statistical method used to evaluate the importance of a word to a document set or a document in a corpus. The importance of a word is directly proportional to its frequency in the document and inversely proportional to its frequency in the corpus.

The TF-IDF method includes two indicators.

Term frequency (TF) is the frequency of the word (keyword) in the text, and it is calculated using the given formula:

1. Term Frequency (TF) refers to the frequency at which a term (or keyword) appears in a text

$$\text{TF} = \frac{\text{number of times the term } w \text{ appears in a category}}{\text{total number of terms in that category}} \tag{12.9}$$

2. Inverse document frequency (IDF) inverse document frequency: the total number of documents is divided by the number of documents containing the term, and then the logarithm of the obtained quotient is taken. If fewer documents contain the term w, the IDF is larger, indicating that the term has good category differentiation ability.

$$\text{IDF} = \log \frac{\log \text{total number of documents in the corpus}}{\left(\text{number of documents containing the term } w + 1\right)} \tag{12.10}$$

The reason for adding 1 to the denominator is to prevent the occurrence of a situation where the denominator is 0.

TF-IDF is TF × IDF. A high term frequency of a word in a specific document and a low inverse document frequency of the word in the entire document set can generate a large TF × IDF value. Therefore, the TF-IDF algorithm can filter out common words and retain important words. The higher the importance of a word to a document, the larger its TF × IDF value. Therefore, the words with the highest TF × IDF values are the document's keywords.

The TF-IDF algorithm is mainly used in search engines, keyword extraction, text similarity evaluation, text summarization, etc.

12.2.2.2 TextRank Algorithm

The TextRank algorithm introduces the concept of edge weight based on the PageRank algorithm, representing the similarity between two sentences. The general model of the TextRank algorithm can be represented as a directed weighted graph $G = (V, E)$, which consists of a set of vertices V and edges E. For a given vertex V_i, In (V_i) represents the set of vertices pointing to V_i, and Out (V_i) represents the set of vertices that V_i points to.

The steps of the TextRank algorithm are as follows:

1. Divide the given text into complete sentences.
2. For each sentence, perform word segmentation and part-of-speech tagging and filter out stop words, only retaining words of specified parts of speech, TF-IDF is TF × IDF. A high word frequency in a specific document and a low document frequency of the word in the entire document set can produce a high-weight TF-IDF. Therefore, TF-IDF tends to filter out common words and retain important words. The higher the importance of a word to an article, the larger its TF-IDF value. Therefore, the first few words are the keywords of this article. These keywords typically include nouns, verbs, adjectives, and similar parts of speech, which are used as candidate keywords.
3. Construct the candidate keyword graph $G = (V, E)$, where V is the node-set composed of candidate keywords generated by (2), and then use the co-occurrence relationship to construct the edge between any two points. There is an edge between two nodes only when their corresponding words co-occur in a window of length K.
4. Then, the co-occurrence relationship is used to construct the edge between any two nodes, and there is an edge between two nodes if and only if the words corresponding to them co-occur in a window of length K.
5. Initialize the weights of each node according to the importance measurement formula in the PageRank algorithm, then iteratively calculate the weights of each node until convergence. Sort the node weights in reverse order to get the most important T words as candidate keywords.
6. Mark the T words obtained from step (5) in the original text. If two words form adjacent word groups, they are combined into multi-word keywords. For example, in the sentence "Matlab code for plotting ambiguity function," if "Matlab" and "code" are both candidate keywords, they are combined into "Matlab code" and added to the keyword sequence.

The formula for calculating node weight is as follows:

$$\mathrm{WS}(V_i) = (1-d) + d \times \sum_{V_j \in \mathrm{In}(V_i)} \frac{\omega_{ji}}{\sum_{V_k \in \mathrm{Out}(V_j)} \omega_{jk}} \mathrm{WS}(V_j) \tag{12.11}$$

where d is the damping coefficient, ranging from 0 to 1, representing the probability of pointing from a specific point in the graph to any other point, generally taken as 0.85.

12.2.2.3 Word2Vec Algorithm

The Word2Vec tool mainly includes two models: Continuous bag of words (CBOW) model and the skip-gram model. As shown in Fig. 12.10, the blue part on the left represents the CBOW model, and the green part on the right represents the skip-gram model. The difference between them is that CBOW is trained to obtain word vectors by predicting the target word based on the context, such as predicting $W(t)$ based on W_{t-2}, W_{t-1}, W_{t+1}, W_{t+2} these four words to predict W_t, while skip-gram model is trained to obtain word vectors by predicting surrounding words based on the target word, such as predicting W_{t-2}, W_{t-1}, W_{t+1}, W_{t+2} based on W_t. According to experience, CBOW is more suitable for small corpora, while skip-gram performs better on large corpora.

How is the Word2Vec algorithm specifically implemented? The following introduces the implementation details of each step based on the principle of the CBOW model shown in Fig. 12.11.

1. In the input layer, there are three words in the context of the target word; each word is represented by one-hot encoding, which is a $1 \times V$ matrix, where V represents the number of words.
2. All one-hot matrices are multiplied by the input weight matrix W. W is a shared matrix of $V \times N$, where N is the dimension of the output word vector.

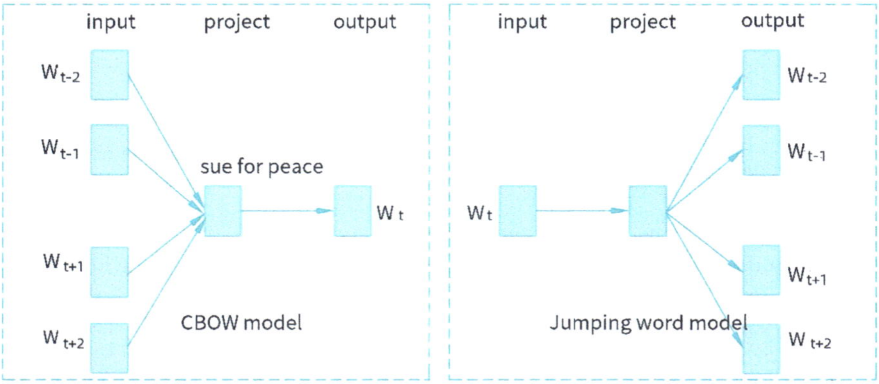

Fig. 12.10 Structure of Word2Vec algorithm models

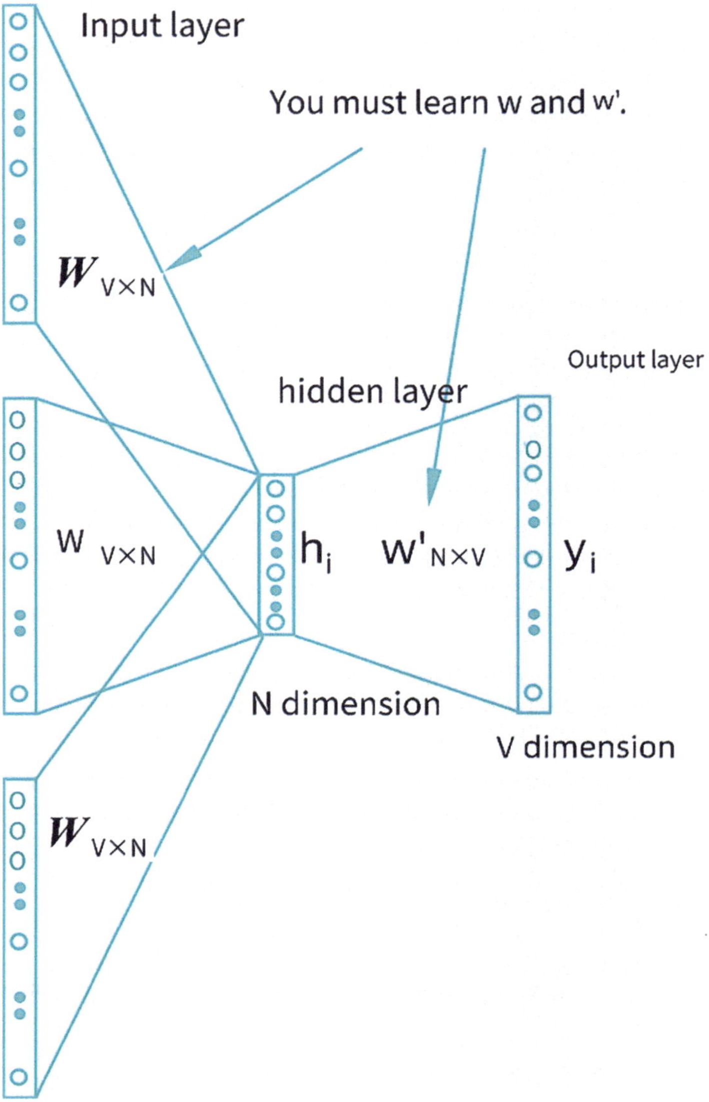

Fig. 12.11 Principle of the CBOW model

3. Multiply the obtained vector's one-hot matrix by the shared matrix of $V \times N$, add the results, take the average, and use it as the hidden layer vector h, which has a dimension of N.
4. Multiply the hidden layer vector h by the output weight matrix W'. W' is a shared matrix of $N \times V$.

5. The resulting vector y, with a dimension of V, is processed by the Softmax activation function to obtain a V-dimensional probability distribution.
6. Since the input is one-hot encoded, each dimension represents a word, so in the V-dimensional probability distribution, the word represented by the index with the highest probability is the predicted middle word.
7. Compare the result with the real label's one-hot encoding, the smaller the error, the better. The error function here generally chooses the cross-entropy cost function.

The above describes the entire process of generating word vectors using the CBOW model. If you only need to extract the vector for each word, you can simply obtain the vector y. However, during the training process, predictions must be made, and errors must be calculated to derive the input weight matrix W and the output weight matrix W'.

12.2.2.4 Topic Model Algorithm Based on LDA

How does the author conceive an article? Simply put, if you want to write an article, you first need to determine several themes, choose words that can describe the theme well, and form sentences, paragraphs, and chapters according to the grammatical rules. In fact, in a statistical sense, the words that are used more frequently for this theme are the keywords of this theme.

The topic model is a probabilistic language model that simulates the author's thinking when writing an article. Its main ideas include two points: (1) the document is a mixed distribution of several topics and (2) each topic is a probability distribution of words. The earliest idea was the probability latent semantic analysis (PLSA) model proposed by Thomas Hofmann. The basic idea of this model is that a document can be mixed from multiple topics, and each topic is a probability distribution of words. Each word in each document is obtained through such a process: first, select a topic by probability, and then select a word from this topic by probability. The formula for calculating probability for each word in a document is as follows:

$$P(\text{word} / \text{document}) = \sum_{\text{topic}} P(\text{word} / \text{topic}) \times P(\text{topic} / \text{document}). \tag{12.12}$$

The main topic model is the latent dirichlet allocation (LDA) model. Its proposer, David Blei, added a Dirichlet prior distribution based on the PLSA model, a breakthrough extension of the PLSA model. The PLSA model does not use a unified probability topic model to calculate the probability distribution of the document corresponding to the topic. When there are many parameters, it will cause overfitting, and calculating the probability distribution of documents outside the training dataset is relatively difficult. On this basis, Blei introduced hyperparameters in the LDA model. No matter how many external documents are changed, only one hyperparameter exists. Then, the model is derived by probabilistic methods to find the semantic structure of the document set and mine the topics of the document. The principle of the LDA model is shown in Fig. 12.12.

In the LDA model, the document–topic and topic–word distributions are multinomial distributions, and their prior probabilities are Dirichlet distribution.

Assuming the total vocabulary of the document set is V, in a document composed of n words, each word is generated according to a multinomial distribution. By sampling n times, a document can be generated, corresponding to the φ distribution (where β is the hyperparameter) in Fig. 12.12. Similarly, different documents have different 'dice' (multinomial distributions). Each time a topic 'die' is selected for a document, this process follows a multinomial distribution, corresponding to the θ distribution (where α is the hyperparameter) in Fig. 12.12. In summary, the LDA model consists of two stages.

1. $\alpha \rightarrow \theta_m \rightarrow Z_{m,n}$; select a document–topic distribution with parameter θ_m, and then generate the topic of the nth word of the mth document, generating the number of $Z_{m,n}$. Sample a multinomial distribution θ_m from a Dirichlet distribution with parameter α as the distribution of this document on k topics;
2. $\beta \rightarrow \varphi_k \rightarrow W_{m,n\mid k} = Z_{m,n}$, generate the nth word of the mth document. For each word in this document, sample a topic number according to the θ_m distribution in the previous step, and then sample a multinomial distribution from the Dirichlet distribution with parameter β corresponding to the topic–word distribution as the distribution of words under this topic.

The parameters α and β are given empirically, and $Z_{m,n}$, θ_m, φ_k are hidden variables that need to be derived. There are two methods for deriving LDA: an exact derivation, such as using expected maximum (EM) calculation, and an approximate derivation, commonly used in actual engineering. The simplest of these is the Gibbs sampling method.

Obviously, the higher the topic feature of the word in the document, the stronger its ability to represent a certain topic. When calculating the topic weight of a word using the topic model, select the training corpus, preprocess the text, and finally, train the LDA model. After obtaining the LDA model, there are two methods to calculate the topic feature values of each word using the distribution of words under each topic in the model: one method assumes that each word can only represent one topic, and the top k words with high weight under each topic in the model are taken as the words of that topic; the other method assumes that words with high topic

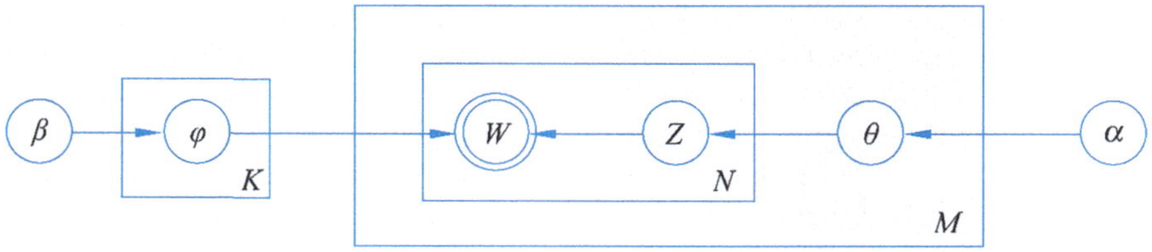

Fig. 12.12 Principle of the LDA model

differentiation should be those with high weight under a certain topic and low frequency in different topics, and each word only represents its most representative topic. After calculating the topic feature values of the words, it can be used as a basis for judging keywords, or the topic feature values of words can be used as a feature of words in other automatic keyword extraction methods.

The topic model is the core of semantic mining, and the LDA topic model is representative. In the topic model, a topic represents an aspect or a concept, manifested as a collection of related words, which is the conditional probability of these words. The advantages of the topic model are as follows:

1. It can obtain the similarity between texts. According to the topic model, the probability distribution of the topic can be obtained, and the similarity between texts can be calculated through the probability distribution.
2. It can solve the problem of polysemy. For example, "apple" may be a fruit or a mobile phone brand or company name. Through the *P*(word|topic) of the topic model, it can be concluded which topic "apple" is related to, and from the perspective of the generative model, the similarity between other words under its belonging topic can be obtained.
3. It can remove the noise in the document. In the topic distribution, the noise of the text often exists in the secondary topics. Therefore, noise can be removed by sorting the topics.
4. Unsupervised and fully automated, the topic model does not require manual annotation. The probability distribution can be directly obtained through the model.
5. Language-independent. The topic model has no language restrictions; any language can obtain topic distribution through the topic model.

The keyword extraction methods for text are divided into supervised and unsupervised types, with more detailed classifications shown in Fig. 12.13.

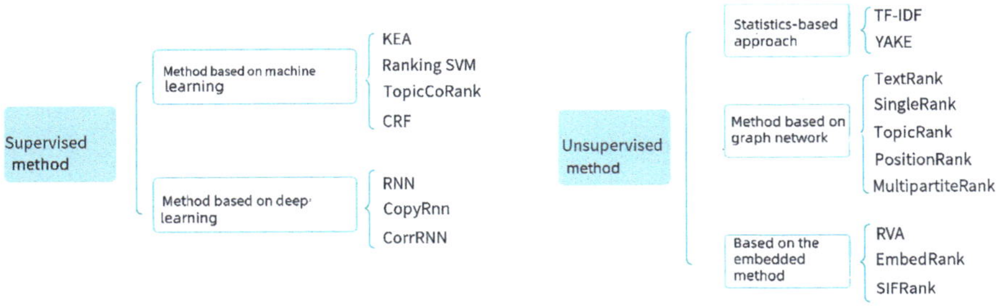

Fig. 12.13 Classification of keyword extraction methods

12.2.3 Algorithm Classification

The methods for extracting keywords from text are divided into supervised and unsupervised. A more detailed classification framework is shown in Fig. 12.13:

12.2.3.1 Supervised Methods

Keyword extraction is viewed as a binary classification problem, determining whether a word or phrase in a document is a keyword. Since it is a classification problem, it requires providing an already annotated training corpus, which is used to train the keyword extraction model, and then the keywords are extracted from the document according to the model. Supervised methods for keyword extraction can be classified into machine learning-based methods (also known as traditional methods) and deep learning-based methods.

Machine Learning-Based Methods

Keyphrase Extraction Algorithm (KEA) [10] is an early algorithm that uses feature vectors (such as TF-IDF values and the position information of the word first appearing in the article) to represent candidate words, uses the Naive Bayes method as the classification method, and scores and classifies candidate words. On this basis, many improved algorithm versions have also been proposed. For example, Hulth [11] introduced linguistic knowledge and proposed an improved version. Caragea and others [12] introduced more feature information through the citation relationship of academic papers when extracting keywords, thereby further improving the keyword extraction effect.

Xin and others [13] used the learning-to-rank method to model this problem and abstracted the training process into fitting the ranking function.

Bougouin is a supervised extension of the unsupervised method TopicRank. This method combines a second graph network with the basic topic graph.

Deep Learning-Based Methods

Zhang [14] used a double-layer RNN structure, represented information through two hidden layers, and used the sequence labeling method to output the final result.

Meng Rui used the encoder–decoder structure for keyword extraction. First, the training data is converted into text–keyword pairs, then, the RNN-based encoder–decoder network is trained, learning the mapping relationship from source data (sentences) to target data (keywords).

Yang Xitong [15] also uses the encoder–decoder structure but introduces two additional constraints:

1. Keywords should cover as many different topics of the article as possible.
2. Keywords should be as different from each other as possible to ensure diversity.

12.2.3.2 Unsupervised Methods

Unsupervised methods target corpora that do not require manual annotation, using the language features of the text to discover important words as keywords for keyword extraction. Unsupervised methods are divided into the following three categories.

Statistics-Based Methods

The TF-IDF-based method is the most basic statistics-based method. After obtaining a set of candidate words (such as using POS tags to extract noun phrases), word frequency and inverse document frequency are used to score the candidate words, selecting high-scoring words as keywords.

Graph Network-Based Methods

TextRank is the first keyword extraction algorithm based on graph networks. This method first extracts candidate words based on POS tags and then uses the candidate words as nodes to create a graph network. If two candidate words co-occur within a certain window, a link is created between the nodes, establishing a connection between the nodes. The PageRank algorithm is used to update this graph network until it converges. Subsequently, various improved algorithms based on graph networks have been continuously proposed, and this type of algorithm has gradually become the most widely used in unsupervised keyword extraction.

Wan and others [16] introduced weights for the edges between nodes based on the TextRank algorithm. Florescu et al. [17] proposed the Biased Weighted Page Rank algorithm by incorporating the positional information of words, thereby enhancing the effectiveness of keyword extraction.

Embedding-Based Methods

Embedding-based methods use embeddings to express the information of articles and words at various levels (characters, grammar, semantics, etc.).

Bennani-Smires first uses POS tags to extract candidate words, then calculates the cosine similarity between the candidate word embeddings and the article embeddings, uses the cosine similarity to sort the candidate words, and obtains the keywords.

Because supervised text keyword extraction algorithms require manually annotated training samples, which are costly, commonly used text keyword extraction mainly uses unsupervised keyword extraction with strong applicability.

12.3 Applications and Analysis

12.3.1 Named Entity Recognition Example

12.3.1.1 Laboratory Work

1. A method of idiomatic expression extraction based on BiLSTM and CRF models

The NLPIR Big Data Search and Mining Laboratory applies the Bert-BiLSTM-CRF model to related projects, proposing a small sample entity extraction method based on BiLSTM and CRF models.

Through data augmentation and mixed annotation methods, a dataset of 1000 small-sample Chinese corpora was created.

Expanded to 10,000, the model's P, R, $F1$ values are 93.05%, 94.30%, 93.67%.

2. NLPIR-ICTCLAS-DocExtractor

NLPIR-ICTCLAS-DocExtractor is a named entity extraction component of the Chinese analysis system.

It can extract names, place names, organization names, time, and various types of custom information from the text.

The algorithm flow of this component is shown in Fig. 12.14.

12.3.1.2 Experimental Examples and Analysis

Evaluation Indicators

Judging whether a named entity is correctly recognized includes two aspects:

1. Whether the boundary of the entity is correct.
2. Whether the type of entity is correctly labeled.

The correctness of named entity recognition can generally be divided into the following two evaluation methods.

Precise Match Evaluation

For the two main subtasks of named entity recognition—boundary detection and type recognitionprecise match evaluation requires the system to correctly identify the boundary and type of the instance at the same time. The definitions of true

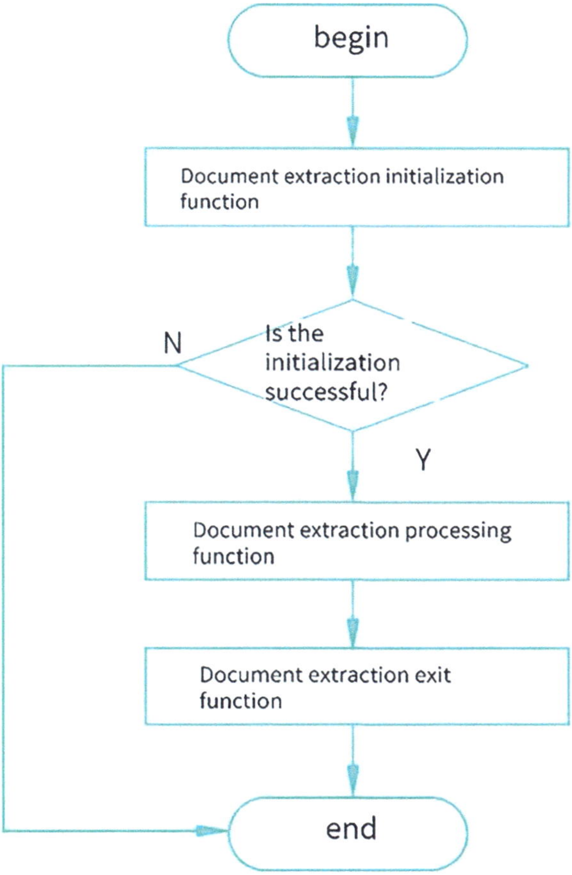

Fig. 12.14 DocExtractor component algorithm flow

positive (TP), false positive (FP), and false negative (FN) used to calculate the precision, recall, and $F1$-score are as follows:

- True Positive (TP): The named entity recognition system recognizes and marks the entity as this type, and it matches the ground truth.
- False Positive (FP): The entity is recognized and marked as this type by the named entity recognition system, but it does not match the ground truth.
- False Negative (FN): The entity is not recognized and marked as this type by the named entity recognition system. The ground truth is this type, but it is not correctly marked.

Precision refers to the percentage of entities correctly identified in the model results, and recall refers to the percentage of entities correctly identified by the system. The formulas for calculating these two indicators are as follows:

$$\text{Precision} = \frac{\text{TP}}{\text{TP} + \text{FP}} \tag{12.13}$$

$$\text{Recall} = \frac{\text{TP}}{\text{TP} + \text{FN}} \tag{12.14}$$

The calculation of $F1$-score is as follows:

$$F1 = 2 \times \frac{\text{Precision} \times \text{Recall}}{\text{Precision} + \text{Recall}} \tag{12.15}$$

Loose Match Evaluation

Some researchers proposed a loose match evaluation scheme in the early development of named entity recognition tasks. This scheme is considered correct or partially correct when the entity position overlaps partially or the entity position is correct, but the category is wrong. Because this scheme is not intuitive and not friendly to error analysis, it has yet to be widely adopted.

Experimental Analysis

Conduct a sample experiment on the Chinese-named entity recognition dataset using the MSRA dataset, as shown in the example below:

当/o 希望工程/o 救助/o 的/o 百万/o 儿童/o 成长/o 起来/o ，/o 科教/o 兴/o 国/o 蔚然成风/o 时/o ，/o 今天/o 有/o 收藏/o 价值/o 的/o 书/o 你/o 没/o 买/o ，/o 明日/o 就/o 叫/o 你/o 悔不当初/o ！/o

藏书/o 本来/o 就/o 是/o 所有/o 传统/o 收藏/o 门类/o 中/o 的/o 第一/o 大户/o ，/o 只是/o 我们/o 结束/o 温饱/o 的/o 时间/o 太/o 短/o 而已/o 。/o

因/o 有关/o 日/ns 寇/o 在/o 京/ns 掠夺/o 文物/o 详情/o ，/o 藏/o 界/o 较为/o 重视/o ，/o 也是/o 我们/o 收藏/o 北京/ns 史料/o 中/o 的/o 要件/o 之一/o 。/o

我们/o 藏/o 有/o 一/o 册/o 1945 年 6 月/o 油印/o 的/o《/o 北京/ns 文物/o 保存/o 保管/o 状态/o 之/o 调查/o 报告/o 》/o ，/o 调查/o 范围/o 涉及/o 故宫/ns 、/o 历博/ns 、/o 古研所/nt 、/o 北大清华图书馆/ns 、/o 北图/ns 、/o 日/ns 伪/o 资料/o 库/o 等/o 二十几家/o ，/o 言/o 及/o 文物/o 二十万件/o 以上/o ，/o 洋洋/o 三万/o 余/o 言/o ，/o 是/o 珍贵/o 的/o 北京/ns 史料/o 。/o

以/o 家乡/o 的/o 历史/o 文献/o 、/o 特定/o 历史/o 时期/o 书刊/o 、/o 某/o 一/o 名家/o 或/o 名著/o 的/o 多种/o 出版物/o 为/o 专题/o ，/o 注意/o 精品/o 、/o 非卖品/o 、/o 纪念品/o ，/o 集成/o 系列/o ，/o 那/o 收藏/o 的/o 过程/o 就/o 已经/o 够/o 您/o 玩味/o 无穷/o 了/o 。/o

When /o Hope Project /o Rescue /o Millions /o Children /o Grow up /o Get up /o, /o Science and Education /o Prosperity /o Country /o Become a common practice /o, /o Today /o Have /o Collection /o Value /o Books /o You /o Did not /o Buy /o /o

Book collection /o is /o all /o tradition /o collection /o category /o middle /o first /o big family /o, /o is only /o we /o end /o have enough food and clothing /o time /o too long /o short /o just/o. /o

Because /o is related to /o Japan /ns Kou /o in /o Beijing /ns plunder /o cultural relics /o details /o, /o Tibet /o boundary /o pays more attention to /o, /o is also one of the /o elements /o of /o us /o collection /o Beijing /ns historical materials/o. /o

We /o have /o a /o book /o June /o1945 /o mimeograph /o /o "/o Beijing /ns cultural relics/o preservation /o custody /o status /o investigation/O report/O"/O, /o Survey /o Scope /o Covers /o Forbidden City /ns, /o Libo /ns, /o Ancient Research Institute /nt, /o Peking University Tsinghua Library /ns, /o North Map /ns, /o Japan /ns Pseudo /o Data /o Library /o, etc. /o More than twenty /o, /o words /o and/o. /o

Take /o's /o history /o literature /o, /o specific /o history /o period /o books /o, /o a /o famous /o or /o famous /o /o various /o publications /o as /o topic /o, /o attention /o boutique /o,/o. /o

The experiment uses the BiLSTM-CRF model, and its embedding part is implemented using the BERT Chinese pretrained language model. The model is defined as follows:

```
  def __init__(self, bert_config, tagset_size, embedding_dim,
hidden_dim, rnn_layers, dropout_ratio, dropout1,
use_cuda=False):
  super(BERT_LSTM_CRF, self).__init__()
  self.embedding_dim = embedding_dim
  self.hidden_dim = hidden_dim
  self.word_embeds = BertModel.from_pretrained(bert_config)
  self.lstm = nn.LSTM(embedding_dim, hidden_dim, num_layers=rnn_
layers, bidirectional=True, dropout=dropout_ratio,
batch_first=True)
  self.rnn_layers = rnn_layers
  self.dropout1 = nn.Dropout(p=dropout1)
  self.crf = CRF(target_size=tagset_size, average_batch=True,
use_cuda=use_cuda)
  self.liner = nn.Linear(hidden_dim * 2, tagset_size + 2)
  self.tagset_size = tagset_size
```

The model structure is as follows:

```
  Bert_LSTM_CRF(
  (word_embeds): BertModel()
  (lstm): LSTM(768, 500, batch_
first=True, dropout=0.5, bidirectional=True)
  (dropout1):Dropout(p=0.5, inplace=False)
  (crf):CRF
  (liner):Linear(in_features=100, out_features=15, bias=True)
  )
```

The experimental sentences are shown in Table 12.3.
The entity recognition effect of the model is as follows:

On January 24, the Xinhua News Agency released the guidance of the Central government on Xiongan New Area, with more than 12,000 words, 17 mentions of Beijing and 4 mentions of Tianjin, which is a large amount of information, and in fact, it has answered many questions that people are concerned about.

```
['O','O','O','O','O','O',"B-ORG','B-ORG',"B-ORG','O','O','O','O','O','O',
'O','O','B-LOC','B-LOC,'B-LOC','B-LOC','O','O','O','O','O','O','O','O','O',
'O','O','O','O','O','O','O','O','O','O','O','O','O','B-LOC','B-LOC','O','O',
'O','O','O','B-LOC','B-LOC','O','O','O','O','O','O','O','O','O','O','O','O',
'O','O','O','O','O','O','O','O','O','O']
```

In the early morning of February 17, Beijing time, the 69th Berlin International Film Festival announced the list of winners of the main competition unit, Wang Jingchun and Yong Mei won the best actor and actress double Silver Bear Award with the Chinese film "Everlasting" directed by Wang Xiaoshuai, which is the first time that Chinese actors swept the best actor and actress award at the Berlin Film Festival, setting a new record for Chinese films. At the same time, the film "The First Separation", directed by young director Wang Lina, also won the best film of the international jury of the New generation of the Berlin Film Festival. It can be said that after several award-winning years, Chinese films once again ushered in a highlight moment at the Berlin Film Festival

```
['O','O','O','O','O','O','O','O','O','O','O','O','O','O','O','O','B-LOC',
'B-LOC','O','O','O','O','O','O','O','O','O','O','O','O','O','O','O','O','O',
'B-PER','B-PER','B-PER','O','O','B-PER','B-PER','B-PER','O','O','O','B-LOC',
'B-LOC','O','O','O','O','O','O','O','O','O','O','O','O','O','O','O','O','O',
'O','O','O','O','O','O','O','B-LOC','B-LOC','O','O','O','O','O','O','B-LOC',
'B-LOC','O','O','O','O','O','O','O','O','O','O','O','O','O','O','O','O','O',
'O','O','O','O','O','O','O','O','O','O','O','O','O','O','B-PER','B-PER',
'B-PER','O','O','O','O','O','O','O','O','O','O','O','O','O','O','O','O','O',
'O','O','B-ORG','B-ORG','B-ORG','B-ORG','B-ORG','B-ORG','B-ORG','B-ORG',
'B-ORG','B-ORG','B-ORG','B-ORG','B-ORG','B-ORG','B-ORG','O','O','O','O','O',
'O','O','O','O','O','O','O','O','O','O','O','O','O','O','O','O','B-LOC','B-LOC',
'O','O','O','B-LOC','B-LOC','O','O','O','O','O','O','O','O','O','O','O','O']
```

Beijing News News According to the State Administration for Market Regulation, in response to media reports that pork products such as dumplings were detected positive for African swine fever virus nucleic acid, the State Administration for Market Regulation and the Ministry of Agriculture and Rural Affairs have required enterprises to immediately trace the source of pork raw materials and dispose of pork products

```
['B-ORG','R-ORG','B-ORG','O','O','O','O','B-ORG','B-ORG','B-ORG','B-ORG',
'B-ORG','B-ORG','B-ORG','B-ORG','O','O','O','O','O','O','O','O','O','O','O',
'O','O','O','O','O','O','O','B-LOC','B-LOC','O','O','O','O','O','O','O','O',
'O','O','O','B-ORG','B-ORG','B-ORG','B-ORG','B-ORG','B-ORG','O','B-ORG',
'B-ORG','B-ORG','B-ORG','B-ORG','O','O','O','O','O','O','O','O','O','O','O',
'O','O','O','O','O','O','O','O','O','O','O','O','O','O','O','O']
```

Table 12.3 Experimental example sentences

No.	Example sentence
1	On January 24, Xinhua News Agency released the central government's guidance on Xiongan New Area, which contained over 12,000 words. The document mentioned Beijing 17 times and Tianjin 4 times, providing a large amount of information. In fact, it also addressed many questions that people were concerned about.
2	In the early morning of February 17, Beijing time, the 69th Berlin International Film Festival announced the list of winners of the main competition unit. Wang Jingchun and Yong Mei won the best actor and actress double Silver Bear Awards for the Chinese film "Everlasting" directed by Wang Xiaoshuai, which is the first time that Chinese actors swept the best actor and actress awards at the Berlin Film Festival, setting a new record for Chinese films. At the same time, the film "The First Separation," directed by young director Wang Lina, also won the best film of the international jury of the New Generation of the Berlin Film Festival. It can be said that after several award-winning years, Chinese films once again ushered in a highlight moment at the Berlin Film Festival

12.3.2 Keyword Extraction Experiment

12.3.2.1 Evaluation Indicators

Like most natural language processing tasks, keyword extraction is evaluated from three dimensions: precision, recall, and $F1$-score. The $F1$-score is generally the most focused.

12.3.2.2 Experiment Example One

Example one uses the TF-IDF algorithm based on jieba segmentation to extract keywords from documents.

The initialization function is as follows:

```
def __init__(self, idf_path=None):
self.tokenizer = jieba.dt
self.postokenizer = jieba.posseg.dt
self.stop_words = self.STOP_WORDS.copy()
self.idf_loader = IDFLoader(idf_path or DEFAULT_IDF)
self.idf_freq, self.median_idf = self.idf_loader.get_idf()
```

The main function is as follows:

```
def extract_tags(self, sentence, topK=20, withWeight=False,
allowPOS=(), withFlag=False):
if allowPOS:
allowPOS = frozenset(allowPOS)
words = self.postokenizer.cut(sentence)
else:
words = self.tokenizer.cut(sentence)
freq = {}
for w in words:
if allowPOS:
if w.flag not in allowPOS:
continue
elif not withFlag:
w = w.word
wc = w.word if allowPOS and withFlag else w
if len(wc.strip()) < 2 or wc.lower() in self.stop_words:
continue
freq[w] = freq.get(w, 0.0) + 1.0
total = sum(freq.values())
for k in freq:
```

```
kw = k.word if allowPOS and withFlag else k
freq[k] *= self.idf_freq.get(kw, self.median_idf) / total
if withWeight:
tags = sorted(freq.items(), key=itemgetter(1), reverse=True)
else:
tags = sorted(freq, key=freq.__getitem__, reverse=True)
if topK:
return tags[:topK]
else:
return tags
```

The sentences of example one are shown in Table 12.4.

The effect of keyword extraction is as follows:

```
origin text:
  A thread is the smallest unit of program execution, it is an execution flow of a process, and the basic unit of
CU scheduling and dispatch. A process can be composed of many threads, and all resources of the process are
shared among threads. Each thread has its own stack and local variables. Threads are independently scheduled
and executed by the CPU, allowing multiple threads to run simultaneously in a multi CPU environment.
Similarly, multithreading can also achieve concurrent operations, with each request assigned a thread to
handle/person
keywords by tfidf:
  [Thread, CPU, Process, Scheduling, Multi threading, Program Execution, Each, Execution, Stack, Local
  Variables, Units, Concurrency, Dispatch, One, Share, Request, Minimum, Can, Allow, Allocate]

origin text:
  Dior new, autumn and winter new doll style sweet round neck with fur collar woolen coat coat, size: SM, P339
keywords by tfidf:
  [New fur collar, size, Dior, SM, P330, woolen coat, round neck, sweet, coat, autumn/winter, doll]
```

12.3.2.3 Experiment Example Two

Example two uses the TextRank algorithm based on jieba segmentation to extract keywords from documents.

The initialization function is as follows:

Table 12.4 Keyword extraction sentences

No.	Example sentence
1	A thread is the smallest unit of program execution; it is an execution flow of a process and the basic unit of CU scheduling and dispatch. A process can be composed of many threads, and all resources of the process are shared among threads. Each thread has its own stack and local variables. Threads are independently scheduled and executed by the CPU, allowing multiple threads to run simultaneously in a multi-CPU environment. Similarly, multithreading can also achieve concurrent operations, with each request assigned a thread to handle
2	Dior new autumn and winter new doll style sweet round neck with fur collar woolen coal, size: SM, P339

```
def __init__(self):
self.tokenizer = self.postokenizer = jieba.posseg.dt
self.stop_words = self.STOP_WORDS.copy()
self.pos_filt = frozenset(('ns', 'n', 'vn', 'v'))
self.span = 5
```

The main calling function is as follows:

```
def textrank(self, sentence, topK=20, withWeight=False,
allowPOS=('ns', 'n', 'vn', 'v'), withFlag=False):
self.pos_filt = frozenset(allowPOS)
g = UndirectWeightedGraph()
cm = defaultdict(int)
words = tuple(self.tokenizer.cut(sentence))
for i, wp in enumerate(words):
if self.pairfilter(wp):
for j in xrange(i + 1, i + self.span):
if j >= len(words):
break
if not self.pairfilter(words[j]):
continue
if allowPOS and withFlag:
cm[(wp, words[j])] += 1
else:
cm[(wp.word, words[j].word)] += 1

for terms, w in cm.items():
g.addEdge(terms[0], terms[1], w)
nodes_rank = g.rank()
if withWeight:
tags = sorted(nodes_rank.items(), key=itemgetter(1),
reverse=True)
else:
tags = sorted(nodes_rank, key=nodes_rank.__getitem__,
reverse=True)

if topK:
return tags[:topK]
else:
return tags
```

For the example sentence in example one, the effect of keyword extraction by TextRank method is as follows:

```
origin text:
```

A thread is the smallest unit of program execution, it is an execution flow of a process, and is the basic unit for cU scheduling and dispatch. A process can be composed of many threads, and all resources of the process are shared among threads

Each thread has its own stack and local variables. Threads are independently scheduled and executed by the CPU, allowing multiple threads to run simultaneously in a multi CPU environment. Similarly, multithreading can also achieve concurrent operations, with each request

Assign a thread to handle.

```
keywords by textrank:
```

[Thread, process, schedule, unit, operation, request, allocation, allow, basic, shared, concurrent, stack, independent, execute, dispatch, composition, resource, implementation, run, process]

```
origin text:
```

Dior new, autumn and winter new doll style sweet round neck with fur collar woolen coat coat, size: SM P339 keywords by textrank:

[New style, size, jacket, doll, round neck]

References

1. Grishman R, Sundheim B, Message Understanding Conference-6: A Brief History. 1996.
2. Krallinger M, Leitner F, Rabal O, et al. Overview of the Chemical Compound and Drug Name Recognition (CHEMDNER) task. Proceedings of the Fourth BioCreative Challenge Evaluation Workshop vol. 2. 2021: 6–37.
3. Zong Chengqing. Statistical Natural Language Processing [M]. 2nd Edition. Beijing: Tsinghua University Press, 2013.
4. Nadeau D, Sekine S. A Survey of Named Entity Recognition and Classification [J]. Lingvisticae Investigationes, 2007, 30(1): 3–26.
5. Rau LF. Extracting Company Names from Text [C]. Artificial Intelligence Applications. IEEE, 1991: 29–32.
6. Humphreys RG, Azzam S, Huyck C, et al. Description of the LaSIE-II System as Used for MUC7 [J]. 1998.
7. Eddy SR. Hidden Markov Models [J]. Current Opinion in Structural Biology, 1996, 6(3): 361–365.
8. Ratnaparkhi A. Maximum Entropy Model for Part-of-Speech Tagging [J]. Proc. Empirical Method for Natural Language Processing, 1996.
9. Lafferty JD, Mccallum AK, Pereira FCN. Conditional Random Fields: Probabilistic Models for Segmenting and Labeling Sequence Data [C]. Proceedings of the 18th International Conference on Machine Learning. 2001: 282–289.
10. Witten, Ian H., et al. KEA: Practical Automated Keyphrase Extraction [C]. Design and Usability of Digital Libraries: Case Studies in the Asia Pacific. IGI global, 2005, 129–152.
11. Hulth A. Improved Automatic Keyword Extraction Given More Linguistic Knowledge [C]. Proceedings of the 2003 Conference on Empirical Methods in Natural Language Processing. 2003.
12. Caragea C, et al. Citation-enhanced Keyphrase Extraction from Research Papers: A Supervised Approach [C]. Proceedings of the 2014 conference on empirical methods in natural language processing (EMNLP). 2014.
13. Xin J, Hu Y H, Li H. A Ranking Approach to Keyphrase Extraction [C]. Proceedings of the 32nd International ACM SIGIR Conference on Research and Development in Information Retrieval. 2009.

14. Zhang Q. Keyphrase extraction using deep recurrent neural networks on twitter. Proceedings of the 2016 conference on empirical methods in natural language processing. 2016.
15. Yang XT. Deep multimodal representation learning from temporal data. Proceedings of the IEEE conference on computer vision and pattern recognition. 2017.
16. Wan XJ, Xiao JG. Single Document Keyphrase Extraction Using Neighborhood Knowledge [J]. AAAI, 2008, 8.
17. Florescu C, Caragea C. Positionrank: An Unsupervised Approach to Keyphrase Extraction from Scholarly Documents [C]. Proceedings of the 55th Annual Meeting of the Association for Computational Linguistics (Volume 1: Long Papers). 2017.

Chapter 13
Big Data Automatic Construction and Application of Knowledge Graph

This chapter explores knowledge graphs, focusing on their development, construction methods, and key technologies such as entity and relationship extraction. It highlights the integration of data from large-scale knowledge bases, web link data, and knowledge fusion from diverse sources. Applications of knowledge graphs in intelligent search, machine learning, automatic question answering, and document representation are also discussed.

13.1 Overview of Knowledge Graph

In the rapidly developing Internet era, people usually obtain the information and knowledge they need through search engines. When querying information related to a certain keyword, after the search engine gets the results, it ranks the web pages based on their importance and relevance to the specified keyword. It returns them as search results to the user. People then search for the content they need from the search results, but the search results often need improvement, requiring some time to find the necessary answers. Google proposed the concept of knowledge graphs to solve this problem, mainly to improve the quality and experience of user searches. A knowledge graph, also known as a scientific knowledge graph, refers to the result of extracting, understanding, and integrating key points of existing knowledge through scientific calculation and describing the relationships between knowledge resources using visualization technology.

In the Internet era, online information is both complex and vast. A large amount of useful information is contained in the text, requiring individuals to extract from a large amount of text information. Even if people obtain useful information, it is difficult to remember it when the volume of information needed is large. However, using a graph to represent it can help people remember information intuitively. As shown in Fig. 13.1, representing animal classification with a tree-like graph can

H. Zhang, J. Shang, *Natural Language Processing and Applications*,
https://doi.org/10.1007/978-981-97-9739-4_13

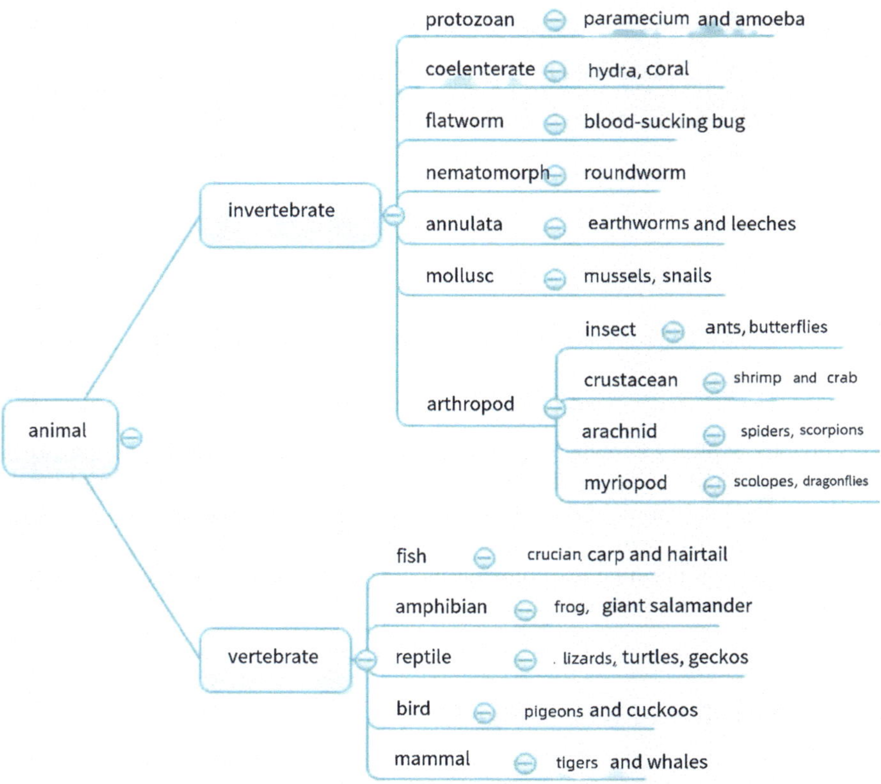

Fig. 13.1 Animal classification chart

Chinese name	Chicago bulls	Current head coach	Billy Donovan
foreign name	Chicago Bulls	Famous person	Michael Jordan, scottie pippen, dennis rodman, Derek.
founding time	January 16(th), 1966		Ross
Belonging area	Chicago, Illinois, USA	Main honor	Six NBA championships
Sports	basketball		The record in the regular season is the second in history (72 wins in 1995-96).
Compete for an event	NBA		9-time division champion
Main venue	united center	galleryful	21,711 people
owner	Jerry Leinsdorf	Jersey color	Home-scarlet letter on white background; Away-black letters on red background

Fig. 13.2 Basic information about the Chicago Bulls

clearly reveal the categories to which different animals belong and more intuitively reflect the relationships between different animals compared to text descriptions [1].

The knowledge graph's output result is displayed in the form of a graph. Here, using the Chicago Bulls as an example, Fig. 13.2 shows the basic information of the Chicago Bulls in the Baidu Encyclopedia, including the establishment time, region,

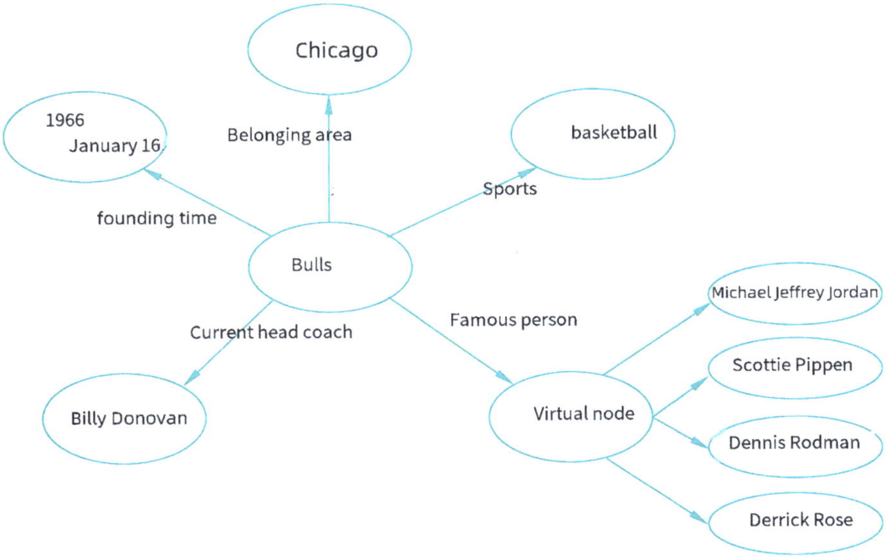

Fig. 13.3 Bulls information knowledge graph

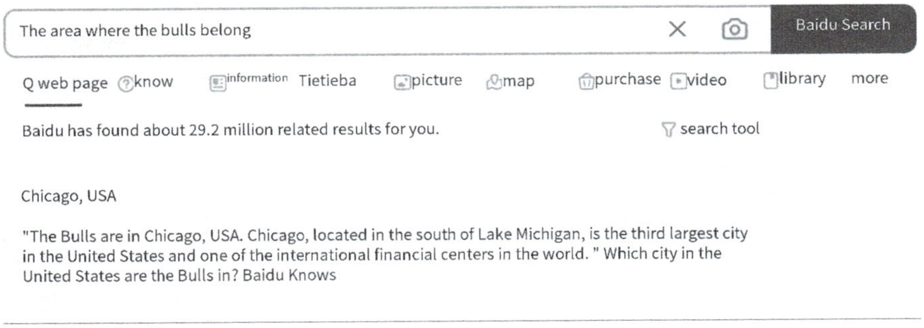

Fig. 13.4 Search results

famous figures, etc. Using "Bulls" as a named entity to build a knowledge graph, the result is shown in Fig. 13.3.

After the knowledge graph is built, when the user searches for the related information, they can get accurate answers, such as the result obtained when searching for "the region of the Bulls," as shown in Fig. 13.4.

Traditional search engines mechanically match web pages related to the query words and cannot understand the user's request. For example, when querying 'the region of the Bulls,' the search engine may list articles containing this information but will not provide a direct answer.

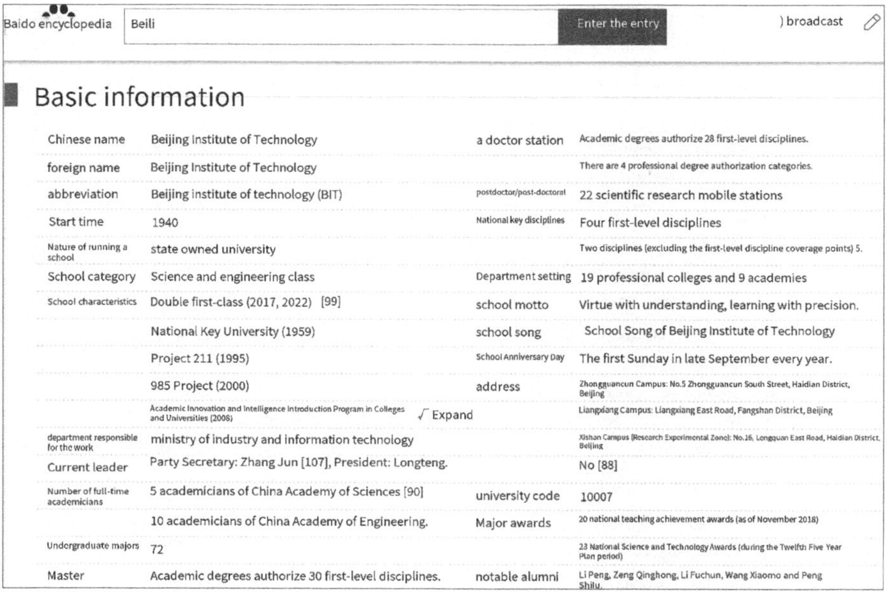

Fig. 13.5 Entry information for "Beijing Institute of Technology"

Currently, Wikipedia is the largest knowledge base, while Baidu Encyclopedia is the largest Chinese large-scale knowledge base. As of 2022, Baidu Encyclopedia's knowledge base contains more than 27 million entries written by over 7.6 million people. The entry information for the search term "Beijing Institute of Technology" is shown in Fig. 13.5.

Wanfang, CNKI, and other literature search websites have collected professional papers from various fields, which helps in knowledge extraction.

13.2 Data Sources for Knowledge Graphs

The construction of a knowledge graph is based on large-scale knowledge bases, such as Wikipedia, Baidu Encyclopedia, Sogou Encyclopedia, etc. These knowledge bases contain much-structured knowledge that can be efficiently transformed into a knowledge graph. In addition, the internet contains vast knowledge hidden in numerous web pages. This knowledge is usually unstructured and disorganized, making it difficult to convert it into useful information for the knowledge graph directly. However, through certain techniques, it can be extracted to build a knowledge graph.

13.2.1 Large-Scale Knowledge Bases

Large-scale knowledge bases generally use entries as the basic unit, with each entry corresponding to a concept in the real world. Editors compile these entries from worldwide to form the knowledge base. Currently, Wikipedia is the largest knowledge base globally, while Baidu Encyclopedia is the largest Chinese-language knowledge base, containing more than 15 million entries compiled by over 6 million contributors.

In addition to the Baidu Encyclopedia, China National Knowledge Infrastructure (CNKI) also covers professional papers in various fields, which helps in knowledge extraction.

13.2.2 Internet Link Data

The Internet contains a vast amount of information, so people often use tools like web crawlers (detailed introduction in Chap. 5) to extract knowledge from the massive number of web pages. The data in Internet web pages is basically unstructured data, which makes the accuracy of the information extracted from the web pages relatively low because the forms of web pages are diverse, and there is a lot of noise information. You can also directly extract structured web page table data with a significant limitation and small coverage. People hope to obtain data from web pages with a good structure and large coverage. Driven by this demand, Internet link data was born.

The World Wide Web Consortium (W3C) launched the Linked Open Data (LOD) project in 2007.

The project aims to expand the World Wide Web, which is composed of documents, into a knowledge space that is composed of data. It can be seen as a database spanning the entire network. The part of the data cloud map released by the Linked Open Data project is shown in Fig. 13.6. DBpedia is a semantic web example composed of structured data extracted from well-known Wikipedia entries. It is also one of the largest multi-domain knowledge ontologies in the world and is part of linked data, linking datasets such as Linked MDB and LinkedCT.

Description framework (RDF) is a form of various open data sets published on the Web. RDF is a framework for describing structured knowledge, representing the relationship between entities as (entity 1, relationship, entity 2) triples. The nodes in the data cloud map represent the published datasets, the larger the node area, the more triples are contained in the data set. The arrows between the nodes in the data cloud map represent more than 50 RDF links between two datasets, the thicker the arrow, the more links between the datasets, and the bidirectional arrows represent the two datasets using identifiers mutually. Linked Open Data also enables RDF links to be established between data items from different sources, thereby creating the semantic Web knowledge base.

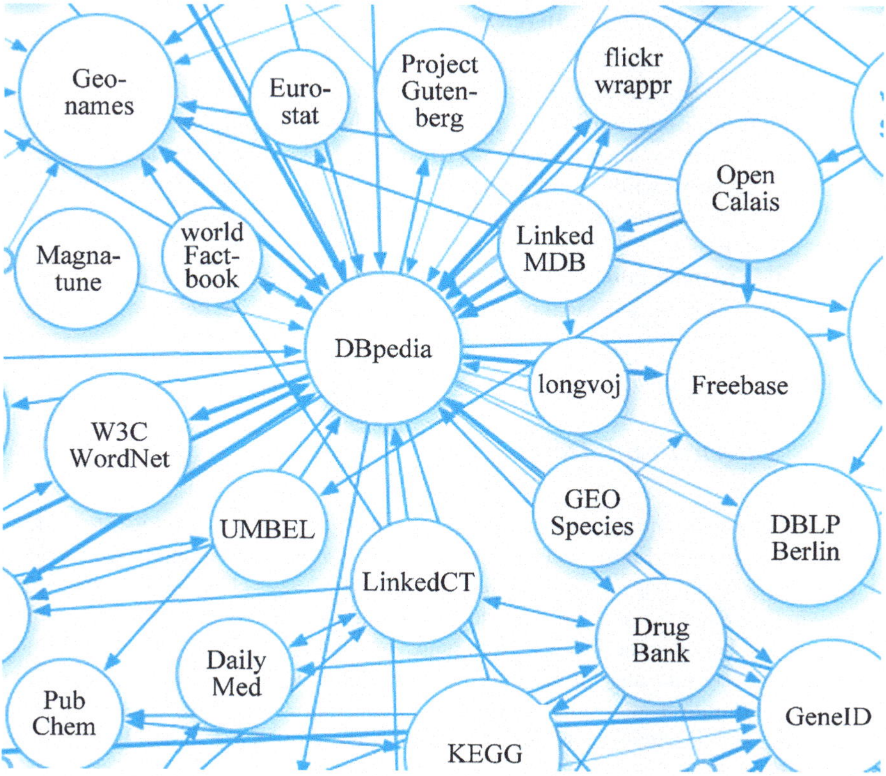

Fig. 13.6 Part of the data cloud map released by the Open Linked Data project

13.2.3 Knowledge Fusion from Multiple Data Sources

Different data sources were used in the construction process of the knowledge graph, and there is no strong correlation between these data; different semantics may have different expressions, which may lead to the problem of knowledge duplication. Knowledge fusion aims to process these data from other sources under a unified data model, thereby eliminating the problems caused by different source data and completing knowledge fusion.

Knowledge fusion generally needs to deal with two levels of problems: the fusion of the pattern layer and the fusion of the data layer. The fusion of the pattern layer mainly unifies different ontologies under the same framework. The fusion of the data layer mainly solves conflicts between entities in terms of references, attributes, relationships, etc.

The pattern layer's fusion methods can be divided into two categories: ontology integration and ontology mapping. The former unifies multiple ontologies under the same framework, thereby eliminating the heterogeneity problem of different ontologies; the latter establishes mapping relationships between ontologies using methods such as neural network-based methods, rule-based methods, etc.

Ontology integration generally requires a lot of manpower and time, so the more commonly used method now is ontology mapping. For example, Stanford University's AnchorPrompt [2] can automatically align similar semantic terms, achieving ontology mapping.

The fusion of the data layer is mainly the problem of entity alignment, and there are three points to note for entity alignment: entity disambiguation, entity unification, and anaphora resolution.

Entity disambiguation can be achieved using the bag-of-words model. In Fig. 13.7, there are two MichaelJordans waiting for disambiguation. Each time MichaelJordan appears in a sentence, if you want to know if this is the same person, you can use the words around the entity to be disambiguated to construct a vector, then use the vector space model (VSM) to calculate the similarity of the two entity references, perform clustering, and thus complete entity disambiguation. This is the method used by the bag-of-words model. As early as 1998, Bagga proposed using the vector space model to complete entity disambiguation.

If entity disambiguation distinguishes the actual different references of entities with the same appearance, then entity unification judges whether the references of multiple entities with different appearances are the same.

To complete entity unification, you need to calculate the attribute similarity of the entity; the commonly used method here is to calculate the Levenshtein distance. Also called the minimum edit distance, as you can see in Fig. 13.8, from "Beijing Institute of Technology" to "BIT," the minimum number of modifications is three, so the minimum edit distance between them is three; therefore, it can be inferred that they are synonyms.

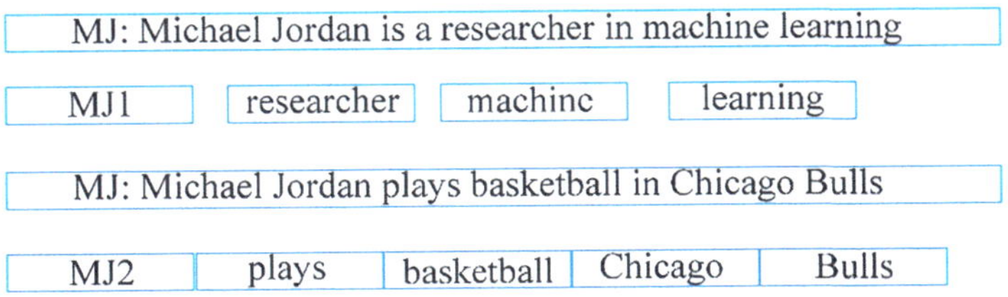

Fig. 13.7 Bag-of-words model example

Fig. 13.8 Levenshtein distance

In addition, you can also calculate the set similarity. For example, Dice distance can be used to measure the similarity of two sets because you can understand a string as a kind of set, so the Dice distance can also be used to measure the similarity of strings.

The entity's attributes can also be introduced to achieve entity unification, and different weights can be assigned to them. Then, weighted summation is performed to calculate the entities' similarity. The algorithm mainly uses TF-IDF to assign weights to each component of the entity vector and build an index, and it calculates the entity similarity for discrimination through cosine similarity.

In the following text, pronouns often replace the entities that appear in the previous text, and semantically, the entities in the last text and the pronouns in the subsequent text belong to the same equivalence set. The problem to be solved by coreference resolution is correlating the pronouns in the following text with the entities in the previous text. There are generally three models for coreference resolution: one is the mention pair model, which forms pairs of pronouns and antecedents and judges whether the two are equivalent for each mention pair; the second is the mention ranking model, which finds the candidate antecedents for each pronoun, ranks them, and the highest scoring one is the actual antecedent; the third is the entity-mention model, which clusters all entities based on their context, thereby dividing equivalence sets.

13.3 Construction of Knowledge Graph

Knowledge graphs can process and display the intricate interaction relationships between subjects or knowledge units with modern visualization big data technologies such as nodes and links, allowing people to clearly and intuitively understand the knowledge structure and research trends in a subject or a field's development process.

Knowledge graphs can effectively extract knowledge from a large amount of data, and they are currently an effective method for people to extract knowledge from the vast sea of data.

Knowledge graphs include entities and relationships, where nodes represent entities and edges between nodes represent relationships. A relationship between entities can be called a fact, usually represented by a triplet (head, relation, tail), where the head represents the subject of the relationship, the tail represents the object of the relationship, and the relation represents the relationship between the two. The triplet is the knowledge graph's basic unit and its core.

As shown in Fig. 13.9, there are eight entities in this knowledge graph, namely, "electric energy metering system," "real-time monitoring system," "transformer," "current transformer," "power station equipment," "smart grid," "station control layer" and "process layer;" the relationships are "belongs to" and "exists in." Starting from the entities "electric energy metering system," "real-time monitoring system, "transformer," and "current transformer," pointing to the entity "power

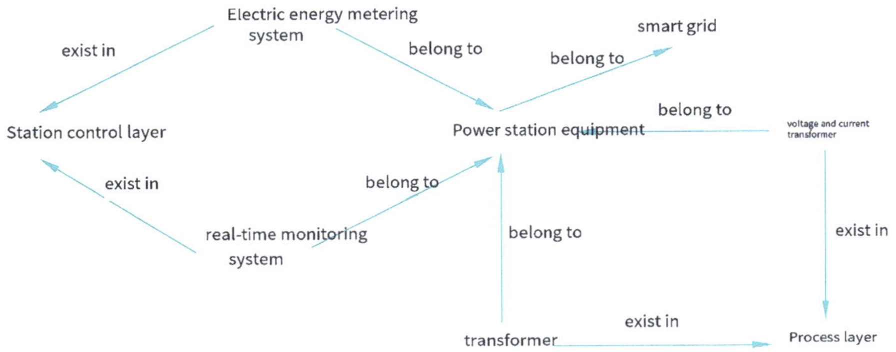

Fig. 13.9 Knowledge graph of power station equipment

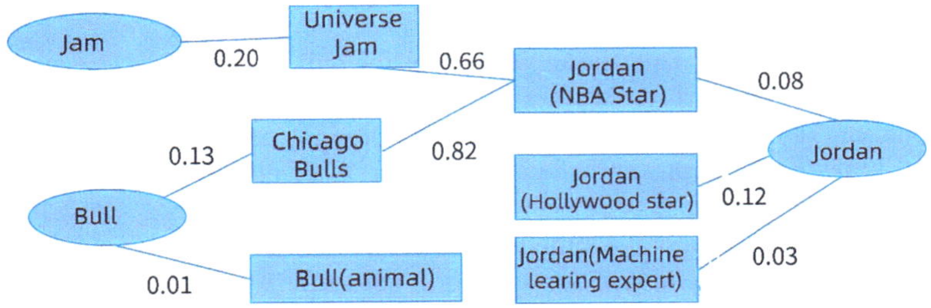

Fig. 13.10 Example of semantic association

station equipment" and accompanied by the relationship "belongs to," it is known that these entities are subclasses of "power station equipment." Similarly, it is known that the "electric energy metering system" and "real-time monitoring system" exist in the "station control layer." The above relationships can be represented by triple relationships (electric energy metering system belongs to power station equipment) and (electric energy metering system exists in station control layer), etc. According to the knowledge graph, the relationships between various entities can be more precise, intuitive, and detailed without reading an enormous paragraph of text and then through complex thinking to deduce their connections.

After extracting the subject from the text, it can sometimes be difficult for computers to discern the exact meaning of the subject. For example, in a conversation between two boys, there is a question, "Do you like bulls?" For individuals, it is clear that "bulls" here refer to the Chicago Bulls team in the NBA, not the animal bull. As shown in Fig. 13.10, semantic association refers to constructing a knowledge graph for related entities, determining the semantic ontology based on the probability relationship between entities. Establishing semantic ontology requires the support of the following three key technologies:

1. XML is used at the syntax layer.
2. RDF is used for the resource management framework.

3. Ontology is used at the ontology layer.

Currently, search engines can answer some of the questions posed by people. Compared to previously returning many web pages directly, they can now understand the user's question, extract facts from the network, and display the most suitable answer. This form is called automatic question answering, and the knowledge graph can serve as the knowledge base for automatic question answering, allowing you to select the best answer from it. For example, people often need to find answers to questions such as "What are the opening hours of the Big Wild Goose Pagoda?" "Where was the Bulls' owner born?" and "How many districts are there in Shanghai?" These questions require reasoning from the complex relationships between entities in the knowledge graph. Whether it's understanding user query intent or exploring new search forms, semantic understanding, and knowledge reasoning is invariably required, and this requires the strong support of large-scale, structured knowledge graphs, so knowledge graphs have become a battleground for major internet companies.

The following uses the question "Where was the bull's owner born?" as an example. As shown in Fig. 13.11a, the question is semantically parsed into a language the computer can understand. The knowledge is filtered according to the already constructed knowledge graph, and the best answer is displayed after being found. Figure 13.11b shows the answer found through a Baidu search.

Next, the construction of the knowledge graph is introduced. Because knowledge extraction is based on Chinese processing, a semantic web must be constructed before building the knowledge graph.

Usually, the text is first segmented, and key points, sentences, phrases, vocabulary, and knowledge points are extracted from it, the relevance between these knowledge points is calculated, and then semantic analysis, syntactic analysis, and lexical analysis, as shown in Fig. 13.12, are performed to construct a semantic web.

As shown in Fig. 13.13, the automatic semantic web construction process has four steps: concept discovery, association calculation, relationship extraction, and integration verification.

13.3.1 Concept Discovery

Concept discovery, entity extraction, or named entity recognition automatically identifies named entities from a text dataset. The most typical entities include names of people, places, organizations, etc. In recent years, attempts have been made to identify a broader range of entity types, such as movie names, product names, etc. In addition, since knowledge graphs involve entities and many concepts, some researchers have proposed to identify these concepts. Detailed information about named entity recognition has been discussed in Chap. 12, so it will not be repeated here.

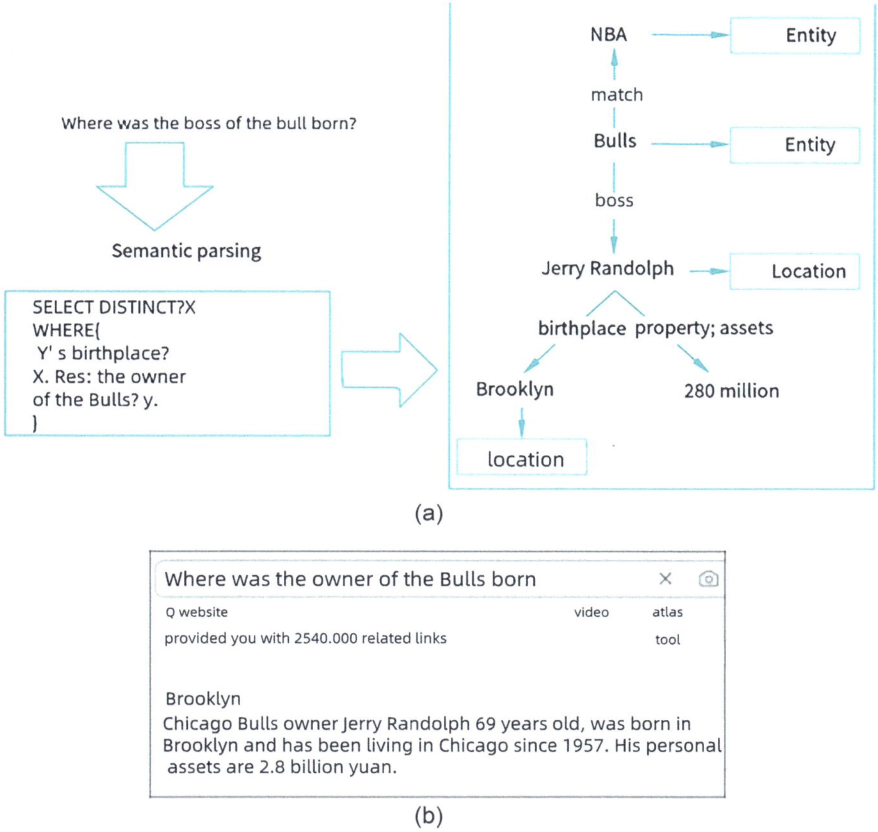

Fig. 13.11 Example of automatic question answering. (**a**) Find the best answer. (**b**) Answer found through a Baidu search

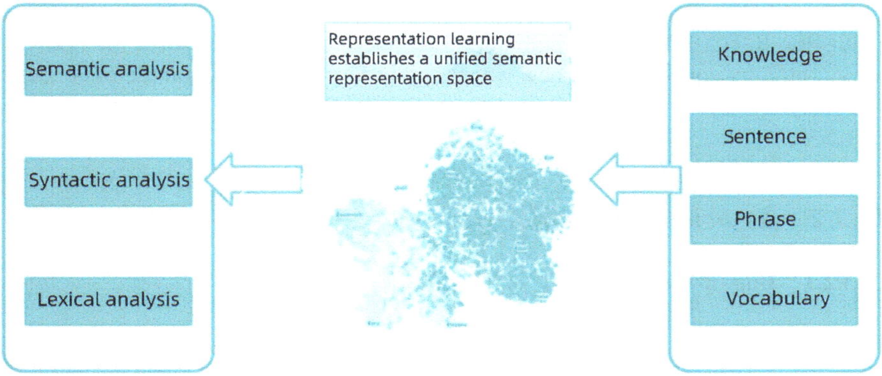

Fig. 13.12 Preparation for semantic web construction

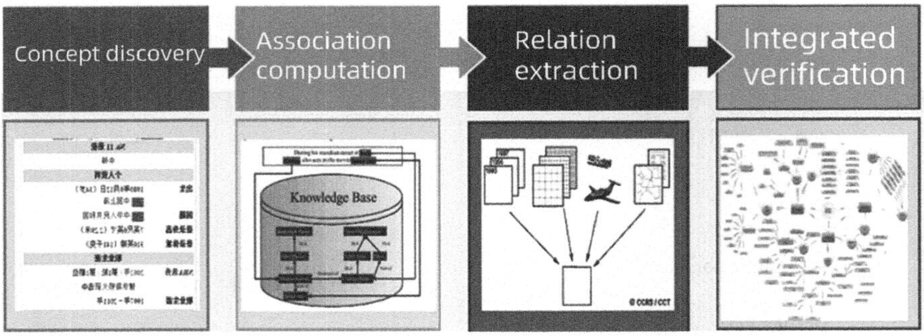

Fig. 13.13 Automatic semantic web construction process

Concept discovery can be completed in four steps: format parsing, word segmentation and tagging, new word discovery, and concept discovery.

1. Format Parsing
   Format parsing refers to extracting structured text information from mainstream documents such as PDF, Word, XML, etc., using the information extraction component of NLPIR-ICTCLAS.
2. Word Segmentation and Tagging
   The NLPIR-ICTCLAS word segmentation system can integrate existing ontology libraries to achieve professional field word segmentation and tagging. It has been successfully applied to Huawei, People's Network, China Post, the Central Bank, and the Central Cyberspace Affairs Commission.
3. New Word Discovery
   NLPIR-ICTCLAS can directly discover new words and new concepts from the original corpus. Using the NLPIR-Parser application, new words and new concepts can be extracted from the original corpus of Chinese files about electricity and displayed in the form of a word cloud, as shown in Fig. 13.14.
4. Concept Discovery
   NLPIR-ICTCLAS uses a combination of rule-based and statistical methods to select ontology concepts from new words. Table 13.1 shows an example of concept discovery results.

13.3.2 Association Calculation

13.3.2.1 Bag of Words Model

The Bag of Words model is a commonly used document representation method in information retrieval. In information retrieval, the Bag of Words model has the following assumptions: for a document, it ignores the order of words and elements

Fig. 13.14 Example of new word discovery

Table 13.1 Example of concept discovery results

Data format	PDF	PDF (add professional words)	XML (add professional words)	Remark
Number of files	351	351	351	Number of file provided
Total size	About 7.1 GB	About 7.1 Gb	About 185 MB	Size of the file itself
Corpus size	About 160 MB	About 237 MB	About 125 MB	Text size after extraction, washing, and archiving
Number of specialized words	None	1865	1865	Number of existing professional words
Number of specialized words (indicating new words)	4000	5865	5865	The sum of the discovered new words and the number of existing specialized words
Part-of-speech continuous bag of words model (POS-CBOW model)	About 81.8 MB	About 62 MB	About 30 MB	The size of generated model file
Word number	70,180	53,013	25,379	The final number of words

such as grammar and syntax, treating each word as an independently occurring object, not dependent on whether other words appear. Thus, it views the document merely as a collection of several words. For example, there are the following two documents:

Document 1: "I like to jog alone, my roommate also likes it."
Document 2: "I also like reading."

After segmenting the two documents, we can get {"I," "like," "alone," "jog," "roommate," "also," "like"} and {"I," "also," "like," "reading"}.

After the documents are segmented, construct a dictionary composed of words in the document:

This dictionary contains a total of seven different words. Using the index number of the dictionary, the above two documents can be represented by a seven-dimensional vector, with integers 0~n (n is a positive integer) representing the number of times a word appears in the document. In this way, based on the number of times each word appears in each document, the above two documents can be represented in vector form:

Document 1: [1 2 1 1 1 1 0].
Document 2: [1 1 0 0 0 1 1].

This vector format is called one-hot encoding.

13.3.2.2 CBOW Model

The task of the continuous bag-of-words (CBOW) model is: given the words w_{t-c}... w_{t-1}, w_{t+1} ... w_{t+c} within a certain neighborhood radius (such as c) of the center word w_t, predict the probability of the output word being the center word wt. Because it does not consider the order of words, it is called the continuous bag of words model. The structure of the CBOW model is shown in Fig. 13.15. The CBOW model includes three layers, namely, the input layer, mapping layer, and output layer. The input layer is the context words of the center word w_t; all are one-hot encoded. If the input layer matrix is W_{in}, then

$$w(i)^T W_{\text{in}} = v_i \left(i = t-c,\ldots,t-1,t+1,\ldots,t+c\right). \tag{13.1}$$

The mapping layer is the sum of these $2c$ word vectors. The average value is taken as the output of the projection layer. The output layer then calculates the probability of w_{t-c} ... w_{t-1}, w_{t+1} ... w_{t+c} generating w_t, that is,

$$P\left(w_i | ,w_{t-c} |,\ldots |,w_{t-1} |,w_{t+1} |,\ldots |,w_{t+c}\right) = P\left(w_i | v_{\text{PROJECTION}}\right) \tag{13.2}$$

13.3.2.3 POS-CBOW Model

The part-of-speech continuous bag of words (POS-CBOW) model can be used for deep computation to determine word similarity. As shown in Fig. 13.16, the POS-CBOW model is divided into five layers: the input layer, filter layer, projection layer, tagging layer, and output layer. Compared with the CBOW model, the POS-CBOW model adds filter and tagging layers. The main purpose of the filter layer is to correct the sequence of the input text corpus because some texts may contain some special symbols, which can change the normal sentence structure of the training corpus and interfere with the training results. The filter layer can optimize the

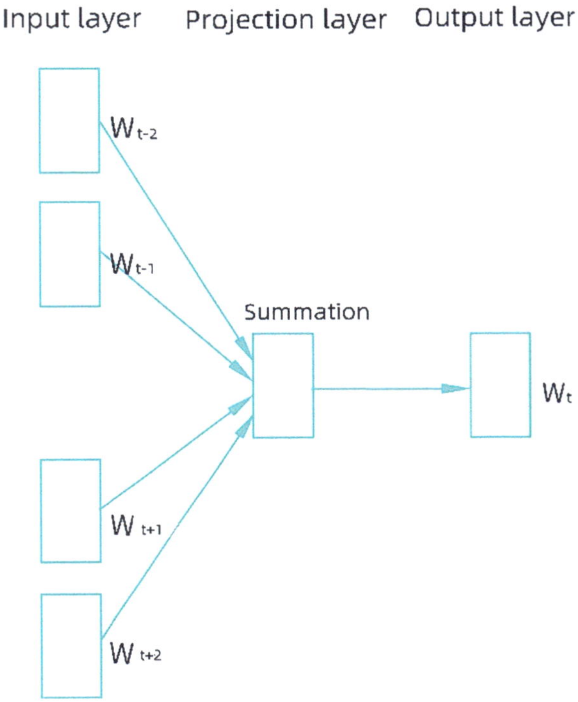

Fig. 13.15 Structure of the CBOW model

word vector space, making the POS-CBOW model more independent than that in the previous stage of work. The tagging layer is located between the projection layer and the output layer. Its main function is to tag all word vectors, establish syntactic relationships between spatial word vectors based on latent semantic relationships, and thus improve the accuracy of word similarity. The generated word vectors can be tagged with part-of-speech using the Chinese word segmentation tool NLPIR. The part-of-speech used comes from the Chinese part-of-speech tag set of the Institute of Computing Technology, Chinese Academy of Sciences.

13.3.3 Relationship Extraction

Relationship extraction is actually the extraction of entities and relationships, generally realized through the iterative triple method mentioned above. For example:

1. Extract (China, capital, Beijing), (USA, capital, Washington) and other triple instances through the "X is the capital of Y" template.
2. Based on these triple instances, more can be discovered from the two entities "China-Beijing" and "USA-Washington."

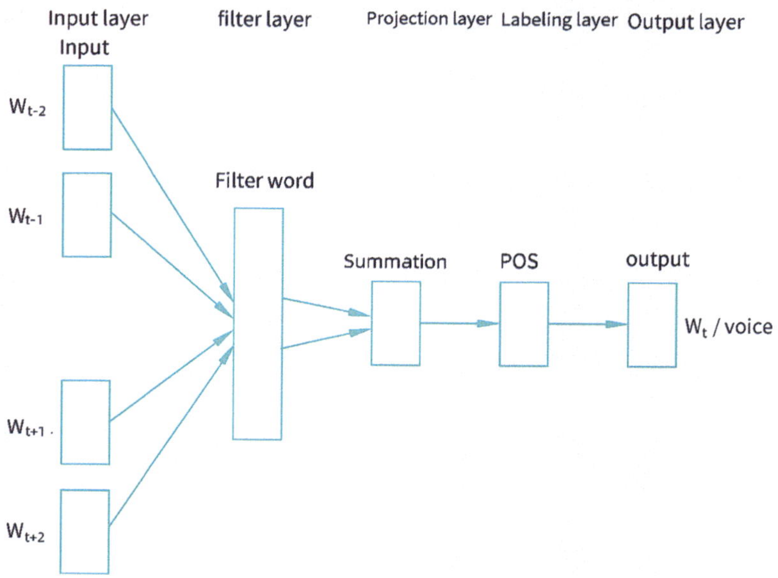

Fig. 13.16 Structure of the POS-CBOW model

Matching templates can be found, such as "The capital of Y is X" and "X is the political center of Y," etc.

3. Use the newly discovered templates to extract more new triple instances. New templates and instances are continuously discovered through repeated iterations.

13.3.3.1 Phrase Extraction

Phrase extraction is the process of extracting relationships between entities by identifying phrases expressing semantic relationships. For example, there are three phrases: (Huawei, headquartered in Shenzhen), (Huawei, headquarters set up in Shenzhen), and (Huawei, built its headquarters in Shenzhen). They express the same meaning, but for the computer, they are different combinations of words. At this time, syntax and statistical data need to be used to filter the extracted triples, where the relationship phrase should be verb-centered.

As mentioned above, relational phrases like 'located in,' 'set in,' and 'built in' should match multiple entity pairs, such as "Huawei" and "Shenzhen."

13.3.3.2 Dependency Syntax Analysis

Dependency syntax analysis refers to determining the relationship between entities based on the context of a phrase. As shown in Fig. 13.17, in the original semantic web, Huawei has relationships with entities such as "mobile phone," "communication infrastructure," and "HarmonyOS 3." At this time, for the sentence "Huawei releases HarmonyOS3 operating system, the number of Harmony devices exceeds 300 million," based on the known semantics.

Associated Knowledge Graph: Through the information in the context, such as "HarmonyOS3" and "operating system," it can be known that "Huawei" is likely to refer to "Huawei Company," not a person's name. According to the dependency syntax analysis, it is known that "Huawei" has a relationship with "HarmonyOS3" and "operating system," and it is added to the semantically associated knowledge graph.

13.3.3.3 Discourse Relation Extraction

Discourse relation extraction refers to extracting the required information from the content of the entire article. Take, for example, the paragraph "Lu Xun, originally named Zhou Zhangshou, later renamed Zhou Shuren, styled Yushan, later changed to Yucai, a native of Shaoxing, Zhejiang, He was a famous writer, thinker, revolutionary, educator, democratic fighter, an important participant in the New Culture Movement, and one of the founders of modern Chinese literature." According to the large amount of information existing in the search engine, basic personal information such as "name," "birthplace," and "occupation" is extracted from it, and the extraction result is shown in Fig. 13.18. When it is necessary to extract the relationship between different entity objects, the required information elements are extracted according to the known extraction rules.

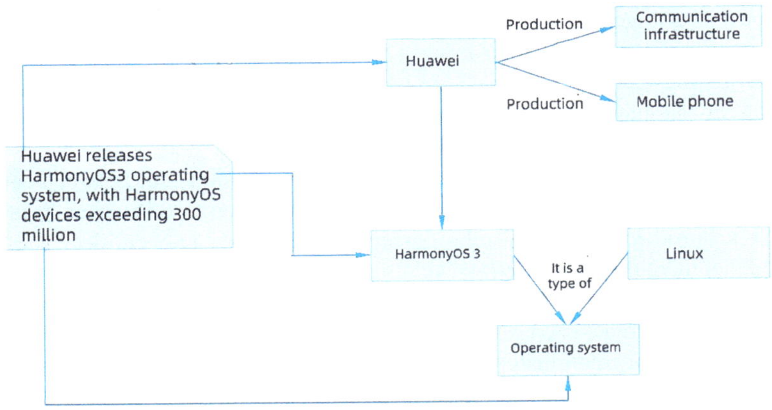

Fig. 13.17 Example of dependency syntax analysis

Name: Lu Xun/Zhou Zhangshou/Zhou Shuren

Place of birth: Shaoxing, Zhejiang

Occupation: writer, thinker, revolutionary, educator, democratic fighter.

Fig. 13.18 Example of discourse relation extraction result

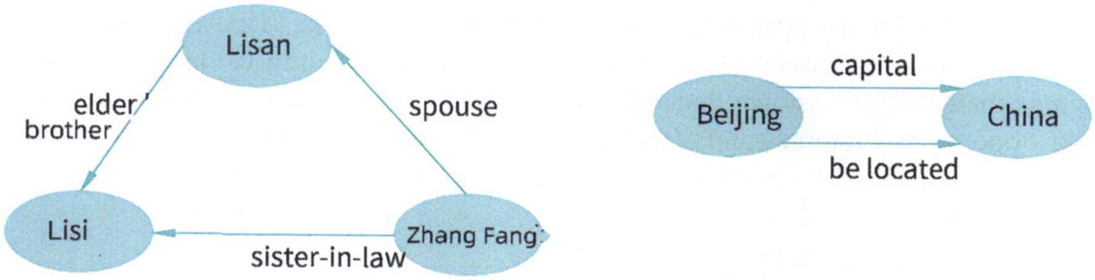

Fig. 13.19 Example of knowledge inference

After extracting the relationship, further inferences can be made on this basis of, for example, "brother" + "spouse" can be inferred to be "sister-in-law," "female" + "parents" can be inferred to be "mother," and the total price can also be inferred based on the unit price and quantity of items.

All the above inferences require the support of relevant rules. These rules can be summarized manually, but finding all the inference rules is time-consuming and difficult. It mainly depends on the co-occurrence between relationships and uses association mining technology to discover inference rules automatically.

Entity relationships contain rich co-occurrence information. As shown in Fig. 13.19, among Li Si, Li San, and Zhang Fang, there are three instances: (Zhang Fang, spouse, Li San), (Li San, brother, Li Si), and (Zhang Fang, sister-in-law, Li Si). Based on a large number of similar instances between entities X, Y, and Z, such as (X, spouse, Y), (Y, brother, Z), and (X, sister-in-law, Z), the inference rule of "brother + spouse ⇒ sister-in-law" can be statistically derived. Similarly, the inference rule of "capital ⇒ located in" can be statistically derived from a large number of (X, capital, Y) and (X, located in, Y) instances.

Knowledge inference can be used to discover new relationships between entities. For example, according to the inference rule "brother + spouse ⇒ sister-in-law," if there is a "brother + spouse" relationship path between two entities, it can be inferred that there is a "sister-in-law" relationship between them. The classic method of implementing relationship extraction using inference rules is the path ranking algorithm, which treats each relationship path as a one-dimensional feature, builds a feature vector for relationship classification by counting a large number of relationship paths in the knowledge graph, establishes a relationship classifier, and performs relationship extraction. This method can achieve good extraction results and has become one of the most representative methods of relationship extraction in recent years. However, this method, based on relationship co-occurrence statistics, faces severe data-sparsity problems.

13.4 Applications and Analysis

13.4.1 Intelligent Search

Intelligent search engines combine artificial intelligence and knowledge graph technology. In addition to fast retrieval and relevance ranking, they can also perform user role registration, automatic interest recognition, content semantic understanding, information filtering, and precise push functions. The primary goal of building a large-scale knowledge graph is to understand the query words input by users better and find accurate answers based on the query words. Usually, the query words input by users are phrases formed by connecting several keywords. Traditional keyword-matching technology matches content based on relevance, and the semantic information behind the query words cannot be understood, resulting in unsatisfactory search results. However, based on the established relationship knowledge, the answer can be quickly and accurately displayed in the search results after using the knowledge graph.

For example, for the query word "calories in rice," if only keyword matching is used, the search engine will mechanically return all web pages containing the two keywords "calories in rice," as shown in Fig. 13.20. You should also open the webpage to find it yourself to get accurate information. But by using the knowledge graph to identify the entities and their attributes in the query words, the search engine can better understand the user's search intent.

As shown in Fig. 13.21, when using the knowledge graph to query "calories in rice," you will find that the search engine directly displays the information "the calories in every 100 grams of fragrant rice are 346 calories" at the top and pushes some information about weight loss, which is more intelligent and humanized.

13.4.2 Robot Learning Machine

The robot learning machine searches for keywords based on user questions, finds the results, and then answers the questions people raise. This is based on an automatic question-answering mechanism. The robot learning machine understands the user's question, extracts facts from the local database or network information, and finally selects a correct answer to tell the user. The robot learning machine is generally suitable for primary school students because the knowledge involved is relatively simple, which can ensure that the robot learning machine gives the correct answer.

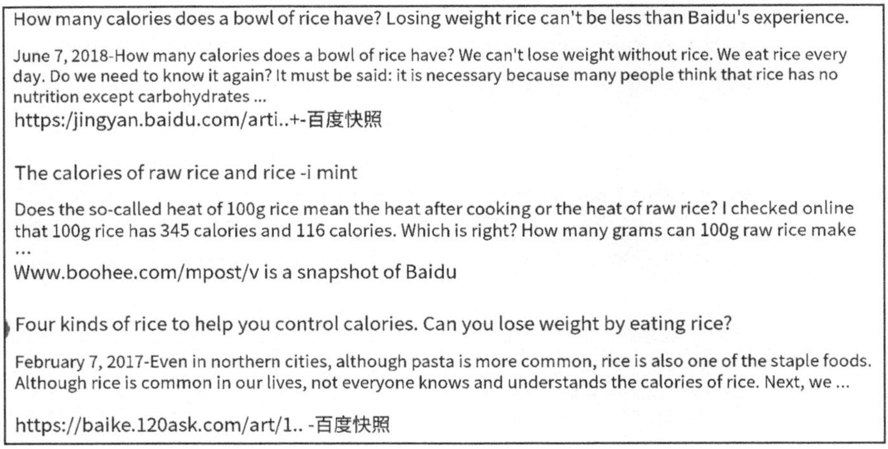

Fig. 13.20 Results of searching "calories in rice" using keyword matching

Fig. 13.21 Results of searching "calories in rice" using the knowledge graph

The robot learning machine uses a knowledge graph and establishes an automatic question-answering knowledge base based on local and network data. For example, when a user queries "the next line of 'the moonlight in front of the bed'," the robot learning machine queries the knowledge graph based on the entity "the moonlight in front of the bed" and the relationship "the next line" and gives the answer "I suspect it is frost on the ground." The robot learning machine has strong learning ability and can add new information to the knowledge graph. For example, when a user tells the robot learning machine, "Mom likes to eat apples," the triplet information ("mom," "likes to eat," "apples") will be added to the knowledge graph. When the user asks, "What fruit does mom like to eat?" next time, the robot learning machine will answer, "Apples."

The robot learning machine is built on an automatic question-answering mechanism. It needs to perform semantic understanding and knowledge reasoning for the questions raised by users. These functions all require the strong support of large-scale, structured knowledge graphs. Therefore, knowledge graphs have become a battleground for major Internet companies.

13.4.3 Document Representation

The classic document representation scheme is the space vector model, which represents the document as a vector of words and adopts the bag-of-words assumption, that is, it does not consider the order information of words in the document. This document representation scheme matches the abovementioned search based on keyword matching. Because of its simple representation and high efficiency, it is the technology mainstream search engines adopt. Document representation is the basis of many natural language processing tasks, such as document classification, document summarization, keyword extraction, etc.

The classic document representation scheme has exposed many inherent serious defects in practical applications, such as the inability to consider complex semantic relationships between words and handle the sparsity problem of short texts (such as query words). People have been trying to solve these problems, and the emergence and development of knowledge graphs have brought new hope to document representation, that is, the knowledge-based document representation scheme. An article is no longer represented by a string of representative words but by the entities in the article and their complex semantic relationships. This document representation scheme has realized the deep semantic representation of documents and laid the foundation for a deep understanding of documents. Some researchers have proposed a straightforward document representation scheme based on knowledge graphs, which can represent documents as a subgraph of a knowledge graph. The document is represented by a graph composed of the entities and their relationships that appear in or are involved in this document. The subgraph of this knowledge graph has a richer representation space than word vectors. It provides richer information for computation and comparison for applications such as document classification, document summarization, and keyword extraction.

Knowledge graphs provide a vast reservoir of knowledge and support for intelligent information processing by computers, which will elevate current technology from a level based on string matching to a level of knowledge understanding. The few applications introduced above can only be a glimpse of the leopard. The construction and application of knowledge graphs is a vast system engineering project, and their potential and possible applications will continue to emerge with the maturity of related technologies.

References

1. Zhang Huaping, Wu Linfang, Zhang Xinming, et al. Construction and Application of Small Sample Domain Knowledge Graph [J]. Artificial Intelligence, 2020(1): 113–124.
2. Noy N F, Musen M A. Anchor-PROMPT: Using Non-local Context for Semantic Matching [C]. International Joint Conference on Artificial Intelligence, 2001.

Part IV
Text Mining

Chapter 14
Information Filtering

This chapter first introduces the background and development of information filtering and then analyzes the reasons for the emergence of information filtering technology. Based on applications, it presents focused information filtering, information filtering recommendations, and their cutting-edge applications according to different application scenarios, analyzes focused information filtering mainly based on text and images, explains the two classic focused information filtering algorithms based on black and white list filtering and content filtering, introduces the three methods commonly used in video recommendation systems: content filtering, collaborative filtering, and hybrid filtering, mainly focusing on algorithm standards and feature descriptions, and introduces the application range and cutting-edge technology of the above filtering technologies and algorithms, and finally presents specific experimental examples.

14.1 Overview of Information Filtering

Information filtering, also known as selective information dissemination, has different definitions and corresponding concepts in many fields, such as routing configuration and radar filtering. Nicholas Belkin and Bruce Croft defined information filtering as a series of processes that deliver information to users who need it. This kind of information filtering refers to network information filtering, which effectively supplements search engines. It can remove harmful information for users and recommend information to users; thus, it has been widely used in fields such as e-commerce and firewalls.

In 1958, Hans Peter Luhn proposed the concept of a business intelligence machine. In this conceptual framework, library staff establish corresponding query models based on the different needs of each user, then generate a new text list that can meet their query needs through precise text selection methods and record the

H. Zhang, J. Shang, *Natural Language Processing and Applications*,
https://doi.org/10.1007/978-981-97-9739-4_14

texts subscribed by the user to update the user's query model. Luhn's work involves every aspect of the information filtering system and lays the foundation for the development of information filtering. In 1969, the selective dissemination of information (SDI) system aroused widespread interest. Most of the systems then followed the Luhn model, and only a few could automatically update user query models. Other systems still rely on professional technicians to maintain the system. The two main reasons for the rise of SDI are the availability of real-time electronic text and the feasibility of calculating the match between the user query model and the text. In 1982, Peter Denning proposed the concept of information filtering. He described an example of the need for information filtering for real-time emails, using the filtering mechanism to identify urgent emails and routine emails. In 1986, Thomas Malone and others published an influential paper and developed the Information Lens system, proposing three information selection models: cognitive, economic, and social.

Information filtering is different from information retrieval. Information filtering focuses on the user's long-term needs (referring to relatively fixed needs over some time). It is designed for unstructured and semi-structured data and is mainly used to process text information. Its goal is to help users handle a large amount of information, filter dynamic information streams, focus on excluding information that users do not want to receive, and filter data from the input information stream based on the user's profile. In information filtering, the user's needs are represented as a profile. A profile is a data structure used to describe a group of topics the user is interested in. The system evaluates (ranks) the articles entering the system based on the profile, and the user provides relevant feedback and updates the profile in time when browsing the results. Due to feedback, machine learning methods have received widespread attention in information filtering, and the main methods used are Bayesian learning methods, neural network methods, support vector machine methods, etc.

When faced with the rich information on the Internet, users hope that the literature they retrieve is what they need. However, the vast ocean of information is overwhelming, and the hyperlinking method often leads people astray, contrary to the original intention. Information filtering allows users to choose service items and content according to their needs actively and quickly find the required information through the filtering mechanism. In addition, user-initiated filtering can save bandwidth, reduce blind information flow, avoid traffic jams, and make network transmission smoother. With the massive increase in machine-readable information, the nature of information retrieval has also changed dramatically. Users of information retrieval systems face two problems: the rapid increase in the number of documents and the huge differences in quality among the papers. The increasingly serious imbalance of documents (quality, type, recording methods, etc.) means that users now need tools to filter and help them select relevant documents more than ever. Therefore, the significance of information filtering technology lies in the following points:

1. Information filtering is a need to improve Internet information query technology. With users' increasing demand for information utilization efficiency, the existing network query technology, mainly based on search engines, has been challenged, and the contradiction between the information needs of network users and the existing information query technology is becoming increasingly acute.

   When using a search engine, as long as the keywords used are the same, the results obtained are the same. The search engine needs to consider the user's information preferences and differences, treat experts and beginners equally, and return thousands of results of varying quality, making it difficult for users to find the information they like. Network information is dynamically changing, and users often care about such changes. However, in search engines, users can only constantly query the same content on the Internet to get changing information, which costs users a lot of time. Therefore, under the current circumstances, traditional information query technology can hardly meet users' information needs. Research on information filtering technology is increasingly valued, and applying information filtering technology to Internet information queries has become a significant research direction.
2. Information filtering is the basis of personalized services. The essence of personalization is targeted, that is, adopting different service strategies for other users and providing different service content. Customized services will allow users to get the best service at the lowest cost. In the field of information services, the goal is to achieve the goal of "information finds people, serves as needed." Since "information finds people," what information finds who is the key? Each user has their own specific, long-term, fixed information needs. Using these information needs as filtering conditions to filter the resource flow, you can extract the content that meets the needs of the resource flow for service. This practice is called information filtering, the basis of personalized, proactive services.
3. Information filtering is urgently needed to maintain national information security. The network has brought great convenience for transmitting information, but it has also made it possible for confidential information to leak out and harmful information about our country's politics, economy, and culture to flow in. Some countries infiltrate politics and instill values and lifestyles through the network, and some criminals use computer networks to copy, disseminate, and read harmful information. Therefore, the issue of national information security is imminent and must be approached with high vigilance and importance. Information filtering serves as an effective preventive measure. Currently, filtering software and grading systems are primarily used to monitor incoming and outgoing information, particularly cross-border data flows. These systems ensure the protection of confidential or valuable information resources that should not be leaked, while blocking information that does not align with national conditions or harmful content from outside the country. The most commonly used tools include Internet reception control software and content selection platforms (Platform for Internet Content Selection, PICS).
4. Information filtering is a means for information intermediaries to carry out network value-added services. The development of the information intermediary

industry has to go through three stages: establishing the initial customer database, enriching the standard archive content, and using the customer archives to gain value. The service focus of the first and third stages involves information-filtering services. Filtering services filter out unwanted sales messages for users. Information intermediaries will set up a filter to check incoming commercial emails and automatically remove unwanted information that does not match the user's needs and preferences. Users can specify the information they want to get through the filtering service or any dealer or product they want to exclude from the filtering service. Information intermediaries will filter unwanted spam out before users get it. In the network environment, reducing invalid data transmission is very important for saving network resources and improving network transmission efficiency. Through information filtering, unnecessary information can be reduced, costs can be saved, and economic benefits can be improved.

14.1.1 Latest Developments in Information Filtering Recommendations

At present, deep learning is widely used in recommendation systems. This can be traced back to a paper published by Hinton and his students in 2007, which applied restricted Boltzmann machines to recommendation systems. Deep learning continues to succeed in computer vision, speech recognition, and natural language processing. More and more researchers and industry professionals are applying it to recommendation services. The application of deep learning in recommendations is burgeoning. Various deep learning algorithms are used for multiple product forms. The current deep learning algorithms are mainly divided into the following three categories:

1. A Recommendation Model Based on Deep Neural Networks. Huang and others [1] proposed a DNN-based deep hybrid recommendation model, which applies deep learning to hybrid recommendations—inputs user and item information into an improved machine learning model for training. The model inputs user and item information into an improved machine learning framework for training, enabling it to learn the interactions between users and recommended items more deeply across multiple dimensions. These advanced deep recommendation models integrate deep learning and machine learning techniques, helping to more comprehensively reflect user preferences and enhance the model's generalization ability. Google's Wide and Deep Learning Recommendation Model [2] was proposed in 2016 and applied to the application (APP) recommendation on the Google Play app store. The model achieved good results in online A/B (AB) testing. This article is also an early application of deep learning in the industry, and it is also a precious article that has a tremendous positive promotion effect on the entire deep learning recommendation system. Many other models are derived based on this model and have succeeded wildly in the industry. The wide and deep model is divided into wide and deep parts. The wide part is a linear model

that learns simple interactions between features, can remember user behavior, and recommends content of interest to users. Still, manual feature engineering requires a lot of time-consuming and laborious work. The deep part is a feedforward deep neural network model that, through the low-dimensional embedding of sparse features, can learn complex cross-combinations of invisible features in training samples, improving the model's generalization ability and effectively avoiding complex manual feature engineering and combining these two parts for joint training results in the final memory and generalization advantages.

2. A Recommendation Model Based on Recurrent Neural Networks. Tencent's WeChat team proposed a deep learning model based on attention mechanisms, RALM [3], which is a deep learning transformation of the traditional look-alike model in the advertising industry. Through user representation learning, look-alike learning captures the local and global information of seed users, while learning user groups and similarity representation of target users helps better mine the audience of long-tail content and applies the selected recommendation in WeChat's "Take a Look." Alibaba Group proposed a deep learning model [4] that represents a user's interests using multiple vectors. Online AB testing significantly improves click-through rates and the diversity of recommended results.
3. A Recommendation Model Based on a Graph Neural Network. In response to the problems in the e-commerce field, Li and others [5] proposed a hierarchical bipartite graph neural network model, HiGNN. By stacking multiple GNN modules and alternately using deterministic clustering algorithms, HiGNN can efficiently obtain hierarchical user and product embeddings and effectively predict user preferences on a larger scale, improving the accuracy of recommendations. Google's neural collaborative filtering (NCF) [6] is a neural network collaborative filtering model based on deep learning recommendation algorithms, embedding user and product vectors in the user behavior matrix into multiple layers of MLP neural network models. The output layer outputs the predicted results through the identity activation function to indicate the user's actual rating, and the model is trained using the square loss function. In 2021, Google published a search pattern centered on large models. Given a search request (query), the model can automatically return results. The model can also complete various knowledge acquisition tasks, including generation and translation tasks.

Deep learning recommendation algorithms have the following characteristics:

1. More Accurate Recommendations. Deep learning models have strong expressive power, therefore, by using deep learning technology to build recommendation algorithms, models can learn deep interactions between features, achieving more accurate recommendation effects than traditional matrix decomposition, factorization machines, and other models.
2. Can Reduce the Investment in Manual Feature Engineering. You need to pour the original data into the model through simple vectorization, and the model can automatically learn features. Building recommendation algorithms through deep learning can significantly reduce the cost of manual feature engineering.
3. Can Conveniently Integrate Additional Information (Side Information).

Deep learning models are highly scalable and can conveniently integrate additional information into the model. However, this model also has the following shortcomings:

1. A large amount of sample data is needed to train a usable deep learning model.
2. A large amount of hardware resources are needed for training.
3. Adapting to the team's existing software architecture is somewhat difficult in engineering implementation. In introducing deep learning models, the organic integration of deep learning-related technology components with the team's existing architecture and components is also an important issue facing the team.
4. Deep learning models are not highly interpretable. Deep learning models are essentially black box models, making it difficult to provide valuable recommendation explanations to users.
5. The parameter tuning process is lengthy and complex. Deep learning models contain a large number of parameters and hyperparameters. Training deep learning models is a complex process that requires following up and observing parameter changes.

Deep learning has driven the arrival of the third wave of artificial intelligence. Within a few years, deep learning has swept across the globe. Almost all technology companies hope to introduce deep learning into real business scenarios, leveraging deep learning to generate tremendous commercial value. The value of deep learning in recommendation systems is gradually becoming apparent.

The future product form will develop in the direction of real time, better adapting to user needs through information flow recommendations. This requires integrating users' real-time interests into the model very conveniently. If the existing deep learning recommendation model can be incrementally optimized and adjusted to reflect changes in user interests, it can better serve users. A deep learning model that can be incrementally learned should be a commercially valuable research topic in the future.

The current deep learning recommendation models still mainly use models built with a single data source. In the future, with the development of 5G technology and the popularity of various sensors, it will be easier to collect multi-source data. How to fully and effectively use this data to build a deep learning recommendation model that integrates multiple data types is a valuable and challenging research direction that researchers must face. At the same time, balancing between protecting user privacy and the data required to train the model is also a problem that should be considered.

Good recommendation products not only recommend precise items, but also consider the visual presentation, visual effects, and interaction methods, which are also essential in determining whether users are willing to use and accept the recommendations. Future deep learning recommendation technologies may model multi-dimensional goals to improve the user experience of recommendation products.

14.1.2 Pay Close Attention to the Latest Developments in Information Filtering

The latest research on information filtering focuses mainly on deep learning methods, such as data augmentation, feature extraction at multiple granularity levels, attention mechanisms, pre-trained language models, and joint models of images and text.

Aiwan and others [7] use data augmentation and improved convolutional neural network model methods to detect image format spam. In terms of datasets, to address the problem of insufficient spam datasets involving privacy, they use k-means clustering to automatically cut suitable parts from a small number of sample images to form new images, effectively expanding the original dataset. In terms of the model, a commonly used CNN model, Spp-Net, is improved to a DSP model, assigning a weight to each grid at each layer and then concatenating them together. The DSP model uses a weighted pool in the convolution-pooling layer, quantifying the amount of information in the pooling area for the first time using information theory. The DSP model also has a module for dynamically updating network parameters. The DSP model can maintain high performance when filtering spam image emails.

14.2 Classic Algorithms for Information Filtering and Recommendation

The information filtering methods are primarily technical in text and image-based spam filtering, collaborative filtering of video recommendation systems, and key focus information.

The key focus here includes various types of harmful information, such as those violating morality and compromising security.

14.2.1 Content Filtering

Content filtering describes user information needs through a user model, calculates the similarity between newly acquired data and the user model, and sends highly similar information to the user corresponding to the user model. The algorithmic process is shown in Fig. 14.1.

1. Constructing Feature Space
   Constructing a feature space establishes measurement standards, which are similar to coordinate axes. For example, a person has many dimensions or attributes, such as gender, age, height, weight, educational level, professional skills, etc.

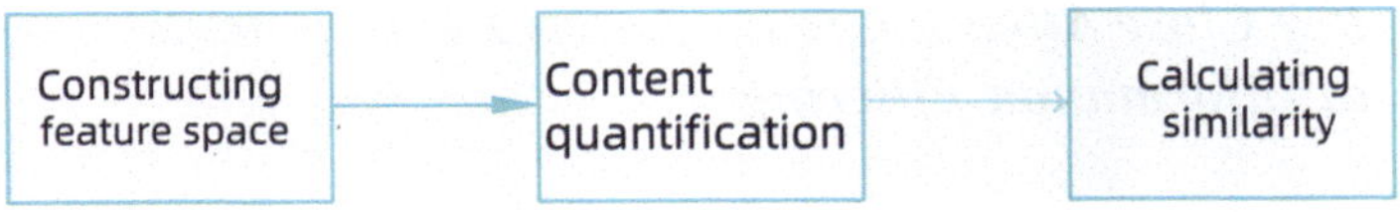

Fig. 14.1 Content filtering algorithm process

These collectively constitute a multidimensional space. Each person will have a specific value on each dimension, thus achieving a quantitative representation of a particular person and realizing the mapping from a person to an N-dimensional vector. Because the demands are different, the feature space constructed may also differ.

2. Content Quantification
   Content quantification refers to evaluating various types of content (such as articles or goods) using the dimensions defined above.
3. Calculating Similarity
   Similarity is the distance of the text in the feature space. The smaller the similarity, the more significant the difference.

Content-based filtering algorithms use different models to evaluate the similarity between texts to generate meaningful recommendation results. They can use vector space models (such as keyword weight calculation algorithms) or probability models (such as Naive Bayes classifiers, decision trees, or neural networks) to simulate the relationship between different text items in the corpus. Then, the basic model is learned through statistical analysis or machine learning techniques to generate recommendation results.

The advantages of the content filtering algorithm are as follows:

1. Users are independent of each other, and each user's recommendation is obtained based on the user's own behavior, unrelated to others.
2. Users can get recommendation results without sharing their personal information, which greatly ensures personal privacy security.
3. It has good interpretability and can explain why these contents are recommended to users.
4. A cold start is quick, allowing newly added items to be directly included in the recommendation results.

The shortcomings of the content filtering algorithm are as follows:

1. The measurement standard is difficult to define, and in most cases, it is difficult to extract features from the project. For example, information is contained in high-dimensional data in videos and other multimedia content, making it difficult to extract features.
2. Unable to mine users' potential interests, the recommended content is only determined based on users' past preferences; therefore, it is similar to users' past preferences.

New users cannot be recommended, as they have no browsing history, making it impossible to obtain their preferences.

14.2.2 Collaborative Filtering

Collaborative filtering analyzes user interests, finds users similar to the specified user in the user group, combines the evaluations of these similar users on a certain piece of information, and predicts the specified user's preference for this information. The algorithmic process is shown in Fig. 14.2.

Taking movie recommendation as an example, the difference between collaborative filtering and content filtering is that the former only needs all users, all movie names, and some scoring information and can guess the scoring situation of all users for all movies, while the latter needs to know the movie's attributes and make recommendations to users based on the analysis of these attributes.

Collaborative filtering can be divided into two types: memory-based collaborative filtering and model-based collaborative filtering.

Memory-based collaborative filtering can be implemented through user-based and item-based techniques. User-based collaborative filtering calculates the similarity between users by comparing users' ratings on the same item, then predicts the active user's rating of the item and uses this prediction as the weighted average of other similar users' ratings. Item-based collaborative filtering uses the similarity between items to predict results, retrieves all items rated by active users from the user–item matrix, establishes an item similarity model, calculates the similarity between items, then selects the top k most similar items, calculates the weighted average of these items, and forms a prediction.

Model-based collaborative filtering uses previous user ratings to model, improving the performance of collaborative filtering. The modeling process can be completed through machine learning or data mining. These techniques include singular value decomposition, latent semantic analysis, regression analysis, cluster analysis, etc.

Compared with content filtering, the advantage of collaborative filtering is that it can recommend items based on the historical information of each user, irrespective of the content attributes of the item itself. Its disadvantages are as follows:

1. Cold Start Problem. When the product is launched, and new users have just arrived, if there is no user behavior data on the application, it is impossible to predict their interests and hobbies. In addition, when new goods are on the

Fig. 14.2 Collaborative filtering algorithm process

shelves, there will also be a cold-start problem. Since no user browsing, clicking, or purchasing behavior data has been collected, it is impossible to recommend the goods.
2. Data Sparsity Problem. This problem occurs when users only rate a small part of the available items in the database. The larger the data scale, the sparser it generally is.
3. Scalability Problem. This is another problem related to the recommendation algorithm because the calculation usually increases linearly with the number of users and items. When the dataset size is limited, the recommendation technology is effective, but when the dataset increases, the generated recommendation results are not very good.

The synonym problem arises when items with different names are very similar. Most recommendation systems struggle to distinguish between such items, such as 'baby clothes' and 'baby fabric.' Collaborative filtering typically cannot establish a match between these terms or calculate their similarity.

14.2.3 Hybrid Filtering

In order to avoid the limitations and problems brought by a single recommendation technique and to achieve better performance, filtering systems combine different recommendation techniques, known as hybrid filtering. Using multiple recommendation techniques can compensate for the defects in a certain technique in the model. There are several methods of hybrid filtering:

1. This refers to the weighted combination of the calculation results of multiple recommendation techniques to produce recommendations.
2. This refers to using different recommendation techniques according to the background of the problem and the actual situation. For example, using a combination of content-based recommendations and collaborative filtering, the system first uses content-based recommendation techniques; if it cannot produce highly credible recommendations, it tries to use collaborative filtering techniques.
3. Cascading techniques build a preference order between items in the iterative refinement process. It is a phased process: the recommendation results of one method are improved by another. One recommendation technique outputs a rough list, which the following recommendation technique improves.
4. Multiple recommendation techniques are used simultaneously to provide multiple user-reference recommendation results. Each recommendation technique has its recommendation results. Each technique has its own recommended results.
5. Feature Combination. The features generated by a specific recommendation technique are input into another recommendation technique.
6. Feature Augmentation. The output of the previous recommendation method serves as the input for the next recommendation method.

Meta-layer mixing involves using the internal model generated by one recommendation technique as the input for another recommendation technique. Compared to a single method, this approach results in a model with richer information.

14.3 Classic Information Filtering Algorithms

14.3.1 Blacklist and Whitelist Filtering

The blacklist and whitelist filtering methods mainly rely on pre-set blacklists and whitelists. Effective blacklists and whitelists are established on the web page side through IP addresses and URLs and on the mobile side through the sender's phone numbers, WeChat numbers, email addresses, and other contact methods. The process is shown in Fig. 14.3.

Take URL filtering as an example. URL is the abbreviation of uniform resource location, commonly known as a web address. Its basic structure is "<URL access method>: //<host >: <port ></path >." Each webpage has a unique URL corresponding to it. URL filtering is based on this unique feature of URLs, which is presetting a URL list to judge whether a user can access the information on a webpage. The URL list is usually stored in the client's firewall or a dedicated URL list server, including both a whitelist and a blacklist. The whitelist consists of safe URLs that are allowed to be accessed, while the blacklist consists of URLs that are prohibited from being accessed. When a user requests to access a webpage, the webpage address is first precisely matched with the addresses in the whitelist. If the match is successful, the access is granted. The address is then matched with the addresses in the blacklist, if successful, access is denied. If not found, i.e., if the URL list does not include this website, it needs to record this website, analyze in the background whether it contains prohibited access information, and allow access to this website this time. Web filtering software based on the URL list method includes Websense, among others. The URL list method has fast filtering speed, high accuracy, and flexibility. It can be embedded in the user's firewall, and users can filter websites according to their needs, but it requires high maintenance costs. Because of the current focus on the concealment of information, many websites can automatically load through hyperlinks or scripts, thereby bypassing URL filtering.

Fig. 14.3 Blacklist and whitelist filtering process

14.3.2 Content-Based Text Filtering

Content-based text filtering technology must analyze whether the content contains key focus information. This filtering technology can match and filter based on keywords and can also, based on the extraction of keywords, apply various natural language processing, artificial intelligence, and data mining methods and technologies, combined with the context, to understand the semantics of the text content, thereby filtering the target information [8]. For example, if the text content appears with the word "Pinduoduo," it needs to comprehensively analyze whether it is advertising information based on the whole sentence or the structure and semantics of the entire text, otherwise, it may mistakenly filter out express delivery and other valid information. These filtering methods require text preprocessing and keyword-based filtering algorithms to achieve filtering targets.

The filtering method based on content understanding needs to extract text feature values and use filtering models for identification and filtering. Its process is shown in Fig. 14.4.

The related algorithms are detailed below.

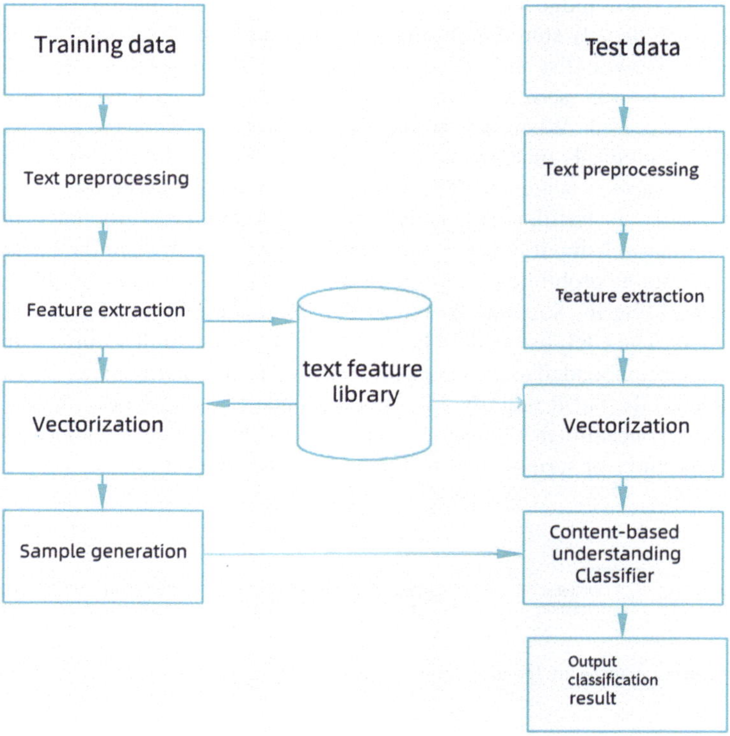

Fig. 14.4 Content-based text filtering process

1. Text Preprocessing

Before extracting and identifying text features, it is necessary to preprocess the text, mainly including removing special symbols (i.e., meaningless symbols to avoid interference), word segmentation, stopping word processing, etc. Due to the current focus on the concealment of information, keywords are usually not used directly. Still, their variants are used, such as split characters, traditional characters, etc., so it is also necessary to recognize and detect variants of keywords.

Automatic word segmentation outputs articles, paragraphs, and sentences written in natural language as words after computer processing, providing a prerequisite for subsequent processing [9]. English uses spaces to split sentences into words, while Chinese requires some algorithms for word segmentation. Common Chinese word segmentation methods include dictionary-based mechanical matching, statistical theory-based, and artificial intelligence word segmentation methods based on neural networks [10]. Commonly used software includes Jieba, word segmentation, etc.

The purpose of stop word processing is to remove the most commonly used functional words (i.e., determiners) in Chinese expressions, such as "的," "一个," "这,""那," etc. These words usually assist in noun description and concept expression, and they do not have much practical meaning. Removing these words can make the main body of the text more prominent, which is beneficial for the subsequent recognition and filtering operations.

Keywords include words that affect the healthy development of culture, and this information is not conducive to the healthy growth of teenagers. Keywords will produce some variants, mainly divided into the following five forms:

1. Replaced with pinyin, for example,"桔子" is written as "ju子" or "juzi."
2. Split characters, for example,"桔子" is written as"木士口子."
3. Homophonic characters or traditional characters, for example,"发票" is written as"蕟票" or "蕟嘌."

Special symbols may appear between words, such as in 'Orange! Son.' Additionally, words may be replaced with English or other languages, for example, 'SM'. According to the associated words and expansion rules of keywords, a keyword library is established, which includes a keyword association word table, a keyword expansion word table, an abbreviation word table, and a synonym table. The Jaccard coefficient can be used to judge the keywords related to the keywords. The keyword expansion word table addresses cases where keywords use pinyin, radicals, homophones, English replacements, or include special symbols. Among these, pinyin and radicals require querying and matching using a Chinese dictionary. An example of a keyword library is shown in Table 14.1.

2. Filtering algorithm based on keywords

The commonly used filtering algorithms based on keywords are as follows:

Table 14.1 Example of keyword library

ID	KeyW	PinYin	Hoph	Anph	Abb	SpC	Synsets	Type
1	Orange	juzi	orange	木士口	null	Ju子	Null	fruit
2.	apple	Pingguo	Pingguo	廾平果	PG	Ping果	Null	fruit

1. Boyer-Moore (BM) Algorithm. It is a complete matching algorithm. Its core idea is reverse comparison, that is, comparison from right to left, and at the same time, it calculates the jump distance through two different heuristic rules and chooses the larger distance to jump, thereby reducing the number of comparisons and improving the comparison efficiency. The two heuristic rules are the bad character rule and the good suffix rule.
2. Aho-Corasick Finite Automaton Algorithm (AC Algorithm). It is a matching algorithm based on pattern trees. Its basic idea is to use the principle of finite automata to merge multiple pattern strings together to form a pattern tree, where each prefix in the pattern tree represents a state, and the search for the string to be matched is completed through state transitions. The matching process mainly involves sequentially reading the text to be matched and comparing it with the pattern string, making judgments and transitions through the transition function and the failure function until the output function is not empty; at this point, the match is completed, and the result is output.
3. WM Algorithm. This algorithm uses a strategy of skipping unmatched characters and a hash method to preprocess the filtered text, building transition tables, hash tables, and prefix tables. This algorithm requires that each string be of the same length when processing strings and only processes the first m characters of each string, where m is the defined shortest length of the string.
4. SWDT-IFA Algorithm. This algorithm first preprocesses the target text by removing HTML, stop words, etc. Second, it uses a keyword decision tree to build the keyword library into a diversion tree to improve efficiency. Third, it passes the preprocessed text through the keyword decision tree in the form of a data stream and records information such as the frequency and regional position of relevant keywords in the text. Finally, it calculates the keyword degree of the text and divides it according to a given threshold [11].

1. Content Understanding-Based Filtering Method

Content-understanding-based filtering methods require text feature extraction operations. Text feature extraction is the extraction of feature values from preprocessed text. Common methods include weight-based feature selection, dependency syntax-based feature selection, and text similarity calculation. Common feature representation methods include vector space, Boolean, and probability models. The vector space model is most commonly used and uses statistical knowledge to model documents. The widely used weight calculation method is the TF-IDF algorithm. Dependency syntax-based feature selection requires constructing a text network structure, analyzing the dependency relationships in the sentences in the text, and building a language network structure based on the words and their frequencies. After converting the text into a language network, the text is mined using parameters such as node degree, clustering coefficient, and betweenness, with the commonly used algorithm being the PageRank algorithm. Text similarity calculations can identify whether two or more texts belong to the same field or category, commonly using Euclidean distance and cosine similarity calculations.

Filtering models match and identify text content based on text feature values, mainly including vector space models, rule-based models, Bayesian decision models, latent semantic indexing (LSI) models, support vector machine models, neural network models, etc.

1. The vector space model is a text calculation model widely used in statistical-based classification systems. It simplifies documents into high-dimensional vectors represented by feature item weights, simplifying the text information filtering process into spatial vector operations, significantly reducing the complexity of the problem.
2. Rule-based models generally use a rule set that contains various constraints to make decisions, thereby performing filtering. They are essentially deterministic deductive reasoning methods, where each rule can represent the user's information needs or information filtering model. According to these rules and patterns, the text to be filtered is matched, and the context determines whether to pass or filter out the text.
3. The Bayesian decision model is usually established using the simple and effective Naive Bayesian theory. Its basic idea is to estimate the probability of a document being relevant or irrelevant based on past judgment experience.
4. The latent semantic indexing model uses mathematical-statistical methods to analyze and infer the properties of some latent semantic structure existing between words, paragraphs, and chapters in the text. This model uses the latent semantic relationships in the document set to construct an index–item document space. Documents with similar topics are very close to each other in this space. Information filtering can be performed by calculating the vector distance between the text to be filtered and the filtering template and setting a threshold.
5. The support vector machine model uses a statistical learning method based on ordered risk minimization induction. Its core idea is to divide the sample space using a simple linear classifier. The division criterion between the two types of topics is obtained by constructing the best hyperplane with the maximum margin in the feature space to minimize the upper bound of the expected risk.

14.3.3 Content-Based Image Filtering

The classic methods used for image filtering mainly include database filtering, tag filtering, optical character recognition (OCR) technology, deep learning-based filtering, etc. Among these, database filtering, also known as black-and-white list filtering, involves creating a blacklist of IP addresses or URLs for websites containing excessive images and blocking access to web pages listed in the blacklist. Tag filtering involves manually or automatically adding tags to images by the uploader, followed by filtering out images with tags such as 'advertising' for specific audiences. Although these two methods are simple to implement and fast in execution, they have significant limitations and are only suitable for basic filtering tasks. They struggle to meet the complex demands of key information filtering. This section introduces the main applications of image filtering.

1. Content-Based Image Filtering

The content-based image filtering method is a complex process of classifying the content of images by establishing a mapping between low-level image features and high-level semantics, including using digital image technology, statistical knowledge, etc., to extract features from images. Through machine learning and other methods for classification and other steps, its process is shown in Fig. 14.5. Feature extraction and classification methods are complementary relationships. Commonly used classification tools include support for simple machine learning algorithms such as support vector machines, k-nearest neighbors, and BP neural networks.

2. Skin feature extraction based on color space

Through actual observation, it can be found that a significant feature of most portrait pictures is a large amount of skin, so large areas of skin can be used as an essential evaluation standard for pictures. Since human skin has good aggregation in certain color spaces, the distribution of skin pixels in the image and the distribution of non-skin pixels have significant separability, so pictures with too many skin pixels can be preliminarily filtered through the proportion of skin pixels.

The HSV color space is also known as the hexagonal cone model, as shown in Fig. 14.6. In this model, H represents hue, measured in degrees, with a range of 0 to 360. Red corresponds to 0°, green to 120°, and blue to 240°. S represents saturation, which indicates how closely a color resembles a pure spectral color. The higher the value, the more saturated the color, with a range of 0% to 100%. V represents value, which indicates the brightness of the color, typically ranging from 0% to 100%, where 0% is black and 100% is white. In addition to the HSV color space, YCbCr and YIQ are also commonly used color spaces.

The model for extracting skin features is typically defined in the HSV space or a mixed space. In this model, the definition of skin pixels is based on prior knowledge, explicitly specifying the boundary conditions for skin pixels. For example, a mixed skin model defines skin pixels as follows:

- The I component in the YIQ color space lies within the range $15 < I < 80$.
- In the YCbCr space, the Y and Cr components lie within the range $20 < Y < 120$ and $130 < \text{Cr} < 150$, respectively.
- In the HSV space, the H, S, and V components lie within the range $0.0095 < H < 0.02$, $S > 0.15$, and $0.2 < V < 0.6$.

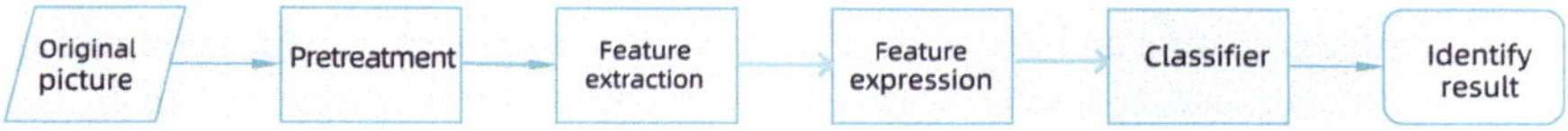

Fig. 14.5 Content-based image filtering process

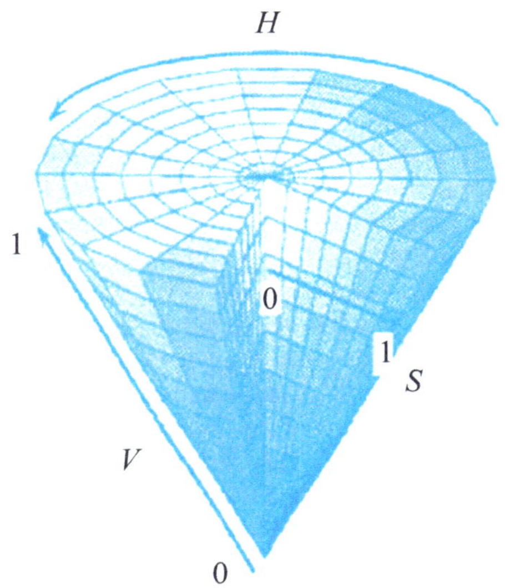

Fig. 14.6 HSV color space model

The definition of skin pixels is simple and computationally efficient, but it requires high accuracy and prior knowledge and has limitations. Moreover, there is overlap between skin pixels and non-skin pixels at the defined boundary, resulting in a high false detection rate. The Gaussian model and Gaussian mixture model (GMM) use statistical methods to match sample distributions, and the skin color probability density formula is:

$$P(x;\phi)=\sum_{i=1}^{K}\pi_i\cdot p_i(x;\theta)=\sum_{i=1}^{K}\pi_i\cdot p_i(x|;i|;\theta)$$

$$=\sum_{i=1}^{K}\pi_i\cdot\frac{1}{(2\pi)^{\frac{d}{2}}|\Sigma_i|^{\frac{1}{2}}}\exp\left[-\frac{1}{2}(x-\mu)^2\Sigma_i^{-1}(x-\mu)\right] \tag{14.1}$$

In the above formula, x is the pixel color vector, μ is the mean vector, Σ is the covariance matrix, and k is the number of Gaussian density functions in the mixture model. The Gaussian model can effectively extract the skin pixels in the picture and judge whether the picture is inappropriate based on this.

Skin feature extraction often also uses texture, shape, and other features to assist in judgment. The method based on skin detection has a certain reliability, but the false detection rate and missed detection rate are relatively high, and the generalization performance is also poor.

3. Face Information and Key Part Monitoring

The face is also highly recognizable, so introducing face detection is essential for determining the characters and attributes in the picture. Face recognition, a key issue in computer image recognition, has been well-researched and applied. The commonly used face detection algorithms include predefined template matching methods, the local binary pattern (LBP) algorithm, the super-resolution algorithm based on linear models, the AdaBoost algorithm based on Haar features, etc.

Taking the LBP algorithm as an example, the LBP feature is a commonly used local feature descriptor that calculates the relationship between adjacent pixels. The calculation process of the LBP feature uses the gray value of the center pixel of the 3×3 window as the threshold, and the gray values of the surrounding eight pixels that are greater than or equal to this threshold are marked as 1, and those less than this threshold are marked as 0. The resulting eight values are used as an 8-bit binary number, which is converted to decimal to get the LBP code of the center pixel. The LBP code can form another picture, the LBP feature spectrum, which can be used for picture classification and recognition.

The key parts of the human body can be used more directly as the basis for judging the picture. The detection methods are similar to face recognition, mainly including predefined feature methods, LBP feature extraction, and methods through AdaBoost ensemble learning. A classifier is established to detect and filter potential key parts of the human body in the picture.

Different detection methods are expected to be combined step by step at the application level to exclude benign pictures and filter out pictures. However, the content features of the picture are complex, and there is often a certain conflict and overlap between different detection methods. The physical cascading method cannot integrate different feature detection methods well, as these feature detection methods all have certain limitations. Therefore, this detection method based on the region of interest often has a high false-positive rate.

4. Picture Filtering Based on Local Features

In addition to the filtering methods based on areas of interest such as skin, face, and key parts, local features are used to characterize local information in the image. The most representative of these is the scale-invariant feature transform (SIFT) feature, which is invariant to the scale and rotation of the image and has been proven to be robust under a wide range of affine transformations, changes in 3D perspective, and changes in lighting conditions. The introduction of SIFT features has effectively improved the efficiency of image filtering. Another classic local feature method is the histogram of oriented gradient (HOG), which divides the image into multiple dense units during calculation, performs gradient direction histogram statistics separately and then merges them to obtain the classification decision result of the entire image. The feature expression part after feature extraction uses sparse coding, a bag-of-words model, etc.

OCR technology recognizes text in images, and it is used in the field of image filtering to filter images with target text. OCR technology can be divided into three categories: template matching, machine learning, and deep learning. The main process of the machine learning OCR method is to input images, process images, extract features, and use SVM or KNN for classification. The basic method of deep learning OCR is to build and train a deep network of convolutional layers, fully connected layers, and classifiers. Then, the trained network is used to recognize characters in the image.

Image filtering based on deep learning uses neural network methods, mainly fully connected layers, convolutional neural networks, activation functions, etc., to complete feature extraction and classification together and automatically extract the required features, avoiding the problem of manual design features not matching reality. The representative model of this method is AlexNet, which is not much different from traditional CNN in structure. It includes five convolutional layers and three fully connected layers, with an input of images of size 224 × 224. The first convolutional layer contains 96 convolution kernels of size 11 × 11, and a maximum pooling layer is placed after the first, second, and fifth convolutional layers for feature-down sampling. AlexNet, as well as later VGGNet and GoogleNet, has achieved great success in the field of image recognition.

14.4 Applications and Analysis

14.4.1 Information Filtering Recommendation Example

There are currently many movie rating datasets. The dataset used in this section is the ml-latest-small dataset, which includes ratings given by 610 users for 9724 movies and user IDs and movie IDs.

1. Collaborative Filtering

Movie Recommendation Based on Users and Products. User-based collaborative filtering is based on finding neighboring users based on user preferences for items and then recommending items that neighboring users like to the current user. In calculation, it is to calculate the similarity between users by taking a user's preference for all items as a vector, and after finding k neighboring users, predict the items that the current user likes based on the similarity weight of neighboring users and their preference for items, and obtain a sorted item list for recommendation through calculation. Item-based collaborative filtering is similar in principle to user-based collaborative filtering, but it uses the item itself (rather than from the user's perspective) when calculating similarity. Similar items are found based on user preferences for items, and then similar items are recommended based on the user's historical preferences. From a computational perspective, it involves taking all user preferences for a particular item as a vector, calculating the similarity between items, and, after obtaining the similarity of an item, predicting items that the current user has not yet shown a preference for based on their historical preferences. Through this calculation, a ranked list of recommended items is obtained. The Pearson coefficient is used to calculate similarity in applications. The results of movie recommendations based on user and item collaborative filtering are shown in Table 14.2.

Table 14.2 Results of movie recommendations based on user and item collaborative filtering

Method	User-Based Collaborative Filtering	Product-Based Collaborative Filtering
Recommended movie ID	1041, 714, 80906, 1235, 3030, 65261, 1178, 1217, 318, 1104	3285, 65088, 41571, 52435, 111113, 261, 290, 1411, 3307
Recommended time/s	18.98s	18.93s

$$\underset{R}{\begin{array}{c|cccc} & \text{item1} & \text{item2} & \text{item3} & \text{item4} \\ \hline \text{user1} & R_{11} & R_{12} & R_{13} & R_{14} \\ \text{user2} & R_{21} & R_{22} & R_{23} & R_{24} \\ \text{user3} & R_{31} & R_{32} & R_{33} & R_{34} \end{array}} = \underset{P}{\begin{array}{c|ccc} & \text{Class1} & \text{Class2} & \text{Class3} \\ \hline \text{user1} & P_{11} & P_{12} & P_{13} \\ \text{user2} & P_{21} & P_{22} & P_{23} \\ \text{user3} & P_{31} & P_{32} & P_{33} \end{array}} \times \underset{Q}{\begin{array}{c|cccc} & \text{item1} & \text{item2} & \text{item3} & \text{item4} \\ \hline \text{user1} & Q_{11} & Q_{12} & Q_{13} & Q_{14} \\ \text{user2} & Q_{21} & Q_{22} & Q_{23} & Q_{24} \\ \text{user3} & Q_{31} & Q_{32} & Q_{33} & Q_{34} \end{array}}$$

Fig. 14.7 Principle of matrix decomposition

1. Movie Recommendation Based on Matrix Decomposition Collaborative Filtering

   The principle of matrix decomposition is shown in Fig. 14.7. For the original rating matrix R, assuming there are three hidden features, the $R_{3\times4}$ matrix is decomposed into the user feature matrix $P_{3\times3}$ and the item feature matrix $Q_{3\times4}$. Considering user1's rating for item1, it can be assumed that user1's interest in the three hidden features class1, class2, and class3 is P_{11}, P_{12}, and P_{13}, respectively, and the relevance of these three hidden features to item1 is Q_{11}, Q_{21}, and Q_{31}, respectively.

   Matrix P is the user–LF matrix, i.e., the user and latent feature matrix.

- Matrix Q is the LF–item matrix, i.e., the latent feature and item matrix.
- Matrix R is the user–item matrix, i.e., the user and item matrix, derived from P × Q, which can handle a sparse rating matrix. The results of the movie matrix based on matrix decomposition and collaborative filtering are shown in Table 14.3.

14.4.2 Spam Filtering Example

This section includes two simple text classification filtering examples: one is using a machine learning model to filter test set SMS data, and the other is to identify whether a single input message is spam.

1. Dataset Filtering

   The original data has 800,000 entries, as shown in Fig. 14.8, and the data distribution is unbalanced. Using undersampling, 20,000 entries are randomly selected (including 10,000 spam and 10,000 non-spam messages).

Table 14.3 Results of movie recommendations based on matrix decomposition collaborative filtering

Method	Collaborative Filtering Results Based on Matrix Decomposition
Recommended movie ID	2122, 3148, 3451, 3783, 3429, 1704, 4993, 1223, 318, 7387
Time required for recommendation	41.22s

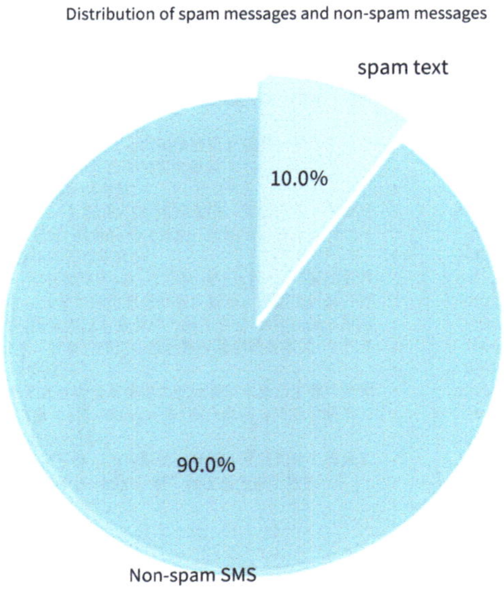

Fig. 14.8 Distribution of original data

First, data preprocessing operations are performed:

1. Data cleaning includes removing spaces, X sequence processing, and text deduplication.
2. Use Jieba segmentation to segment the text messages.
3. Add new words to the dictionary and remove stop words.
4. Draw a word cloud.

Text features are represented by the TD-IDF weight matrix, using the text feature extraction module in scikit-learn.

The key code is as follows:

```
cuntVetorizer = CountVetorizer
data_tr = countVectorizer.fit_transform(data_tr)
X_tr = TfidfTransformer().fit_transform(data_tr.toarray()).toarray()
```

Finally, use the Gaussian Naive Bayes (GNB) model for data classification and recognition. The key code is as follows:

```
model = GaussianNB ()
model .fit(X_tr, labels_tr)
   ..
```

The original data of the test set is shown in Fig. 14.9. Label 1 represents spam messages, and label 0 represents non-spam messages. The recognition accuracy rate is 95%, as shown in Fig. 14.10.

	label	message
0		
197631	1	你好，我们是阳光信贷，办理免担保无抵押信用贷款的，我是客户经理左婷，有资金需要或朋友需要可以...
403647	0	发送x到xxxxxxxxxx兑换xx元
463751	1	更多的乐趣。凡是元宵期间成功认购客户均可享受一票抵万金优惠。此外老带新购房成功新老客户各有x...
466457	0	镇静凉爽的喷雾+粉底+防晒+美白+抗皱5种机能的多功能遮盖产品2
71610	0	是杭州长运运输集团有限公司的一名高级修理技师
325894	1	亲爱的，三月促销，幻活新生 时光礼遇衔一:购买任意幻时和幻时佳产品每满xxxx元，即可获得价...
515863	1	另外伊人坊特以回馈天下女性:?产品优惠?套盒单支?一律八折?仅此三天?x.x.x.号?伊人坊...
398710	0	md大早上睡不着起来看花千骨我真的遭不住我自己了
638755	1	尊敬的客户: x月x日至x月x日，融e购商城开展元宵节、女神节大型促销活动，多款商品全网最低价...
112423	1	活，每月可赎回。更有月添利进取型收益率x.xx供您选择，赶紧备好您的资金抢购吧！详询 理财...
688907	1	端设计品团购会 您在活动中将免费体验到全国连锁大型装饰企业已为您做好的各种设计方案和彩色效果...
96044	1	您好； 我是领秀中原一楼惠达卫浴的任姝丽、我们现在x.xx活动正式开始了、是今年最优惠的活动...
170107	0	飞机特别小跟公交一样居然有单人座
430950	0	明明说着生日快乐可是我想去旅游不让我去我不想办生日非要我办我喜欢做什么都不支持我我不喜欢做什...
19089	1	..亲朋好友们，商场大秘?秘?xxxx年x月x（周五）日，国美电器一年一次的员工内购会即将开...
260568	0	16时15分至16时35分奥委会委员投票
773956	1	本公司（林涛贸易）新到一船辐射松。卸于常熟兴华港，x米 x米 x.x米各种规格。如老板需要。...
201723	0	他想了想还是回到了那一年的一个晚上静静的躺在床上用手机发了一条短信“我们不分手好不好”
32612	0	隋炀帝杨广在江都被部下缢杀
68308	0	我总是和一个群里机器人说晚安

	label	message
0		
197631		Hello, we are Sunshine Credit, which handles unsecured and unsecured credit loans. I am Zuo Ting, the account manager. If you need funds or friends, you can ...
403647	0	Send x to >xxxxxxxxx>Exchange xx yuan
463751		,"trans_result":[{customers who successfully subscribe during the Lantern Festival can enjoy a one-vote discount of 10,000 yuan. In addition, the old and new buyers are successful, and new and old customers have their own X. ..
466457	0	Calm and cool spray+foundation+sunscreen+whitening+wrinkle-resistant multifunctional covering product 2
71610	O	He is a senior repair technician of Hangzhou Changyun Transportation Group Co., Ltd.
325894	one	Dear, March promotion, magic life new life courtesy Street 1: You can get a price for every xxxx yuan you buy any magic time and magic time products. ..
515863	one	In addition, Yirenfang specially gives back to women all over the world: Product discount? Single box? All 20% off? Only three days? X.x.x? Yiren house ...
398710	nine	Md can't sleep in the morning and watch Hua Qiangu. I really can't stand myself.
638755	one	Dear customers: From X to X, Ronge Shopping Mall launched large-scale promotional activities for Lantern Festival and Goddess' Day, with the lowest prices for many commodities in the whole network. ...
112423	one	Live, redeemable every month. There is also a monthly profit-making rate of return x.xx for you to choose from, so get ready for your funds to snap up! Inquire about financial management.
688907	one	At the end of the design group purchase meeting, you will experience all kinds of design schemes and color effects that the national chain large-scale decoration enterprises have done for you for free. ...
96044	one	Hello; I am Ren Meili, who leads the show of Huida Sanitary Ware on the first floor of Zhongyuan. Now our x.xx activity has officially started, which is the most favorable activity this year. ..
170107	0	The plane is so small that it has a single seat like a bus.
430950	0	Obviously, I said happy birthday, but I want to travel and don't let me go. I don't want to have a birthday. I have to do it. Nothing I like to do supports me. What I don't like to do. ..
19089	one	.. friends and family, the secret of the mall? Secret? On Friday, xxxx, Gome's annual employee internal purchase meeting will be held soon. ...
260568	nine	From 16: 15 to 16: 35, IOC members voted.
773956	one	Our company (Lin Tao Trade) has a new shipment of radial pine. Unloaded in Xinghua Port, Changshu, x m x m x m of various specifications. If the boss needs it. ...
201723	nine	After thinking about it, he went back to one night that year and quietly lay in bed and sent a short message on his mobile phone, "Shall we not break up?"
32612	O	Yang Guang, Emperor Yangdi, was killed by his subordinates in Jiangdu.
68308	O	I always say good night to a group of robots.

Fig. 14.9 Original data of the test set

```
The accuracy of test set is 0.9500.
[10101110111100101000]
        label                    message
0
403647      0      Send x to xxxxxxxxxx for xx yuan.
466457      0      Calm and cool spray+foundation+sunscreen+whitening+wrinkle-resistant multifunctional covering product 2
398710      0      Md can't sleep in the morning and watch Hua Qiangu. I really can't stand myself.
170107      0      The plane is so small that it has a single seat like a bus.
430950      0      Obviously, I said happy birthday, but I want to travel and don't let me go. I don't want to have a birthday. I have to do it. Nothing I like to do supports me. What I
                   don't like to do. ..
260568      0      From 16: 15 to 16: 35, IOC members voted.
201723      0      After thinking about it, he went back to one night that year and quietly lay in bed and sent a short message on his mobile phone, "Shall we not break up?"
32612       0      Yang Guang, Emperor Yangdi, was killed by
68308       0      his subordinates in Jiangdu. I always say
                   good night to a group of robots.
```

Fig. 14.10 Test set recognition filter results

Table 14.4 The 4 machine learning algorithms used in this example

Machine Learning Algorithm	Model Implementation
Logistic regression (LR)	Implemented using the logistic regression model in sklearn
Support vector machine (SVM)	Implemented using svm.SVC in sklearn
Decision tree (DT)	Implemented using DecisionTreeClassifier () in sklearn
Gradient boosting decision tree (GBDT)	Implemented using GradientBoostingClassifier in sklearn

2. Single Message Recognition

After preprocessing the data and extracting the TF-IDF feature values, train with the four machine learning algorithms shown in Table 14.4, save the model using joblib, and then recognize the content of the input message to determine whether it is a spam message.

Taking the SVM algorithm as an example, the key code is as follows:

```
def __init__ (self, training_ data, training_ target):
self.training_data = training_data
self.training_target = training_target
self.clf = svm.SVC(C=1, class_weight=None, coef0=0.0,
decision_function_shape=None, degree=3, gamma= "auto",
kernel='linear', max_iter=-1, probability=False,
random_state=None, shrinking=True, tol=0.001, verbose=False )
def train_classifier(self):
self.clf.fit(self.training_data, self.training_target)
joblib.dump(self.clf, 'model/SVM_sklearn.pkl')
training_result = self.clf.predict(self.training_data)
print(metrics.classification_report(self.training_target,
training_ result))
#performance_report(self.training_target, training_result)
```

Table 14.5 shows some test cases and model prediction results. The first and second are the recognition results of spam messages, among which only the second one was correctly recognized by SVM, which is widely used in reality. The recognition result of ordinary text is shown in the third.

Table 14.5 some test cases and model prediction results

Test Examples	Models			
	Logistic regression (LR)	Support Vector machines (SVM)	Decision Tree (DT)	Gradient Booled Decision Tree (GBDT)
Dear VIP,first of all.I wish you a happy new year,Spring Departement store will officer the lowest discount of the year from February 25th,with 30%off on all auntumnand winter products. This is a rare opporunitiy, so huuy up and buy it.	Spam	Spam	Spam	Spam
[Princess and the pea] your 0 yuan free trial product Wild Vegetable Udon Bowl Noodles is about to expire. There is only one step left. Take it home	Not spam	Not spam	Spam	Not spam
The Big Data Analysis and application course is held in Boom 2001,Building 8.	Not spam	Not spam	Not spam	Not spam

14.4.3 Intelligent Filtering System Display

1. System Overview

The intelligent filtering system shown in this section is a content filtering system for complex text with big data. In real time, it can intelligently recognize common variants such as keyword deformation, sound change, and character splitting and has achieved precise semantic disambiguation. The system has built China's latest and most comprehensive knowledge base, suitable for intelligent filtering and discovering content such as spam advertisements.

2. System Features

The intelligent filtering system has the following features:

1. Intelligent Variant Recognition. The intelligent filtering system uses the perfect double array trie (PDAT) management and retrieval method to automatically recognize morphological, phonetic, split, traditional/simplified characters, full-width/half-width characters, and various types of interference noise. The system supports a custom keyword library and can incrementally add millions of words to the library. When recognizing phonetic variants, the system uses a built-in Chinese character pinyin library to automatically convert keywords into phonetic forms, generating full and abbreviated pinyin for keywords to facilitate the recognition of homophonic characters or pinyin variants, significantly increasing the filtering range and hit rate. For example, "fa票" and "fapiao" will automatically be converted to"发票," and "ju子"and "just" will automatically be converted to"桔子." When recognizing morphological variants, the system uses a built-in Chinese character homograph

library to automatically convert keywords into different forms, making all types of split characters, combined characters, and other morphological variants unrecognizable. For example, "弓长"will automatically be converted to "张," and "發嘌" will automatically be converted to "发票."
2. Semantic Disambiguation. Example: "桔子" will automatically be converted to "木士口."
3. Fast and Real-Time Recognition and Filtering. The intelligent filtering system uses a patented algorithm for fast scanning, with a single-machine speed of 20 MB/s. The system supports single-machine multi-threading, multi-machine parallelism, Hadoop cloud services, etc., and can perform parallel, efficient online checks on PB-level information content.
4. Built-in, Latest, and Most Comprehensive Word Library. The system has ten built-in categories of keywords, including fraud, pyramid schemes, online gambling, anti-ethics, spam advertising, etc. The word library covers almost all industries and suits users in different fields. The word library can continue to accumulate and optimize during use, customizing the latest and most comprehensive word library for users in their professional fields. The system supports custom keyword categories and weights and can incrementally add up to millions of words to the library.

3. System Technical Architecture

The technical architecture of the intelligent filtering system is shown in Fig. 14.11. The intelligent filtering system uses the perfect double array trie management and retrieval method. The performance of this method determines the processing speed of the intelligent filtering system so it can achieve efficient online checks in PB-level data scale scenarios with a single-machine speed of up to 20 MB/s.

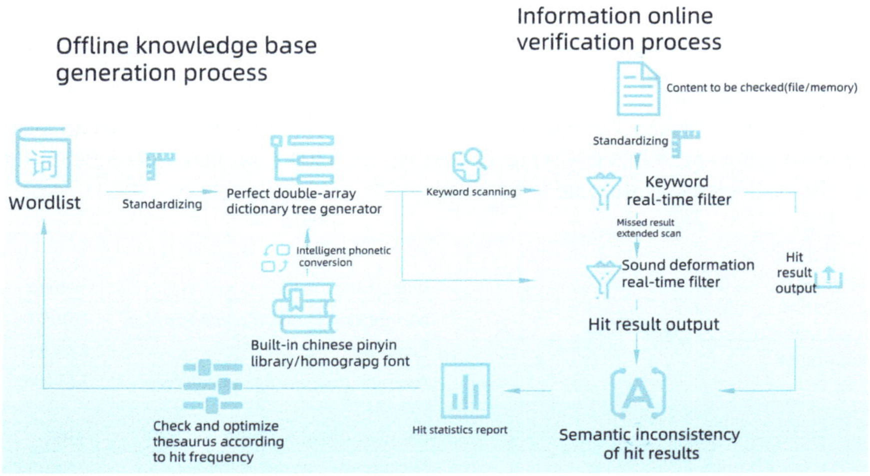

Fig. 14.11 Technical architecture of the intelligent filtering system

In the process of offline knowledge base generation, the system normalizes the keyword table imported by the user (including encoding conversion, traditional/simplified conversion, etc.) into PDAT format. The keywords are formatted as one per line, including the word, category, and weight.

The following are explanations about this system:

1. Keywords and categories are entirely set by the user, with no restrictions on length, format, or encoding.
2. The maximum number of categories currently supported by the system is 255.

The recommended weight is 1~10, with 1 being the smallest and 10 being the largest.

1. The same word can belong to different categories.

4. Application Function

The intelligent filtering system is specifically designed to filter the content of complex text and big data. After users submit the content to be reviewed and keywords (the system has the latest and most comprehensive word list built-in), the system intelligently transforms keyword variants. It checks the target content and then semantically disambiguates the check results. The intelligent filtering system uses NLPIR's precise word segmentation and sentiment analysis technology to understand the semantics of the content and accurately disambiguate it by analyzing the context, excluding harmful information, significantly improving the accuracy of information filtering and reducing the rate of misjudgment. Finally, the system outputs real-time scanning results and hit statistics reports. The system can also adjust the keyword dictionary based on the hit statistics report, and after accumulation, it obtains a complete dictionary, making the output results more and more accurate.

The account verification results of a well-known Internet Q&A community are shown in Table 14.6.

The system scanned 1,006,076 records, taking a total of 129.94 s, with an average processing speed of 1154.96 records/s, hitting 3304 rules and 144,032 records, and the suspected harmful rate is 14.32%.

Table 14.6 Account verification results of a well-known Internet Q&A community

Indicators	Values
Number of scanned records/piece	1,006,076
Scanning time	129.94
Average processing speed/(pieces/second)	1154.96
Number of rules hit/piece	3304
Number of hit records/piece	144,032
Suspected harmful rate%	14.32

References

1. Huang Z, Yu C, Ni J, et al. An Efficient Hybrid Recommendation Model with Deep Neural Networks[J] . IEEE Access, 2019, 7: 137900–137912.
2. Cheng H T, Koc L, Harmsen J, et al. Wide & Deep Learning for Recommender Systems[C] . ACM. The 1st Workshop on Deep Learning for Recommender Systems. 2016: 7–10.
3. Liu Y, Ge K, Zhang X, et al. Real-time Attention Based Look-alike Model for Recommender System[C] . ACM. The 25th ACM SIGKDD International Conference on Knowledge Discovery & Data Mining. 2019: 2765–2773.
4. Li C, Liu Z, Wu M , et al. Multi-interest Network with Dynamic Routing for Recommendation at Tmall[C] . ACM. Proceedings of the 28th ACM Incovery & Data Mining. 2018: 1040–1048.
5. Li Z, Shen X, Jiao Y, et al. Hierarchical Bipartite Graph Neural Networks: Towards Large-scale E-commerce Applications[C] . 2020IEEE36th International Conference on Data Engineering(ICDE) . IEEE, 2020: 1677–1688.
6. He X, Liao L, Zhang H , etal. NeuralCollaborativeFiltering[C] . ACM. the26thInternational-Conference. 2017: 173–182.
7. Aiwan F, Zhao F Y. Image SPAM Filtering Using Convolutional Neural Networks[J] . Personal and Ubiquitous Computing, 2018, 22(5–6): 1029–1037.
8. Elkin-Koren N. Contesting Algorithms: Restoring the Public Interest in Content Filtering by Artificial Intelligence[J] . Big Data& Society, 2020, 7(2): 1–13.
9. Wu F, Liu J, Wu C, et al. Neural Chinese Named Entity Recognition via CNN-LSTM-CRF and Joint Training with Word Segmentation[C]. The Worldwide Web Conference. 2019: 3342–3348.
10. Liu Lifang. Research on the identification of bad information on the Internet based on rules and statistics[D]. Wuhan: Central China Normal University, 2017.
11. Gao Wen, Li Ronghua, Chen Changqi, et al. Research on the system for filtering sensitive words in the content of online virtual community texts [J]. Modern Business Trade Industry, 2017(16): 169–172.

Chapter 15
Text Classification

This chapter covers text classification, starting with methods based on statistical rules, which require manual feature selection. It then discusses machine learning approaches like support vector machines and decision trees, which automate feature selection but struggle with context and semantics. The chapter shifts to deep learning techniques, such as CNNs, RNNs, and graph neural networks (GNNs), which improve text classification by better capturing deep semantic features. Finally, it includes a practical example, detailing the process from data collection to evaluation, demonstrating the real-world application of these classification methods.

15.1 An Overview of Text Classification

With the advent of the big data era, text has become an important medium for information sharing and dissemination, such as news articles, product reviews, and social media posts. Classifying these texts can help quickly analyze the text, discover patterns, and mine data value, so text classification has important practical significance. Early text classification methods mainly used algorithms based on statistical rules, but the disadvantages of this algorithm are also very obvious: it requires a lot of manual intervention and manual setting of extraction rules, and the classification accuracy depends on the level of people. With the in-depth study of text classification, some machine learning algorithms have been proposed, but they still cannot meet various text classification tasks. In recent years, hardware development has allowed hardware facilities to no longer be the bottleneck of deep learning model training. Deep learning methods have been applied to many natural language processing fields. Researchers have tried to use deep neural networks to solve the problem of text classification, and the results prove that deep neural networks can learn useful text features and context information for classification prediction.

H. Zhang, J. Shang, *Natural Language Processing and Applications*,
https://doi.org/10.1007/978-981-97-9739-4_15

15.1.1 Text Classification Based on Statistical Rules

Traditional statistical rule methods mainly classify texts through feature selection, determine specific rules, use selection algorithms, and select feature words from training corpora. The goal is to obtain a set of words representing a certain category in the training corpus. Feature selection can reduce the problem's size and improve the execution of the classification task. The classification results depend on the selected features. Statistical rule methods are easy to understand, and the algorithm implementation is intuitive. Traditional feature selection algorithms based on statistical rules can extract text features well in some aspects, but this method relies on manual work, which is time-consuming and costly, and the classification accuracy depends on the selected features.

15.1.2 Text Classification Based on Machine Learning

Most existing machine learning algorithms are based on feature selection methods. The main process is to perform feature engineering, extract local feature words from the text, and form local text feature representations. Support vector machine is a feature selection algorithm for machine learning that, along with decision trees, dominates current machine learning classification practices. These methods have more advantages than early statistical-based methods. The implementation process of the support vector machine is to find a maximum separation hyperplane, with the aim of dividing the sample dataset into different areas by this hyperplane. Support vector machines can classify samples in many text classification tasks. The main reason for the widespread use of support vector machines is that their principle is intuitive, implementation is simple, and the classification effect is good.

The decision tree algorithm adopts a tree-like structure, a decision tree corresponds to a classifier, and based on a certain discrimination rule, the final classification is obtained using the inference method. The process of constructing a decision tree is a process of learning parameters. Using a decision tree to predict node labels is a judgment process. Based on the specific node attributes, enter a specific branch node according to the decision rule. If the branch node is not a leaf node, continue to judge, if the branch node is a leaf node, stop judging and output the final result. As a commonly used machine learning algorithm, the decision tree has the characteristics of easy implementation, strong interpretability, and conforming to human intuitive thinking. It is still widely used in the field of text classification.

However, these methods still need to be designed according to specific scenarios, and the classification accuracy heavily depends on feature selection results. In addition, they usually ignore the natural order structure or context information in the text data and lack semantic expansion, making it difficult to learn the context semantics of words and effectively capture useful information. They cannot obtain the context semantic information of the text well.

With the development of social networking platforms, a large amount of text data is emerging, supervised classification methods can no longer meet the needs of various classification scenarios, and unsupervised classification methods have become a research hotspot. Among them, as an unsupervised learning method, the latent dirichlet allocation (LDA) topic model has achieved good results in the unsupervised learning of text. The commonly used sampling methods of the LDA topic model are Gibbs sampling and variational inference. As an unsupervised method for learning unlabeled data, the LDA topic model only needs to set the number of topics in the document during training for different corpora. The training results of the LDA topic model are intuitive and convenient for analysis and observation. The topic distribution is uniform, so it is adopted by many natural language processing tasks, especially for text classification.

While the LDA topic model has shown promise in unsupervised learning of text, it is not well-suited for short text classification tasks. The short text classification topic model method, bayesian latent dirichlet allocation (BLDA), has emerged to address the challenges of sparse short text semantics and few text words. The BLDA model assumes that all documents follow the same topic distribution, fully considering the problem of less text semantic information in short text classification tasks. The results are better than the traditional LDA topic model. However, BLDA only considers word co-occurrence and does not consider the relationship between words or the relationship between positions. This limitation underscores the need for further research and innovation in the area of short text classification.

While the methods discussed perform well on specific classification tasks, they also face significant challenges. Among these, the acquisition of text semantic information has become a major bottleneck affecting the accuracy of text classification. The text representation implemented by traditional machine learning methods has a poor ability to express the context of features and has sparsity problems. Decision trees use a tree structure to make predictions based on certain rules. Support vector machines can find the maximum separation hyperplane to achieve node classification. These are all feature engineering methods that cannot extract the context-semantic information of the text. These challenges underscore the urgency and importance of the topic of text classification.

15.1.3 Text Classification Based on Deep Learning

Deep learning methods are widely used in the field of text classification. Text classification algorithms related to deep neural networks are emerging one after another, such as the classification model proposed by Zeng and others [11]. Deep learning methods avoid manual design rules and functions. The convolutional neural network TEXTCNN has achieved good results in text classification. By fixing the field of view of the convolution kernel and using convolution and pooling to automatically extract features from the text, saving a lot of manual work, we can obtain the information that best represents the text, achieve text feature extraction, and then classify. Although the convolutional neural network performs well in local feature extraction, it cannot model longer temporal information, and the convolution operation requires a lot of calculations. FastText is also a commonly used text classifier in the field of natural language processing. It generates word vector representations through training and directly inputs the obtained word vectors, or their combinations, into the classification algorithm for calculation and prediction by the model. However, these models focus more on feature extraction and do not consider the context and semantic information of the text, ignoring the importance of context information in text classification.

The neural variational inference network model based on the neural variational document model (NVDM) introduces a multilayer perceptron and Softmax function and applies the topic model to the neural network. The NVDM is an unsupervised generative model capable of extracting continuous data, divided into variational encoding and decoding processes to capture the latent semantics of documents. The model consists of two main components: the variational encoding process, represented by the inference network, which uses a multilayer perceptron to implement the inference process, and the decoding process, achieved through the Softmax function. While this model can achieve good results in classification tasks, it still has deficiencies in short text classification and deep semantic information acquisition.

Contextual semantic information dramatically impacts the accuracy of text classification tasks. Recurrent neural networks are models that learn the temporal sequence information of text through a memory mechanism. They introduce self-connection and interconnection mechanisms in the hidden layer to learn the context information of the text. They excel at extracting semantic information from the text, thus becoming a commonly used classification method in deep learning. However, early recurrent neural network models used continuous multiplication in the gradient update process. If the eigenvalues are particularly small, it can easily lead to the gradient disappearing during propagation, and if the eigenvalues are too large, it can easily lead to an explosive increase in the gradient during propagation. LSTM networks can handle these types of problems well. In addition, other recurrent networks, such as bidirectional recurrent networks (BILSTM, RCNN, distributed semantic network models, etc.), have all extracted text context semantics. Although these methods can obtain context semantic information, they all focus on acquiring shallow semantics of text and are not good at mining its deep potential semantic information.

The Dense Convolutional Network (DenseNet) proposal provides a good solution for acquiring deep semantic information from text. DenseNet solves the problem of gradient disappearance in the propagation process of deep networks. Information is passed between different network layers in a feed-forward manner. The output of all previous layers serves as the input of the current layer, applying the information flow directly to subsequent layers to maximize information reuse. Some researchers have applied recurrent neural networks to dense connection structures, achieving dense connection recurrent neural networks, and have achieved good results in experimental corpora. This fully proves that mining a text's context semantic information can help enhance text representation ability and improve classification accuracy.

The above classification methods apply to semantically rich texts and perform slightly worse on short text classification tasks. Short texts have problems with short length, sparse semantics, and weak description information. Text semantic information can significantly improve text classification tasks' accuracy and recall rate. It is crucial for short text potential semantic mining and deep semantic information acquisition. However, deep learning methods (such as convolutional neural network models) use convolution kernels to extract features from the text, achieving local feature extraction. They are not good at mining the potential semantic features of short texts. The key to applying deep learning to solve large-scale text classification problems is effectively handling text representation problems. The extraction of word features only targets the local feature representation of the text. Mining the potential semantic representation of short texts to construct global feature representations is currently the most important issue in short text classification research.

In recent years, graph neural networks (GNNs) have been widely used in processing non-Euclidean data. Graph neural networks were first applied in the field of computer vision. Relying on the powerful expression ability of graphs, they model sets and mine the relationships between nodes in the graph. The excellent ability of graph neural networks to handle unstructured data and their high interpretability have made new breakthroughs in network data analysis, recommendation systems, physical modeling, natural language processing, and graph combination optimization problems. The text classification method that applies convolutional neural networks to graph structures—graph convolutional network (GCN)—uses the expressive power of graphs [1] and the feature extraction ability of convolutional neural networks to achieve text classification. GCN is a method that learns the word vector representation of vertices in a graph by combining topological structure and node attribute information. However, due to their model characteristics, graph convolutional neural networks (GCN) require learning the word vector representation of nodes in a specific graph. They cannot directly generalize to nodes that do not appear during the training process. GCN belongs to transductive learning and cannot be extended to large graphs, nor can it achieve incremental text classification. To solve this problem, Hamilton et al. [2] proposed a method to inductively generate the vector representation of unknown nodes using the attribute information of the nodes in the GraphSage model.

This is a graph-based inductive representation learning method. Its core idea is to learn the neighbor nodes of the target vertex, aggregate the information of the neighbor nodes, and generate the representation of the target node. Neural networks based on graph topic models, such as GraphBTM [3], can also achieve high classification accuracy in unsupervised text classification tasks. GraphBTM, based on the BTM topic model idea, proposes a neural topic model with a graph neural network as the encoder. It first obtains the co-occurring word pairs in the dataset, inputs the co-occurring word pairs in the text into the GCN for encoding, and uses the variational autoencoder to output the topic distribution of the entire corpus. This model can also build a decoder to reconstruct the input items. Although this model can enhance the ability to handle the sparsity problem of short text semantics, it cannot generate the topic distribution of a single document.

The classification model of text-level graph structure network takes the words in each text as nodes, builds a graph structure for each text in the corpus, realizes information sharing through global parameter sharing, reduces performance overhead, enhances the expression ability of words in the text, and minimizes the dependence of each text on the entire corpus. However, this model will ignore the difference in length between texts, does not consider the order of words in the text when obtaining text representation, and directly connects the word vector representation of the text. Zhang et al. [4] proposed a sampling method that obtains global word co-occurrence information by constructing heterogeneous text graphs. The nodes in the graph include word nodes and document nodes. The graph has two types of edges: edges between words and edges between words and documents. The output subgraph representation is encoded through heterogeneous encoding and Weisfeiler–Lehman isomorphic encoding, and the aggregation information of the subgraph is output through the Transformer. Due to their powerful expressive ability, graph neural networks can achieve high classification accuracy in natural language processing text classification tasks, especially in obtaining global text semantics. Dense structures can maximize information transmission, helping to obtain deeper levels of text context semantics.

15.2 Text Classification Algorithm

15.2.1 Dense Connection Network

A series of efficient convolutional neural network models (such as GoogLeNet) can be used to increase the depth and width of the network. They use deep networks to enhance the expressive power of features and use VGGNet to improve model performance with deeper network layers and introduce residual network (ResNet).

Residual Connections: These methods have achieved good results in classification tasks. HighWayNet provides an effective training method for end-to-end networks with over 100 layers, aiming to solve the problem of training deep networks. The basic idea of this model is that when using the information transmission mechanism between gated units, the high-speed network can be easily optimized, where the transmitted network path is a key factor in simplifying the training of these deep network structures. Based on the high-speed network, the residual network (ResNet) structure was proposed, which solves the problem of gradient vanishing in the propagation process of deep networks by allowing the network to learn residuals, and it performs well in many fields, such as image recognition. The ResNet structure inputs information directly to the next layer. With the multi-layering of convolutional neural network structures, the gradient vanishing problem is alleviated. However, the identity function in the ResNet structure and the combination of the output of each layer by summation hinder the transmission of information in the network.

DenseNet, or Densely Connected Network, is a revolutionary dense convolutional neural network structure proposed based on the ResNet structure. It not only effectively avoids the phenomenon of gradient vanishing during propagation but also enhances the feature extraction ability. This dense network structure, proposed based on the CNN model, passes information to the next layer in a feed-forward manner, retaining all information from the feed-forward layers as input information for the current layer. By directly passing the information flow to each subsequent layer, DenseNet achieves dense connections, thereby avoiding the problem of gradient vanishing during propagation and achieving information reuse.

Before the advent of DenseNet, there were already some deep learning models based on CNN. However, DenseNet stands out by retaining the output of all previous layers, concatenating the information, and directly passing it to the next layer, thereby achieving dense connections. It can obtain denser network connections than ResNet. DenseNet uses the feature channel propagation mechanism to maximize feature reuse on the channel, reducing the overhead performance of the model. Compared with ResNet, it can achieve better performance with fewer parameters, a testament to its efficiency and effectiveness.

DenseNet is implemented by a dense connection mechanism, which includes two parts: the dense block, which defines the connection method between input information and output information, and the transition layer, which is used to control the number of channels. The structure of DenseNet is shown in Fig. 15.1, which shows a five-layer dense block with a growth rate of four. The main idea of this model is feature reuse; there is no need to learn redundant features, so fewer parameters are needed repeatedly.

For a DenseNet with a network depth of L, the number of connections required is k:

$$k = \frac{L(L+1)}{2} \tag{15.1}$$

DenseNet retains the output of all previous layers, concatenates the information, and directly passes it to the subsequent layers. The input of the last layer, x_l, is represented as

$$x_l = H_l\left(\left[\, x_0, x_1, x_2, \ldots, x_{l-1} \,\right]\right) \tag{15.2}$$

Here, this term represents the nonlinear transformation function, and multiple convolutional layers may exist between different network layers.

DenseNet uses a transition layer structure, connecting dense blocks through transition blocks. Its function is to adjust the size of the feature map. Figure 15.2 shows a dense network with four dense blocks. The transition layer structure exists between two dense blocks, and each transition layer is composed of a convolution or pooling layer.

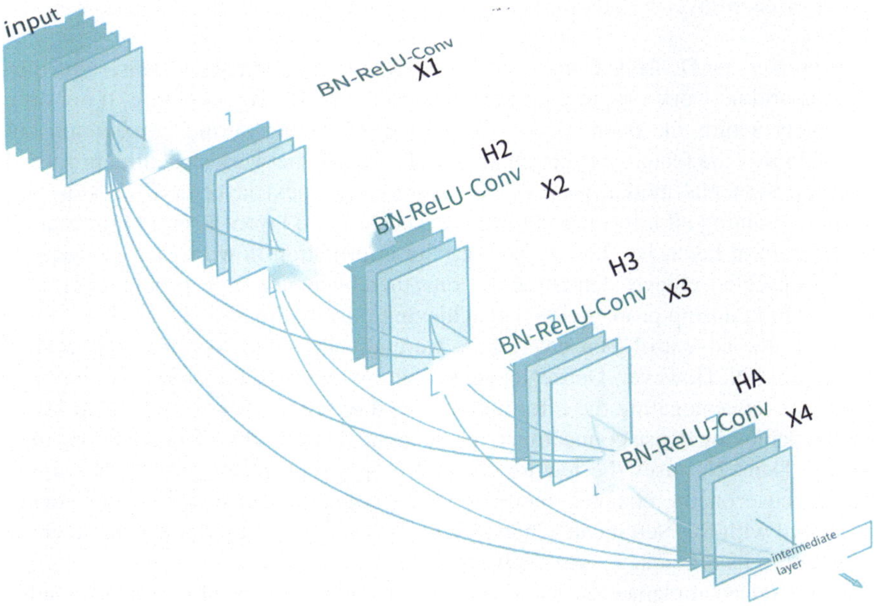

Fig. 15.1 Structure of DenseNet

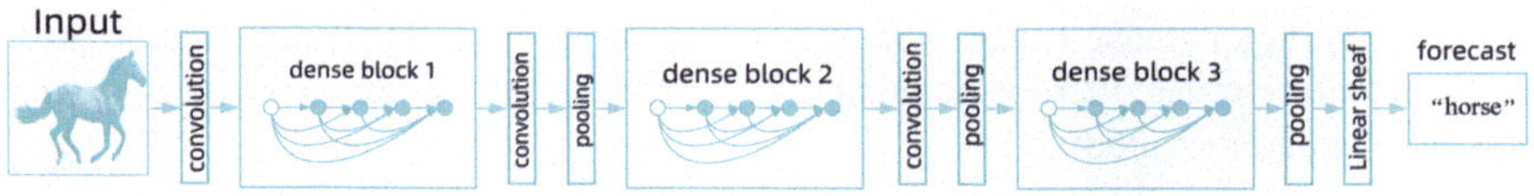

Fig. 15.2 DenseNet with four dense blocks

The dense connection method not only maximizes feature reusability but also enhances the ability of gradient backpropagation, reducing the difficulty of network training. Maximum feature reuse is achieved by retaining the output of all network layers and concatenating them as the input of the last layer. DenseNet directly connects features, making the network parameters smaller and computation more efficient. The network adopts a smaller growth rate to ensure that the feature maps of each network layer maintain a smaller size. DenseNet is widely used, and many networks based on DenseNet structures have been proposed, all of which are based on dense connection structures and have good performance on their respective experimental datasets.

15.2.2 Graph Neural Networks

Graph neural networks were first proposed in 2009 and have achieved good results in dealing with non-Euclidean structured data. Existing deep learning methods can obtain effective representations of these data when dealing with non-graph-structured data. However, the development of network information has led to an increasing amount of data being represented in graph structures. For example, in the Internet era, much of the social network data exist in graphs, such as links between web pages, network subsets, etc. Learning systems based on graphs can mine the relationships between network subsets and improve the performance of search engines. In image processing, special pixels in the image are modeled as graph structures, so it is necessary to learn the modeled graph structures. Before the proposal of graph neural networks, deep learning methods mainly dealt with Euclidean data, which is unsuitable for dealing with irregular graph structure data. In the graph network structure shown in Fig. 15.3, the number of neighbor nodes of a node is uncertain, and the number of neighbor nodes of different nodes may not be the same, which leads to operations similar to convolution that cannot be directly applied to graph data. Therefore, traditional neural networks perform poorly when dealing with non-Euclidean data and cannot even be applied to non-Euclidean data processing. The proposal of graph neural networks is mainly to solve these

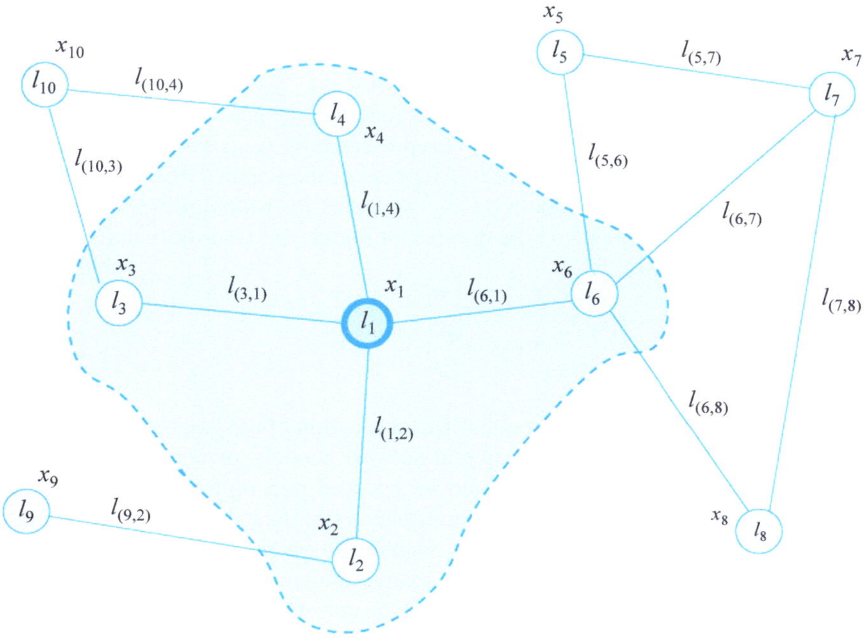

Fig. 15.3 Example of graph network structure

problems. The graph structure has also been proven suitable for mining the structural information and association relationships between data and has made breakthrough progress on Euclidean data.

The graph structure mainly includes nodes and edges, with nodes being associated with edges. Each node in the graph is defined by its own features and the nodes' related features. Graph embedding aims to obtain a low-dimensional vector representation of nodes, mine the association between nodes in the graph, and obtain potential context-semantic information. In the implementation process, the network's topological structure and the nodes' features are preserved. Relevant information, which can utilize the association relationship of nodes in the graph to learn node representation in text classification tasks, is already used in the application of node classification using graph neural networks.

Figure 15.3 shows a graph network structure G with some marked nodes, which are learned and trained, and then the labels of the nodes in the test set are predicted based on the training results. As shown in Fig. 15.3, when dealing with classification problems, the feature of node n is represented as l_{n}, the state of the node is represented as x_{n}, and $l(n,m)$ represents the state of the edge between the node and its neighboring nodes, and each node is associated with a real valid label y_{n}.

The new state of the nth node and the final output of the node are, respectively, represented as:

$$x_{\mathrm{n}} = f_{\mathrm{w}}\left(l_{\mathrm{n}}, l_{\mathrm{co}[n]}, x_{\mathrm{ne}[n]}, l_{\mathrm{ne}[n]}\right) \tag{15.3}$$

$$o_{\mathrm{n}} = g_{\mathrm{w}}\left(x_{\mathrm{n}}, l_{\mathrm{n}}\right) \tag{15.4}$$

where f_{w} is a parameterized local transition function, g_{w} is a label prediction network, $l_{\mathrm{co}[n]}$The features connected to n are represented by $x_{\mathrm{ne}[n]}$, which represents the state of the nodes adjacent to n, and $l_{\mathrm{ne}[n]}$ represents the features of the nodes adjacent to n. Here, f_{w} and g_{w} can both be interpreted as feed-forward fully connected neural networks. Assuming p is the number of nodes, the L1 loss function can be directly expressed as:

$$\text{Loss} = \sum_{i=1}^{p}\left(y_i - o_i\right) \tag{15.5}$$

The process of node classification, a crucial function of the graph neural network structure, is outlined above. Graph neural network models, pivotal in this context, can be categorized into methods based on spectral decomposition and methods based on the spatial domain. These methods play a significant role in feature extraction and dimension reduction. Spectral decomposition methods primarily utilize spectral decomposition operations, while spatial domain methods rely on aggregation operations to obtain the representation of the current node by aggregating information from neighboring nodes in space. The field of graph neural networks is further delineated into graph convolutional networks, graph-structured recurrent networks, and spatiotemporal graph neural networks. The following section will delve into the graph convolutional network model as an illustrative example.

Graph convolutional network, a key component in the field, is a general term for models that apply convolutional neural networks to graphs. Their primary function is to aggregate the features of the node itself and the features of neighboring nodes to generate node representation information. In building a cyclic dependency structure, graph convolutional networks, unlike graph recurrent neural networks that iterate node states, use a fixed number of layers at each layer, each with different weights. In terms of feature extraction implementation, graph convolutional networks can be divided into spectral graph convolutional networks and spatiotemporal domain graph convolutional networks.

Spectral graph convolutional networks, which include GCN [5], ChebNet [6], CayleyNet [7], AGCN [8], and DualGCN [9], have their own set of limitations. The essence of graph convolution, which they employ, is to aggregate signals on the graph through a series of designed filters. Spectral graph convolutional networks are based on spectral decomposition, a method that was used early in the field of graph signal processing. The spectral decomposition of a graph utilizes the properties of the Laplacian matrix, that is, it is a semi-positive, definite, symmetric matrix. Based on the convolution theorem and this Laplacian matrix property, the graph's Fourier transform is constructed. However, spectral graph convolutional networks have their limitations, mainly including any changes in the graph that will lead to changes in the feature base, the learned filters are domain-related, which also means that spectral graph convolutional networks cannot be applied to graphs with different structures, they require a large amount of computation, and the complexity is extremely high.

Spatial domain graph convolutional networks are based on spatial methods and have developed rapidly in recent years. Compared with spectral graph convolutional networks, spatial domain graph convolutional networks are more efficient, more flexible, and more universal. The success of GCN has drawn much attention to methods based on the spatial domain, realizing the transition from spectral decomposition methods to the spatial domain. Typical representatives of graph convolutional networks based on the spatial domain include GraphSage [10], GAT [11], and FastGCN [12]. Similar to traditional convolutional neural networks, methods based on the spatial domain define graph convolution according to the spatial relationship of nodes, where each central node is directly connected to its neighboring nodes, and the representation of the central node and neighboring nodes is derived and updated based on their representations. The core idea of spatial domain graph convolutional networks is to use the association of nodes in the graph structure for information transmission.

GraphSage is an inductive model. Compared to existing graph neural networks, the GraphSage network model has obvious advantages in classification accuracy and graph processing.

The following section will briefly explain how to graph neural networks, taking the GraphSage network model as an example, use the relationships between nodes to obtain low-dimensional vector representations of nodes. The low-dimensional vector representation of nodes applies to various prediction tasks. Although many methods based on graph neural networks have obtained low-dimensional vector representations of nodes, these methods generally directly focus on the nodes in the graph, representing inherent nodes, and cannot be extended to unseen nodes. The GraphSage network model can solve the above problems. It is a general inductive framework that aggregates the feature information of a node's neighboring nodes. Figure 15.4 shows the sampling and aggregation methods of GraphSage.

As shown in Fig. 15.4, first, each node's neighboring nodes in the graph are randomly sampled to reduce computational complexity, then, the aggregation function is used to aggregate the feature information contained in the neighboring nodes, first aggregating the vector representations of the few neighboring nodes closest to the current node, then using the aggregated vector representation to continue aggregating, generating the vector representation of the target node and obtaining the feature information contained in the neighboring nodes in this way. Finally, the low-dimensional vector representation of the node is input into the downstream task.

It predicts the label of the target node. The GraphSage network model provides three types of aggregation functions: LSTM aggregation, recurrent network aggregation, and average and pooling aggregation. The excellent performance of the GraphSage network model in various prediction tasks also fully demonstrates the powerful expressive ability of the graph neural network.

The graph attention (GAT) network applies the attention mechanism to the graph structure network. Like other convolutional networks, the GAT model also aggregates the representation information of neighboring vertices through aggregation and applies the results to the central node to learn the feature representation of new nodes in this way. The difference between the GAT network and GCN is that GCN uses the Laplacian matrix, while the GAT network uses attention coefficients. When aggregating the features of nodes, the GAT network uses attention coefficients to obtain the correlation between nodes and then aggregates the features of the nodes to the central node.

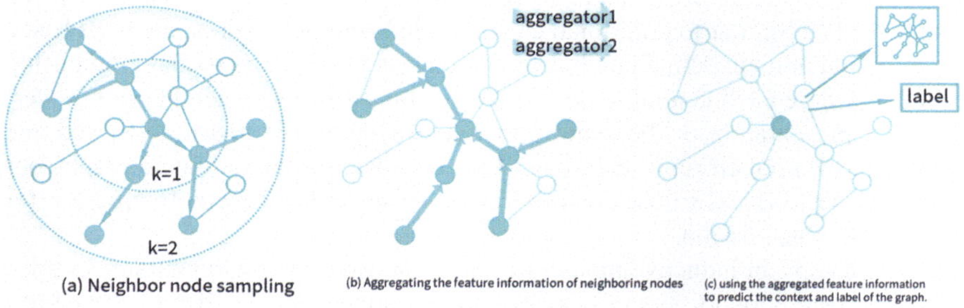

Fig. 15.4 Sampling and aggregation methods of GraphSage

15.2.3 Attention Model

The attention model, initially used to solve problems such as machine translation, has quickly found applications in other tasks. This model, which simulates the way people pay attention to things, is not only easy to understand but also has a certain interpretability. Its outstanding performance in various tasks and its immediate recognition underscore its significant practical value. The attention mechanism, a key component of the model, has not only been widely used in the field of natural language processing but has also found applications in multiple fields such as speech recognition and computer vision. It has become an important concept in neural networks. The attention mechanism's adaptability to different neural network structures demonstrates its superiority and interpretability in a multitude of tasks.

The main reasons for the attention model's high attention and rapid development are as follows:

1. The attention model can be used for various tasks, such as machine translation, question-answering systems, sentiment analysis, partial speech tagging, dialogue systems, etc., and has achieved good results in multiple tasks.
2. The attention model can improve the interpretability of the model. One of the bottlenecks in deep learning model research has always been uninterpretable, and neural networks are considered black box models. People are paying increasing attention to the transparency and interpretability of network models. Because the attention model has a certain interpretability, it has developed rapidly.
3. The attention model helps to deal with some of the drawbacks of recurrent neural networks, such as the problem of model performance degradation when the length of the input time series increases. The problem of low computational efficiency is caused by input order processing.

The attention model draws on how people pay attention to things, selectively focusing on some content, assigning higher weights to the content of attention, lower weights to the content not paid attention to, and even ignoring irrelevant content. This method helps the system improve its ability to discern meaningful information. The attention mechanism aids in identifying important information in classification tasks; for example, it focuses on key feature words in the input text sequence while disregarding other less important words. Additionally, the attention mechanism can address issues of excessive complexity and reduce performance overhead.

The traditional encoder–decoder network has the following problems:

1. The traditional encoder converts all the input information into a fixed-length vector and passes it to the decoder for output. Compressing the input sequence with a fixed-length vector has an apparent problem: if the input information is too long, a lot of information will be lost, resulting in an inaccurate representation of the input. This information loss phenomenon is especially obvious in long-distance dependency tasks.

2. The traditional encoder cannot model the alignment between input and output information, an important aspect of structured output tasks.
3. Intuitively, in sequence-to-sequence tasks, people expect the output word to be greatly influenced by some part of the input sequence. However, the traditional decoder cannot meet people's expectations well because, when getting the output representation of each word, the traditional decoder does not selectively pay attention to the related input words.

The introduction of the attention model significantly improved the above problems by allowing the decoder to access the entire encoded input sequence to minimize the loss of information on the sequence. Calculate the weight of the input sequence, the higher the weight, the higher the attention. First, consider the position set of related information and generate the next output word embedding. An additional feed-forward neural network is introduced in the attention architecture to learn special attention weights. The network model based on the attention mechanism has achieved good performance in its respective fields. The Transformer model is introduced in detail, as shown in Fig. 15.5.

Like most sequence models, the Transformer model structure includes two parts:

1. Encoder, which encodes the input information.
2. Decoder, which decodes and outputs the encoded information.

Each identical layer contains a fully connected feed-forward neural network layer, another network sub-layer is a multi-head attention mechanism layer. The output representation of each subnetwork layer is as follows:

$$\mathrm{LN}\left(x+\mathrm{Sub}\left(x\right)\right) \tag{15.6}$$

where LN represents layer normalization and Sub represents the function of the subnetwork layer.

Let Q be the query vector, and K and V be the key-value pair vectors. The attention function obtains the output vector representation based on the mapping of the query and the key-value pair. Querying the weight of each value with the corresponding key, the attention function is represented as

$$\mathrm{Attention}\left(Q,K,V\right)=\mathrm{softmax}\left(\frac{QK^T}{\sqrt{d_k}}\right)V \tag{15.7}$$

where d_k represents the dimension of the Q vector and the K vector. In order to stabilize the gradient, d_k is used in the model. The results of the calculation are normalized, and the processed output is used as the input to the softmax function, which calculates the weight of each input vector.

Use n linear transformations to map queries and key-value pairs, and finally connect different results, where the mapping is implemented through parameter matrices $\boldsymbol{W}_i^Q$, $\boldsymbol{W}_i^K$, $\boldsymbol{W}_i^V$, $\boldsymbol{W}^O$:

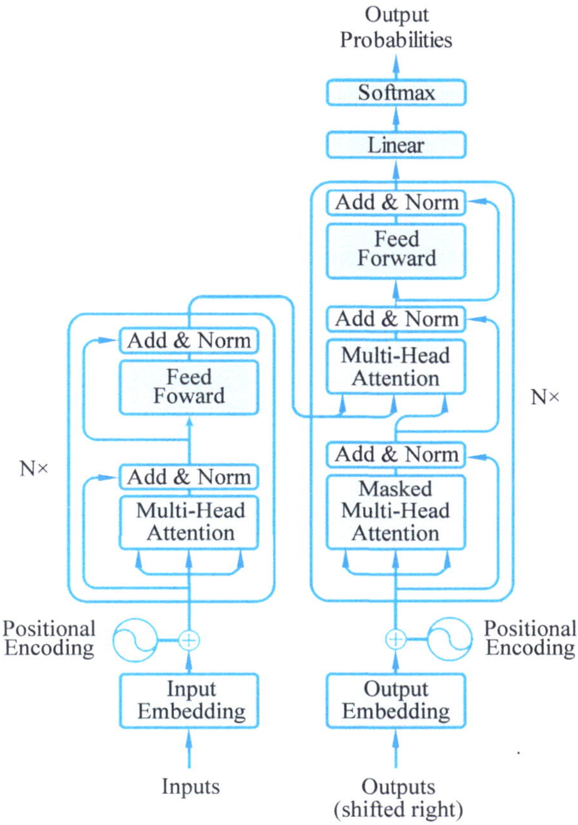

Fig. 15.5 Transformer model structure

$$\text{MultiHead}(Q,K,V) = \text{Concat}(\text{head}_1,\ldots,\text{head}_n)\boldsymbol{W}^{O} \tag{15.8}$$

$$\text{head}_i = \text{Attention}\left(QW_i^{Q}, KW_i^{K}, VW_i^{V}\right) \tag{15.9}$$

The Transformer model not only includes encoder and decoder networks, but another major highlight and breakthrough is the abandonment of the recurrent neural network structure, using position encoding instead of the recurrent neural network that can model data on a time series. Here is a method for encoding positions: by summing the vector representation data and the encoded data, introduce relative position. The model uses the following formula to directly calculate the position encoding:

$$\text{PE}_{(\text{pos},2i)} = \sin\left(\text{pos}/10000^{2i/d_{\text{model}}}\right) \tag{15.10}$$

$$PE_{(pos,2i+1)} = \cos\left(pos / 10000^{2i/d_{model}}\right) \tag{15.11}$$

where Eq. (15.10) is a sine function, Eq. (15.11) is a cosine function, the two functions are of different frequencies. Pos represents the embedding position of each word in the text, i represents the dimension, $2i$ and $2i + 1$ represent whether the position of each word's position vector is even or odd, and dmodel represents the output dimension.

The Transformer model abandons the recurrent neural network structure, uses position encoding to achieve sequential learning, and has achieved good results in various research tasks. Compared with previous deep learning methods, the Transformer model not only has a certain improvement in the results of the model but also has a great improvement in performance. At the same time, the application of the Transformer model is very wide, not only can it be applied in the field of machine translation, but it is also widely used in other directions, such as text classification.

15.3 Application and Analysis

15.3.1 Dataset

R8 is a subset of the Reuters news dataset, which consists of eight categories, namely, ship, money-fx, grain, acq, trade, earn, crude, and interest, including 5485 training data and 2189 test data.

15.3.2 Experiment

Tokenization and lemmatization are performed on the R8 dataset, followed by stop word removal and punctuation removal. A vocabulary is obtained from the cleaned samples. A 300-dimensional Glove [13] pretrained word vector is used as the vector representation of the word. The vector representation of the text is input into TextCNN for training. The model parameters are listed in Table 15.1.

The experimental results show that the accuracy of TextCNN on the R8 dataset is 87.76%.

Table 15.1 Model parameters

Parameters	Values
Dropout	0.5
Epoch	0
Batch size	128
Padding size	2
Learning rate	e–3
Word vector dimension	300
Convolution kernel size	$2 \times 3 \times 4$
Number of convolution kernels	256

References

1. Yao L, Mao CS, Luo Y. Graph Convolutional Networks for Text Classification[C]. Proceedings of the AAAI Conference on Artificial Intelligence, volume 33, 2019: 7370–7377.
2. Hamilton W L, Ying Z, Leskovec J, et al. Inductive Representation Learning on Large Graphs[C]. Neural Information Processing Systems, 2017: 1024–1034.
3. Long QQ, Jin Y L, Song GJ, et al. 2020. Graph Structural-topic Neural Network[C]. Proceedings of the 26th ACM SIGKDD Conference on Knowledge Discovery and Data Mining(KDD '20), 2020.
4. Zhang H, Zhang J: Text Graph Transformer for Document Classification[C]. Proceedings of the 2020 Conference on Empirical Methods in Natural Language Processing (EMNLP). 2020: 8322–8327.
5. Kipf T N, Welling M. Semi-supervised Classification with Graph Convolutional Networks[C] . Proc. of ICLR, 2017.
6. Defferrard M , Bresson X, Vandergheynst P. Convolutional Neural Networks on Graphs with Fast Localized Spectral Filtering[C] . Proc. Of NIPS, 2016.
7. Levier, Montif, Bresson X, et al. Cayleynets: Graph Convolutional Neural Networks with Complex Rational Spectral Filters[J] . IEEE Transactions on Signal Processing, 2017, 67(1) : 97–109.
8. Lir, Wang S, Zhu F, et al. Adaptive Graph Convolutional Neural Networks[C] . Proc. Of AAAI, 2018.
9. Zhuang C, Ma Q. Dual Graph Convolutional Networks for Graph-based Semi-supervised Classification[C] . WWW , 2018.
10. Hamilton W L, Ying Z, Leskovec J, et al. Inductive Representation Learning on Large Graphs[C] . Neural Information Processing Systems, 2017.
11. Velickovicp, Cucurull G, Casanova A, et al. Graph Attention Networks[C] . Proc. of ICLR, 2017.
12. Chen J, Ma T, Xiao C. Fastgcn: Fast Learning with Graph Convolutional Networks via Importance Sampling[C] . Proc. of ICLR, 2018.
13. Pennington J, Socher R, Manning C. Glove: Global Vectors for Word Representation[C]. Proceedings of the 2014 Conference on Empirical Methods in Natural Language Processing (EMNLP), 2014: 1532–1543.

Chapter 16
Text Clustering

This chapter provides an overview of text clustering, its process, key algorithms, semi-supervised methods, and research from the Beijing Institute of Technology's NLPIR lab. It covers five major clustering algorithms and their applications, with a focus on text similarity measurement methods like cosine similarity. The chapter also introduces a new method for detecting Top N hot topics using key feature clustering.

16.1 Overview of Text Clustering

Clustering is a very important concept in data mining. It means dividing a dataset into different categories, making the data in the same category as similar as possible and the data in different categories as dissimilar as possible. Text clustering, also known as document clustering, is based on the following clustering hypothesis: the similarity of documents in the same category is greater, and the similarity of documents in different categories is smaller.

However, unlike text classification with clear supervision information or classification standards, text clustering is an unsupervised, unguided clustering process. The data processing does not know in advance how many categories the clustering result will be. It does not need training samples in the clustering process, and several clusters are completely formed according to the distance of the samples or other calculation standards. The text data in a cluster has a larger similarity and has no or a smaller similarity with the text data in different clusters. The result of text clustering is often the natural division of data.

Text clustering has found extensive application in data mining, information retrieval, and topic detection. It has become an indispensable tool for effectively organizing, summarizing, and navigating text information. Its versatility and utility

H. Zhang, J. Shang, *Natural Language Processing and Applications*,
https://doi.org/10.1007/978-981-97-9739-4_16

have piqued the interest of a growing number of researchers, underscoring its relevance in the field.

So far, text clustering has formed a general process that first requires preprocessing of the text, then uses text representation methods to transform the text into feature vectors, and finally uses clustering algorithms for clustering.

Data preprocessing is a crucial step in text clustering. It involves taking the raw data as input and subjecting it to cleaning, integration, and reduction processes. The output is refined data that is ready for further operations, ensuring a smooth and efficient clustering process.

16.2 Text Clustering Algorithm System

There are currently many clustering algorithms, mainly divided into the following five categories: grid-based clustering algorithms, hierarchical clustering algorithms, partition-based clustering algorithms, density-based clustering algorithms, and model-based clustering algorithms.

16.2.1 Grid-Based Clustering Algorithm

The typical feature of the grid-based clustering algorithm is to transform the processing object from the original data point to the self-divided grid.

Unit. The main process of this type of algorithm is as follows: first, each dimension of the data space is divided into equal-length intervals on average, then the data space is divided into nonoverlapping grid units, and the points in the same grid unit belong to the same class. The possibility is relatively large, so the points that fall into the same grid unit are treated as one object, and all subsequent clustering operations are based on the grid unit. The clustering efficiency is greatly improved because the number of grid units is generally less than the number of original data points. However, the grid-based clustering algorithm also has disadvantages. This algorithm depends more on parameters, and the clustering accuracy decreases as the grid division increases.

16.2.2 Hierarchical Clustering Algorithm

The hierarchical clustering algorithm decomposes the given data set hierarchically until a certain condition is met. The data processor defines this condition and is similar to the threshold set in data analysis. For different data, this type of algorithm uses different similarity calculation algorithms. Hierarchical clustering algorithms can be distance-based, density-based, or connectivity-based, and some extensions of this type of algorithm also consider subspace clustering. In order to reduce computational costs, hierarchical clustering algorithms have a strict rule. Once a step (merging or splitting) is completed, it cannot be revoked, which is also the disadvantage of this type of algorithm.

Hierarchical clustering methods are divided into agglomerative and divisive types. Agglomerative hierarchical clustering is a bottom-up clustering method. At the beginning, each individual data point is a class, and then the two most similar classes are continuously merged to form larger and higher-level classes. The hierarchical clustering ends when it is incorporated into a final class containing all the data or reaches a certain termination condition. Divisive hierarchical clustering is the opposite. In the beginning, all data are considered the same class. Then, each class is split into smaller subclasses based on similarity, so the subclasses have better intra-cluster similarity after splitting. It recursively occurs until each class contains only one data point or reaches a certain termination condition to end the algorithm. Hierarchical clustering algorithms mainly include the BIRCH algorithm, the CURE algorithm, the CHAMELEON algorithm, etc.

16.2.3 Partition-Based Clustering Algorithm

The main content of the partition-based clustering algorithm includes the idea of iteration. This type of algorithm refers to a given dataset with N tuples or records. The partition method will construct k groups, each group represents a class, $k < N$. These K groups satisfy the following conditions:

(1) Each group contains at least one data record.
(2) Each data record belongs to and only belongs to one group.

For a given k, the algorithm first provides an initial grouping scheme. It then continuously improves the grouping situation by iterative methods while keeping the value of k unchanged so that the grouping scheme after each improvement is better than the previous one and finally forms k relatively stable classes. The partition-based clustering algorithm's standard for good is the higher the similarity of data in the same group, the better, and the lower the similarity of data in different groups.

The famous k-means algorithm belongs to the partition-based clustering algorithm. Here, k represents the number of classes in the clustering algorithm and means indicates that this algorithm is a mean algorithm. Clustering the text into k classes using the mean algorithm means clustering the text into k classes. Algorithms like k-medoids, k-modes, and k-medians also belong to the category of partition-based clustering algorithms.

16.2.4 Density-Based Clustering Algorithm

The typical feature of the density-based clustering algorithm is that it is not based on various distances but on density. Based on the distribution density of samples (usually defined by the number of sample points n in the area with radius ε), it determines whether the sample points are "density reachable," classifies the density reachable sample points into the same class, and finally gets the clustering results. A major feature of clustering using distance-based similarity calculation methods is that the data often shows a "circular" shape, and the density-based method breaks this limitation. The fundamental principle of this approach is that, as long as the density of points in a region exceeds a certain threshold, the points are grouped together, regardless of their shape or distribution.

Whatever shape it forms, add it to the class close to it. Its disadvantage is that the results are not good on datasets that do not conform to the Gaussian distribution, and it is difficult to improve by adjusting parameters. Density-based clustering algorithms mainly include DBSCAN, OPTICS, etc.

16.2.5 Model-Based Clustering Algorithm

The model-based clustering algorithm's main idea is first to set a framework or model for each class and then find the dataset that meets this model for filling. A potential assumption of this algorithm is that a series of probability distributions determine the dataset to be processed. Such a model may be the density distribution function of data points in space.

Among them, the basic idea of deep clustering based on the model is to integrate the powerful representation ability of deep learning into the clustering target and optimize the clustering effect through fine-tuning. Common pretrained methods use deep neural networks (such as autoencoders). The original high-dimensional data is mapped to a low-dimensional feature representation and then fine-tuned through KL divergence loss, k-means loss [1], subspace loss, and cross-entropy loss [2], etc., on the pretrained representation, making it more discriminative in the clustering process.

16.3 Semi-supervised Text Clustering

Traditional text clustering algorithms are an unsupervised learning method, the text being processed is unlabeled. However, in practical applications, sometimes a small amount of prior knowledge about the data can be obtained, including class labels, text partitioning constraints (such as pairwise constraint information), etc. However, these small amounts of prior data cannot form a precise classifier, so the text classification method cannot be used. How to use this only prior knowledge to cluster a large amount of text without prior knowledge has become a meaningful and challenging problem. Semi-supervised text clustering has been proposed to solve this problem and has received widespread attention in recent years. Semi-supervised text clustering is the study of using a small amount of data with prior knowledge to assist unsupervised text clustering. Semi-supervised text clustering algorithms can be roughly divided into two categories: one contains constraint-based semi-supervised text clustering algorithms, which use class label data or pairwise constraint information to improve the clustering algorithm itself, the other contains distance-based (metric-based or distance-based) semi-supervised text clustering algorithms.

In the process of natural language processing, the problem of how to calculate the similarity between two texts is often encountered. Text similarity measurement is widely used; it can preprocess large-scale text data corpora for deduplication, and it can also perform fuzzy matching, that is, find other names related to a certain entity name. When performing text clustering analysis, text similarity measurement is indispensable. The methods used for similarity measurement and distance calculation profoundly impact the clustering results, making it particularly important to choose a suitable measurement method for specific data.

1. The simplest word-based similarity calculation method is to directly use the number of common words in two texts divided by the longest text length to measure their text similarity. This method is very simple, but there are obvious deficiencies in the accuracy of the calculation results.
2. The Jaccard similarity coefficient is also a simple similarity calculation method that calculates the ratio of the intersection and union of two sets, A and B, in mathematics. Its calculation formula is as follows:

$$J(A,B) = \frac{|A \cap B|}{|A \cup B|} = \frac{|A \bigcap B|}{|A| + |B|} \tag{16.1}$$

When used for text similarity calculation, the larger the J-value, the more similar the two texts are. The Jaccard similarity coefficient is used to compare the similarity and dispersion of different sample sets. This method has certain efficiency advantages when it comes to large-scale parallel computing.

Below are three important similarity calculation methods.

16.3.1 Cosine Similarity

Cosine similarity is a similarity calculation method based on the cosine function in mathematical trigonometry. Suppose the coordinates of two-dimensional space vectors α and β are (x_1, y_1) and (x_2, y_2), respectively, then the cosine function calculation formula is as follows:

$$\cos\theta = \frac{x_1x_2 + y_1y_2}{L\sqrt{x_1^2 + y_1^2}x\sqrt{x_2^2 + y_2^2}} \tag{16.2}$$

When dealing with text data, there is usually more than one type of feature information, so the actual calculation will expand from two dimensions to multidimensional space. Suppose the coordinates of vectors α and β are $(x_1, x_2 \ldots x_n)$ and $(y_1, y_2 \ldots y_n)$ respectively, the corresponding cosine similarity expansion formula is as follows:

$$\cos\theta = \frac{\sum_{1=1}^{1}(x_i x y_i)}{\sqrt{\sum_{1=1}^{n} x_i^2} x \sqrt{\sum_{1=1}^{n} y_i^2}} \tag{16.3}$$

When calculating similarity, x and y represent the word frequency in two entries, if it appears, it is assigned a value of 1, if it does not appear, it is assigned a value of 0. The closer the calculated result is to 1, the higher the similarity. When using the cosine function to calculate text similarity for text data, vectors α and β represent the two texts to be compared. After segmenting the two texts, list all the words, or words after segmentation, and assign values to vectors α and β based on whether each word appears in each text, thus obtaining their similarity. The calculation method of cosine similarity is detailed and clear and is widely used in similarity calculation, but its efficiency is not high for large amounts of words. At the same time, the traditional cosine similarity method mainly considers whether the word appears and does not consider the relationship between words.

16.3.2 Edit Distance

Edit distance, also known as Levenshtein distance, refers to the minimum number of editing operations required to transform one word into another. The operations specified by the edit distance calculation method include inserting a character, deleting a character, and replacing one character with another. Generally speaking, the smaller the edit distance between two texts, the higher their similarity. Therefore, a text similarity based on edit distance can be defined, and its calculation formula is as follows:

$$\text{Sim} = 1 - \frac{ld}{\max(m,n)} \tag{16.4}$$

where Sim represents text similarity, *ld* is the edit distance between two texts, and *m* and *n* are the lengths of the two texts, respectively. The concept of edit distance is easy to understand and is more suitable for datasets that are highly similar and relatively small. However, the initial method of calculating the edit distance did not consider the order of words. For example, the phrases "Hello tomorrow" and "Tomorrow hello" have a low similarity when calculated using edit distance.

This does not conform to reality. Therefore, there are many improved algorithms for edit distance, among which the most famous one is the DL algorithm based on edit distance. It adds the operation of swapping positions between two adjacent characters, thereby improving the accuracy of the calculation results.

16.3.3 Jaro Distance

Jaro distance is actually a type of edit distance, and the formula for calculating Jaro distance is as follows:

$$d_j = \frac{1}{3}\left(\frac{m}{|s_1|} + \frac{m}{|s_2|} + \frac{m-t}{n}\right) \tag{16.5}$$

where m is the number of matching words, t is the number of transpositions. When the distance between two words from texts s_1 and s_2 does not exceed d as shown in formula (16.6), these two texts are considered to be a match. The value of t is half the number of words that match in different orders in the two texts.

$$d = \left[\frac{ma(|s_1|,|s_2|)}{2}\right] - 1 \tag{16.6}$$

As can be seen, the Jaro distance mainly considers the situation where the two texts have the same words in different positions. The commonly used Jaro-Winkler distance is an extension of the Jaro distance. The Jaro-Winkler distance calculation method gives higher weights to two texts that are the same at the beginning, and its formula is as follows:

$$d_w = d_j + \left(lp\left(l - d_j\right)\right) \tag{16.7}$$

where l is the length of the same initial part of the two texts, generally not exceeding 4, p is a newly defined constant, mainly used to adjust the score to ensure that the result obtained is not greater than 1. Winkler defined the constant p as 1. The Jaro-Winkler distance calculation method is an improvement on the original Jaro distance calculation method, making the measurement of similarity more accurate.

16.4 Research on Top N Hot Topic Detection Method Based on Key Feature Clustering

16.4.1 Research Overview

The Top N hotspot topic detection method based on key feature clustering is a highly efficient approach that aims to study an accurate hot topic discovery algorithm. This algorithm extracts key features from each document in a large-scale corpus, maps the key features of the document to the topic space, forms initial topics, generates subtopics through clustering of initial topics, and extracts key features of subtopics. It then cleans erroneous subtopics and merges similar subtopics based on the coverage of key features, resulting in the final Top N hot topics. The main content of this study is as follows:

1. Automatic extraction of key features from a single text. This stage mainly consists of two steps: The first step is preprocessing for key feature extraction. This step mainly involves text mining-related processing for a single document, including Chinese word segmentation, part-of-speech tagging, stop word removal, etc., and selecting words from specific parts of speech as key feature candidates. The second step is the extraction of key features from a single document. This step uses a document key feature extraction algorithm to extract key features from the document. This study uses three key feature extraction algorithms: the TF-IDF algorithm based on statistical information, the TextRank algorithm based on graph models, and the LDA algorithm based on hidden topic models. To improve the efficiency of the key feature extraction algorithm, this study proposes an equivalent path replacement method to optimize the TextRank algorithm.
2. Detection of Top N Hot Topics This stage mainly consists of three steps: The first step is subtopic clustering based on document key features. This step first deduplicates and integrates all extracted key features and establishes a corresponding relationship with the document. Then, the key features are mapped to the topic space to establish an initial topic space. Again, the initial topic space is dimensionally reduced to reduce the time required for subsequent subtopic clustering. Finally, hierarchical clustering is used to cluster the topics in the initial topic space to form subtopics. The second step is the extraction of key features from subtopics. This step uses meaningful strings and named entities as the key features of subtopics and proposes a subtopic key feature extraction algorithm based on transition probability bidirectional matching. This method can quickly and effectively extract key features from subtopics. The third step is to obtain the top N hot topics. Based on the previous two steps, this step uses the coverage of subtopic key features to clean up erroneous subtopics and clusters similar subtopics based on subtopic key features, finally obtaining the Top N hot topics. This section mainly elaborates on the first step of hot topic detection.

16.4.2 Topic Clustering Based on Document Key Features

16.4.2.1 Topic Definition

In this study, for convenience of research, a topic is extendedly defined as an event, and the events or activities directly related to it are usually represented by H-key descriptors.

A set of related reports centered on this event are formalized as

$$t = (\text{event}, \text{document}). \tag{16.8}$$

For example, the main events of topic A are {employee benefits, total wages, system reform, communication subsidies, collective welfare}, and the related reports are shown in Table 16.1.

This topic can also be expressed as:

$$t = (\text{event}, \text{documents}):$$

$$event = \{\text{职工福利，工资总额制度改革，通讯度改革，集体福利}\}$$

$$\text{document} = \{1,2,3,4,5,6,7,8,9,10\}.$$

In this topic, $H = 5$.

Table 16.1 Related reports of topic A

Document number	Document name
1	The person in charge of the Ministry of Finance answers questions about strengthening the financial management of employee welfare expenses.txt
2	Ministry of Finance: Enterprises shall not purchase and build houses for employees, or bear gift expenses.txt
3	The Ministry of Finance issued a document to standardize the abnormal high welfare of employees in monopoly enterprises.txt
4	Welfare subsidies are included in the total wages, taxpayers' reactions vary.txt
5	The Ministry of Finance answers questions about employee welfare management, clarifying the scope of welfare expenses.txt
6	Subsidies are taxed as part of the total wages; the impact is smallest on low- and middle-income earners.txt
7	Ministry of Finance: Housing and car subsidies and other welfare expenses are included in the total wages for taxation.txt
8	The regulations on corporate subsidies are not tax policies.txt
9	Standardizing welfare is beneficial to adjust income distribution.txt
10	Ministry of Finance: Employees of state-owned enterprises such as telecommunications and electricity should not enjoy internal prices.txt

16.4.2.2 Topic Space

After extracting key features from a single document, a document key feature set $K_i = \{k_{i,1}, k_{i,2}, \cdots, k_{i,m}\}$ can be obtained for each document, where M (this parameter can significantly affect the clustering accuracy of the topic) is the number of features.

$$K_i = f(d_i), d_i \in D \tag{16.9}$$

Similarly, we can get,

$$K = f(D) \tag{16.10}$$

where $K = \{K_1, K_2, K_3, \ldots, K_i\}$, $D = \{d_1, d_2, d_3, \ldots, d_i\}$, $K_i = \{k_{i,1}, k_{i,2}, k_{i,3}, \ldots, k_{i,m}\}$. When extracting key features from documents, the same key feature may be extracted from multiple documents. Therefore, K can be integrated and duplicated to obtain a complete key feature set$' = \{k_1, k_2, k_3, \ldots, k_m\}$, that is, there exists an integrated deduplication filter mapping:

$$K = g(k) = g(f(D)) \tag{16.11}$$

where $D = \{d_1, d_2, d_3, \ldots, d_i\}$. Therefore, there must exist a many-to-many mapping F,

$$F = g(f). \tag{16.12}$$

$$\text{So that}: \{K_1, K_2, K_3, \ldots, K_i\} = F(\{d_1, d_2, d_3, \ldots, d_i\}). \tag{16.13}$$

Figure 16.1 is a many-to-many mapping of key features and document sets, which satisfies Eq. (16.13).

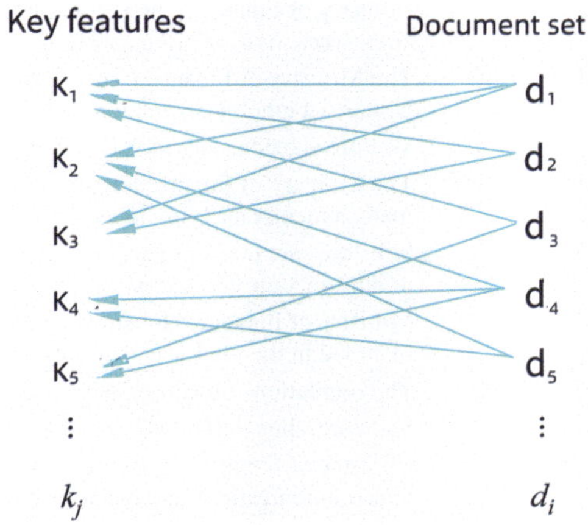

Fig. 16.1 Many-to-many mapping of key features and document set

Table 16.2 Correspondence between key features and documents

Key feature	*Correspondence with the document*
K_1	k_1->$\{d_1, d_2, d_3\}$
K_2	k_2->$\{d_1, d_4, d_5\}$
K_3	k_3->$\{d_1, d_2\}$
	k_4->$\{d_4, d_5\}$
K_5	k_5->$\{d_3, d_4\}$
K_i	k_i->$\{\ldots, d_i, \ldots\}$

Table 16.3 Mapping of initial topics, key features, and documents

Topic	Event	Documents
t_1	$\{k_1\}$	$\{d_1, d_2, d_3\}$
t_2	$\{k_2\}$	$\{d_1, d_4, d_5\}$
t_3	$\{k_3\}$	$\{d_1, d_2,\}$
t_4	$\{k_4\}$	$\{d_4, d_5\}$
t_5	$\{k_5\}$	$\{d_3, d_4\}$
...	...	...
t_j	$\{k_j\}$	$\{\ldots, dj\ldots\}$

At the same time, there must exist a corresponding relationship as shown in Table 16.2.

To proceed with the next step of topic clustering, it is necessary to map the key features to the topic space.

Each key feature can be mapped to an initial topic, the event field of which is the key feature, and the document field of which is the corresponding document.

According to the correspondence between key features and documents in Table 16.3, the mapping of initial topics, key features, and documents can be obtained.

Therefore, the initial topic space can be constructed

$$T = \{t_1, t_2, \ldots, t_j\} \tag{16.14}$$

Since the number of key features may be very large, the corresponding initial topic space will have a high dimension, so it is necessary to reduce its dimension.

This study mainly focuses on how to discover the top N topics, that is, the first N largest topics. In the initial topic space, some initial topics only correspond to a few documents, indicating that only a few documents are distributed on these topics; hence, these topics are sparse. Since the subsequent sub-topic clustering is based on the initial topics, sparse topics cannot aggregate into large topics and cannot appear in the Top N topics; therefore, the topic space can be reduced by removing sparse topics. In this study, a simple method is used to remove sparse topics, that is, all topics in the initial topic space are sorted in descending order according to the number of topic-related documents, and topics that can cover 80% of the document set are retained, thereby reducing the dimension of the topic space.

16.4.2.3 Topic Clustering

In order to perform topic clustering, define topic similarity:

$$\text{Similarity}\left(t_i,t_j\right)=\frac{\text{Count}\left(c\right)}{\text{Count}\left(t_i.\text{documents}\cup t_j.\text{documents}\right)} \tag{16.15}$$

For t_i, t_j, when Similarity(t_i, t_j) is greater than the predetermined topic similarity (Threshold, generally 0.6), the two topics are merged into one topic. That is:

$$\text{Similarity}\left(t_i,t_j\right)>\text{TopicThreshold} \tag{16.16}$$

t_i and t_j can be merged into a new topic t_{ij}, that is:

$$t_{ij}.\text{event}=t_i.\text{event}\cup t_j.\text{event} \tag{16.17}$$

$$t_{ij}.\text{documents}=t_i.\text{documents}\cap t_j.\text{documents} \tag{16.18}$$

$$t_{ij}=\left(t_{ij}.\text{event},t_{ij}.\text{documents}\right). \tag{16.19}$$

When performing topic clustering based on document key features, a hierarchical clustering algorithm is mainly used. The basic idea is: to calculate the similarity between each pair of topics in the initial topics, determine whether two topics are similar, if they are similar, merge them into one topic. The topic clustering algorithm Topic Clustering (T) is as follows:

```
TopicClustering(T)

Input: Originaltopics
Output: Subtopics

flag=false
    do:

        T'= ∅
        for ti in T:

            fortj inT:

                iftі≠tj andSimilarity(ti ,tj )≥TopicThreshold:

                    tij =combine(ti ,tj )
                    T'.add(tij )

                    flag=true

                        T=T'

            while flag
            return T
```

16.4.3 Experimental Results Display

This study has conducted a comparative experiment on the topic detection algorithm proposed (hereinafter referred to as the algorithm in this section) and the traditional topic detection algorithm. The traditional topic detection algorithm uses words as features to represent documents, employs the *k*-means clustering algorithm for topic detection, and sorts the clustering results based on the number of topic documents to produce hot topics. In the algorithm of this section, the *M* value, a significant factor in our experiment, is selected as 10. The single document key feature extraction algorithm uses the TextRank algorithm based on equivalent path replacement. The comparative experiment includes two parts: Experiment 1 compares the accuracy of the traditional algorithm and the algorithm in this section. Experiment 2 compares the detection time of the traditional algorithm and the algorithm in this section.

The results of Experiment 1 are shown in Table 16.4. Among them, p@10 represents Top10, and so on.

Experiment 1 shows that in the task of detecting the top N hot topics, except for the top 20 topics, the accuracy of the algorithm in this section is slightly better than the traditional algorithm in detecting other topic quantities, and Fig. 16.2 is the line chart of the experimental results.

Table 16.4 Comparison of the accuracy of Top N hot topics between the traditional algorithm and the algorithm in this section

Topic detection algorithm	p@10	p@20	p@30	p@40	p@50
Traditional algorithm	0.74	0.72	0.686	0.605	0.56
Algorithm of this section	0.76	0.7	0.686	0.62	0.564

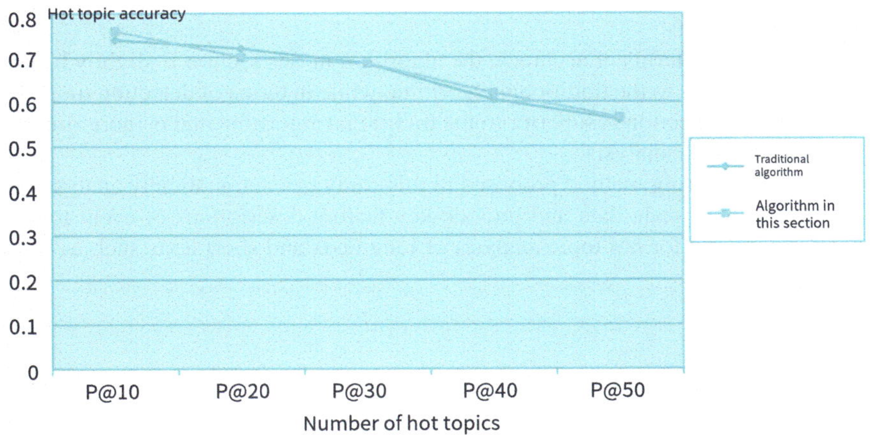

Fig. 16.2 Line chart comparing the accuracy of Top N hot topic detection between the traditional algorithm and the algorithm in this section

Table 16.5 Comparison of detection time between the traditional algorithm and the algorithm in this section. Unit: s.

Topic detection algorithm	1000	2000	5000	10,000
Traditional algorithm	38.457	123.34	517.4	1892.6
Algorithm of this section	17.351	55.67	258.22	884.9

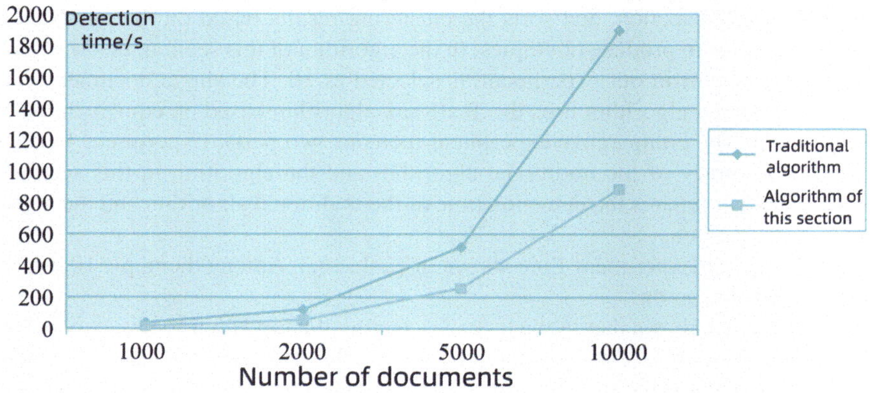

Fig. 16.3 Comparison line graph of the detection time of Top N hot topics between the traditional algorithm and the algorithm in this section

The results of Experiment 2 are shown in Table 16.5.

The experimental results given in Table 16.5 were obtained in this experimental environment. Different results may be obtained in different experimental environments, but the overall trend of the experimental results will not change. The results of Experiment 2 indicate that the detection time of the algorithm in this section is shorter than that of the traditional algorithm. Figure 16.3 is a line graph of the experimental results.

In summary, in terms of accuracy, the algorithm in this section is slightly higher than or comparable to the traditional algorithm, while in terms of detection time, the algorithm in this section is superior to the traditional algorithm and is more suitable for large-scale data analysis.

The text clustering method proposed in this study can automatically analyze hot events from large-scale data and provide key feature descriptions of event topics, making it suitable for hot topic analysis of long texts and short texts such as SMS and microblogs.

References

1. XIE J, GIRSHICK R, FARHADIA. Unsupervised Deep Embedding for Clustering Analysis[J]. Computer Science, 2015.
2. ZHANG T, JIP, HARANDIM, etc. Neural Collaborative Subspace Clustering[J]. 2019.

Chapter 17
Text Proofreading

This chapter discusses text proofreading technologies, covering the significance, current status, and advanced algorithms, including deep learning and pretrained models like BERT. It introduces the NLPIR KDN algorithm, which combines spelling, grammar, and language error correction using phonetic and glyph similarity. A case study showcases its application in an automated proofreading system, improving text accuracy and efficiency in real-world use cases like online platforms and office plugins.

17.1 Overview of Text Proofreading

The position and role of the cultural industry in the development of our national economy are becoming increasingly prominent. According to the survey data of 58,000 large-scale cultural and related industry enterprises by the National Bureau of Statistics, in 2019, China's large-scale cultural and related industry enterprises achieved a business income of 8.6624 trillion yuan, an increase of 7% over 2018 on a comparable basis. The vigorous development of the cultural industry has also brought a problem: text errors are inevitable, and the increasing amount of text information brings a larger proofreading workload and higher proofreading efficiency requirements, and traditional manual proofreading methods can no longer meet the demand.

In the traditional publishing field, with the rapid development of the publishing industry's business volume and electronization in recent years, the workload of the proofreading link has greatly increased, and the error rate of traditional manual proofreading is high. The year 2013 was the "Year of Publication Quality Assurance," and the China Book Quality Spot Check activity inspected 2300 kinds of books, of which 90 were unqualified, and some had a proofreading error rate as high as 0.0626% [1]. In 2019, the National Press and Publication Administration issued a

H. Zhang, J. Shang, *Natural Language Processing and Applications*,
https://doi.org/10.1007/978-981-97-9739-4_17

notice announcing the results of the special work quality inspection of "Quality Management 2019" and determining that 35 kinds of books were unqualified in proofreading quality, involving 29 book publishing units. The quality of book proofreading urgently needs to be improved.

Taking WeChat self-media (We Media) as an example, as of December 2016, the emerging WeChat public account platform had 10 million accounts [2]. The rapid growth of the market size in the self-media field has led to a surge in the number of texts on the Internet, and the demand for text proofreading is constantly increasing. At the same time, due to the characteristics of the self-media industry, anyone can become an uploader of text. Currently, the quality of self-media operators varies greatly, with some having poor writing skills. In order to keep up with the news hotspots, many self-media articles have a short writing cycle, and the authors simply check or even do not check at all before publishing their manuscripts, leading to frequent occurrences of errors such as wrong words, missing words, and improper usage. Taking some WeChat public account articles as an example, not only is the quality worrying but there are also many ridiculous errors. For instance, 年轻产妇剩(生)下双胞胎后；在好奇心的趋势(驱使)下；由小清晰(新)变成了男子汉；一只变(表)现得很乖的猫咪；以领土换经援对俄不舍(合)算 (A young mother left (gave birth to) twins, under the trend (drive) of curiosity, from a small, clear young (new) to a man, a very changed (well-behaved) cat, exchanging territory for economic aid to Russia is not lent (worth) it [3]). The frequency of textual errors in the self-media field is even more frequent than in the traditional publishing industry. As the influence of self-media expands year by year, frequent typos will significantly impact people's cultural quality and reading experience. From the perspective of text input methods, in addition to keyboard input, current computer text input methods implemented using artificial intelligence technology, such as optical character recognition (OCR), handwriting recognition, voice recognition, etc., all have certain errors. Take OCR as an example; the recognition rate of domestic high-level Tsinghua Wenhong, Hanwang, and foreign ABBYY, IRIS, etc., is around 95%, and the recognition rate of scanned clear manuscripts can reach more than 99%. However, no matter which OCR product is used, it cannot reach a 100% recognition rate, and text errors are inevitable. Automatic text proofreading technology also has a high application demand in this field.

In July 2017, our country issued the "New Generation Artificial Intelligence Development Plan," setting a "three-step" strategic goal for the development of a new generation of artificial intelligence. Applying artificial intelligence to text proofreading and researching automatic proofreading technology can achieve higher efficiency and shorter cycles than manual proofreading, freeing the workforce from tedious and monotonous proofreading work.

17.2 Text Proofreading Algorithm

17.2.1 Text Proofreading Methods Based on Statistical Machine Learning

As early as the 1960s, foreign countries began to study automatic proofreading of English texts [4], and in the 1990s, China gradually began to study automatic proofreading of Chinese texts [5]. Early text proofreading was based on the analysis of sentences according to grammatical rules. The sentence will be proofread if it does not comply with the sentence generation rules [6]. With statistical natural language processing development, the n-gram statistical model based on words is widely used in automatic text proofreading [7]. The traditional rule-based and statistical models have certain limitations when applied to text correction. Some deep learning language models, such as RNN, LSTM, ConvS2S, etc. [8–10], can fully understand the semantics and context of words and can find more complex error types. After more than a decade, some pretrained language models, such as ELMo [11], BERT [12], etc., have quickly become popular due to their powerful language representation capabilities and the convenience of training and transfer applications. Looking at the text proofreading methods from ancient to modern times, they can be roughly divided into three categories: text proofreading methods based on statistical machine learning, text proofreading methods based on deep learning, and text proofreading methods based on pretrained language models.

17.2.2 Text Proofreading Methods Based on Statistical Machine Learning

In the early days, text proofreading methods based on semantic rules were initially applied, but they have certain limitations and heavily depend on expert knowledge, which is not highly scalable. Text proofreading methods based on statistical machine learning can alleviate the dependence on expert knowledge, such as the n-gram method and the minimum edit distance method.

17.2.2.1 The n-gram Method

Assume wi is the symbol that is about to appear, when predicting its probability of occurrence [13], we need to consider the influence of several symbols that appeared before it. If we only consider the $n-1$ language symbols that appeared before $(w_{i-n+1}, w_{i-n+2}, \ldots, w_{i-1})$, $(w_{i-n+1}, w_{i-n+2}, \ldots, w_{i-1})$ $(w_{i-n+1}, w_{i-n+2}, \ldots, w_{i-1})$, the occurrence probability of, w_{i-1}, w_i) can be estimated by the conditional probability $P(w_i|w_{i-n+1}, w_{i-n+2}, \ldots, w_{i-1})$. The larger the value of n, the closer the model reflects the real syntactic pattern, but the actual calculation of $P(S)$ is very difficult. In the field of natural language processing, the most commonly used are bi- or tri-gram frequency statistics.

The n-gram method performs bi- or tri-gram frequency statistics on existing corpora to obtain an n-gram co-occurrence table. Combined with the n-gram training results, each n-gram string in the input string is looked up in the table to get the occurrence frequency of the n-gram string. If it cannot be found in the n-gram co-occurrence table, and the occurrence frequency is very low, it is considered that this n-gram string may be erroneous.

The n-gram method is the most widely used method in traditional machine learning. It captures all the information provided by the previous $n-1$ words but can easily lead to error propagation in cases where there is no semantic association between words. Additionally, due to the use of one-hot encoding, the parameter space becomes excessively large, resulting in issues with data sparsity.

17.2.2.2 Minimum Edit Distance Method

The edit distance refers to the number of editing operations required to transform one word into another. The minimum edit distance refers to the minimum number of editing operations necessary for the transformation. This method determines the candidate words for error correction by calculating the minimum edit distance between the misspelled string and a word in the dictionary.

The reverse minimum edit distance method is also commonly used for word-level error correction. This method first generates a candidate set by swapping each possible single error, then checks the dictionary to determine which words are valid and uses these valid words as error correction suggestions for the misspelled string.

The minimum edit distance method exhibits extremely high noise resistance and effectively describes the differences between sequences. It addresses the issue of error propagation caused by the lack of semantic association between words in the n-gram method. However, the minimum edit distance method may still result in semantic errors.

17.2.3 Deep Learning-Based Text Proofreading Method

In view of the limitations of text proofreading methods based on statistical machine learning, researchers have proposed deep learning-based text proofreading methods by combining various deep learning methods, such as neural network models, based on traditional methods.

The traditional language model is a unidirectional model from left to right, only using the context and masking the predicted word, without the information of the predicted word, directly predicting the probability distribution of all words in the word table, most of which is useless information. It is easy to cause dimension disaster. In response to the problems that arise in traditional methods, the improvement measures after adding the neural network language model are transforming the traditional language model into a bidirectional model, thereby utilizing the information of the context, at the same time, introducing the confusion information of the current word (such as words that are similar in sound and shape), restricting the predicted word in the near-sound, near-shape, and confusion word table, thereby improving efficiency and the distinction between correct and incorrect words.

17.2.3.1 NPLM

NPLM [14] is a classic neural probabilistic language model that follows the idea of the n-gram model. In this model, the previous $n-1$ words of the word w are taken as the context of w. The context in different neural language models may vary.

First, each word in the context is mapped to a word vector of length m as the model's input. The word vectors are random at the beginning of training and participate in the training process. All context word vectors are concatenated into a long vector as the feature vector of w, with a dimension of m $(n-1)$. The concatenated vector goes through a hidden layer of size h and finally through a SoftMax output layer of size N to obtain the probability distribution of each word in the vocabulary as the predicted next word. This model also considers the case where edges connect the projection layer and the output layer, so there is an additional weight matrix, but it is essentially the same as the n-gram model. The computational load of this model is mainly concentrated on the matrix calculation of the hidden layer, the output layer, and the normalization calculation of SoftMax. Subsequent related research mainly optimized this part, including the work of Word2Vec. Compared with the n-gram model, the advantage of NPLM is that the similarity between words can be reflected through word vectors, and it comes with smoothing processing.

17.2.3.2 BiLSTM Model

In Chinese text proofreading, the language model constructed by LSTM and the language model constructed by n-gram is essentially the same. They obtain the probability of the next word or character through the previous text. However, n-gram is based on statistics and can only artificially set a limited n-value. In contrast, LSTM can calculate probabilities by integrating information from all preceding words.

For example, Wang and others [15] used the BiLSTM model, in which the forward LSTM learns $hk-1$ from left to right, and the backward LSTM learns $hk + 1$ from right to left, then merges the two results to get hk. First, hk performs an attention operation with the input character vector to get ck, then concatenates ck and hk to get pk, and then performs an attention operation with pk and the candidate character vector, using the obtained score as the predicted probability distribution. This model can effectively solve the problem of neighboring characters also being miswritten.

In the neural network model for character and word-level detection, n-gram and LSTM each have advantages and disadvantages. N-gram has lower requirements for the same distribution of training and testing, strong recognition ability on scattered strings, and strong interpretability, and the principle is clear. The artificial participation in LSTM is less than that in n-gram, and the requirement for the number of corpora is not high. The model selection should be determined according to the specific situation.

17.2.4 Text Proofreading Methods Based on Pretrained Language Models

Pre-trained technology is divided into two categories: static pre-trained and dynamic pre-trained. Static pre-trained technology includes Neural Network Language Model (NNLM), Word2Vec, and FastText, while dynamic pre-trained technology includes Embeddings from Language Models (ELMo), Generative Pre-trained Transformer (GPT), and Bidirectional Encoder Representations from Transformers (BERT) [16]. In 2003, Bengio proposed NNLM, which uses neural networks to build language models and treats the feature mapping matrix of the first layer of the model as the word's text representation. This approach paved the way for representing words with vectors. Later, Mikolov and others [17] built on the idea of NNLM and proposed Word2Vec, which simplified the model's structure and utilized context information to generate word vectors, unlike NNLM. FastText achieves pre-training using supervised text classification data, and its most significant advantage is its extremely fast prediction speed.

In 2018, Peters and others [18] proposed a text representation method related to context and built the ELMo model. The ELMo model can be trained with a large-scale unsupervised corpus and conveniently and flexibly transferred to specific tasks in downstream applications. Although the ELMo model has strong representation capabilities, it is a unidirectional language model. To better utilize the semantic information of the context, Google proposed the BERT pretrained language model. BERT stands out from other pretrained language models due to its unique bidirectional language model, which is achieved by adding a masked language model to the Transformer sub-layer. This allows BERT to predict the next sentence based on the previous one, enhancing its ability to handle the relationship between sentences. BERT shows unparalleled language representation capabilities and transfer application capabilities compared to other pretrained language models, but the number of parameters and the computational resources required by BERT are much greater than those of other pretrained language models.

Chinese spell check (CSC) is a task that detects and corrects spelling errors in Chinese text. Most of the latest research on the CSC task uses the masked language model (MLM) and has achieved good results. The training content of MLM is to complete the cloze task, mask a certain percentage of words in a given sentence, and predict the masked words based on the preceding and following words.

17.2.4.1 FASPell

In 2019, FASPell [19] was introduced at the EMNLP-IJCNLP conference. It is a Chinese spell checker that operates on the Denoising AutoEncoder (DAE)-decoder structure, as depicted in Fig. 17.1. Denoising AutoEncoder is the masked language model in BERT, which leverages unsupervised pretrained methods to circumvent the need for a large-scale, manually annotated corpus required for supervised learning. Using the decoder overcomes the deficiency in the similarity of Chinese characters caused by the use of confusion sets. FASPell, compared to the previous SOTA model, boasts higher computational efficiency, making it suitable for various scenarios of simplified, traditional, artificial, or machine-generated Chinese text. Its simpler structure and the fact that it has achieved the accuracy of the SOTA model in error detection and correction in experiments make FASPell a promising tool in the field of Chinese spell check.

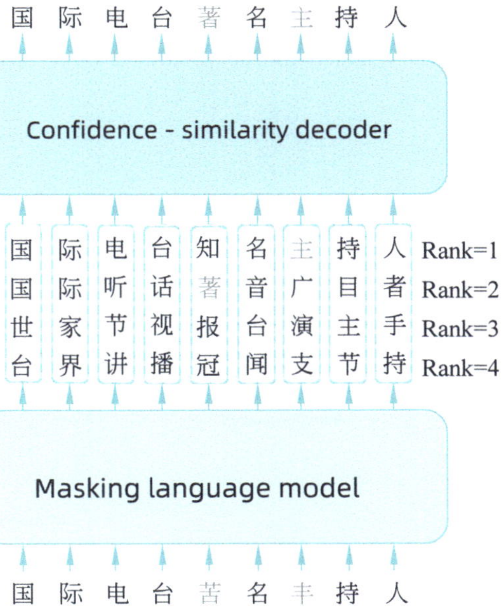

Fig. 17.1 Structure of the FASPell model

17.2.4.2 SoftMasked BERT

SoftMasked BERT, a variant of BERT, uses a masked language model for pretraining, which means it does not have enough ability to detect every position. However, it holds tremendous potential and is an area of interest for further exploration and research.

The presence of errors is determined. Hence, a new neural structure, SoftMasked BERT [20], is proposed. SoftMasked BERT, a Chinese spell check model, is shown in Fig. 17.2. It consists of two networks: a detection network and a BERT-based correction network. The detection network is a Bi-GRU network that can predict the probability of errors at each character position and then embed the mask at the corresponding position according to the probability. The correction network is similar to the corresponding structure in BERT. Experimental results show that all evaluation indicators of SoftMasked BERT are superior to BERT.

17.2.4.3 SpellGCN

The Ant Financial team proposed a new correction model, SpellGCN [21], at ACL 2020, as shown in Fig. 17.3. The main idea of SpellGCN is to use GCN to fuse the embedding vectors of characters with similar pronunciation and shape, then use BERT as the base model for character-level classification, and finally, use Softmax for target character prediction. The model consists of two modules: module one is SpellGCN, and module two extracts character features from the extraction model, using the BERT model for character feature extraction in this section. The SpellGCN model completes detection and correction tasks at once. When predicting the target character for the input character in the last layer, the character with the highest probability is selected as the predicted target character. When the input and target

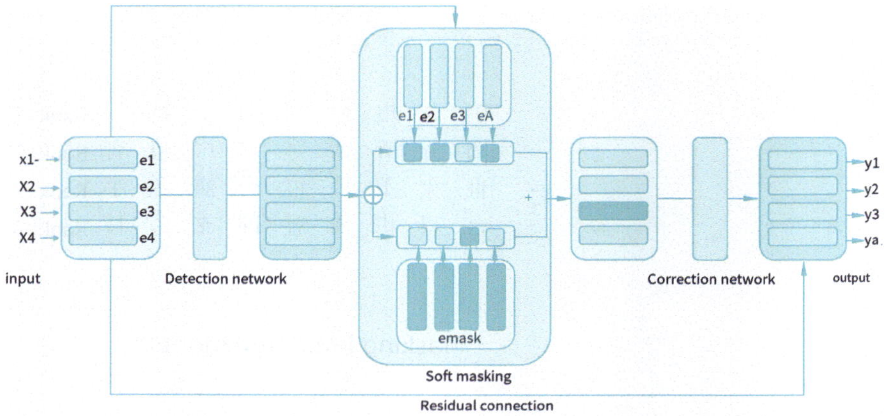

Fig. 17.2 Structure of the SoftMasked BERT model

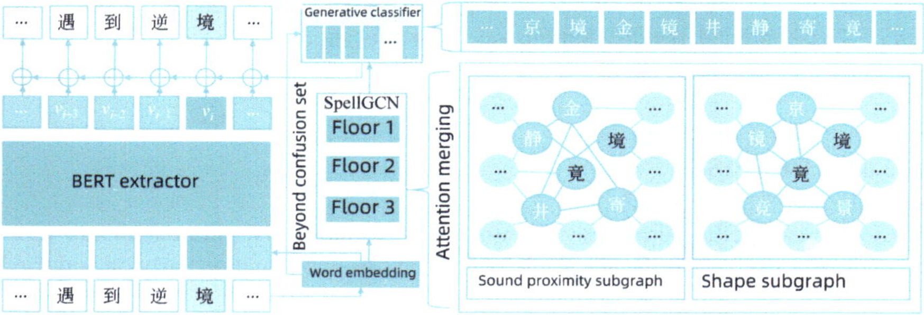

Fig. 17.3 Structure of the SpellGCN model

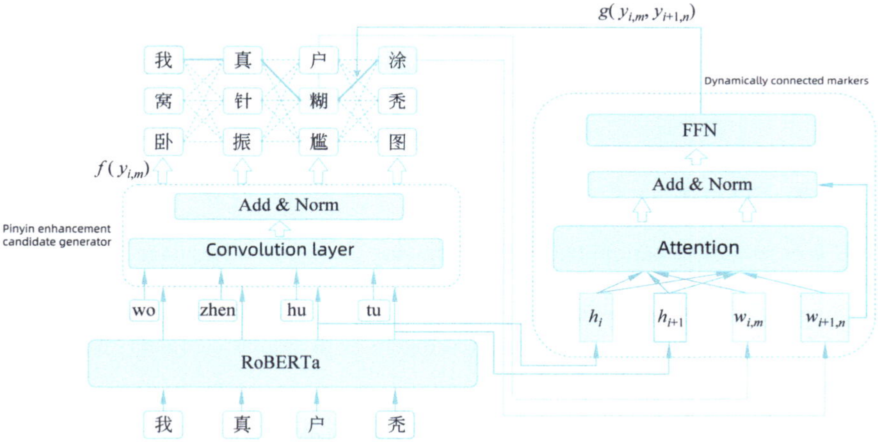

Fig. 17.4 Structure of the DCN model

characters are the same, the text is error-free; if they are not the same, it means that the text is misspelled, and the corrected value is the predicted character.

17.2.4.4 DCN

Most of the latest research on CSC tasks uses BERT-based non-autoregressive language models, which rely on the output independence assumption. The inappropriate independence assumption prevents BERT-based models from learning between target tokens.

Dependency issues often lead to incoherence. To address the above problem, Wang et al. [22] proposed a new model called the Dynamic Connected Network (DCN), as illustrated in Fig. 17.4. This model generates candidate Chinese characters using a pinyin-enhanced candidate generator and then employs an attention-based network to model the dependency between adjacent Chinese characters. Experimental results demonstrate that this model performs well on three manually annotated datasets.

17.2.4.5 REALISE

Most Chinese spelling errors involve misuse of semantically, phonetically, or morphologically similar words. Researchers noticed this phenomenon early on and tried to use these similarities to complete the task.

Based on this, Xu et al. [23] proposed REALISE, a Chinese spell checker directly using multimodal information about Chinese characters. Its model structure is shown in Fig. 17.5. REALISE's approach to the CSC task is: first, use three encoders to capture the semantic, phonetic, and morphological information of the input characters, then selectively fuse this information into the corresponding modality to predict the correct output. Experiments on the SIGHAN benchmark test show that this model performs much better than the strong baseline model.

17.2.4.6 PLOME

CSC is essentially a natural language processing problem, so language understanding ability is very important in this task. Liu et al. [24] proposed a pretrained masked language model with spelling error knowledge for CSC PLOME, as shown in Fig. 17.6. This model masks the selected tokens with similar characters according

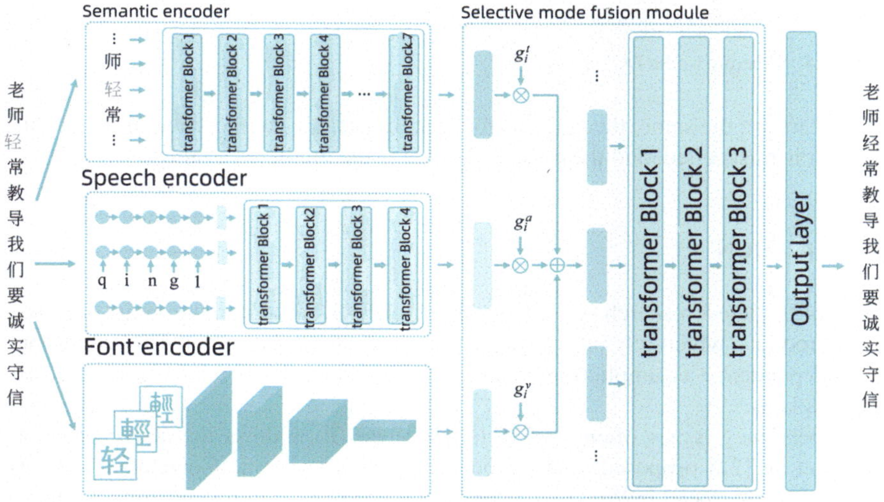

Fig. 17.5 Structure of the REALISE model

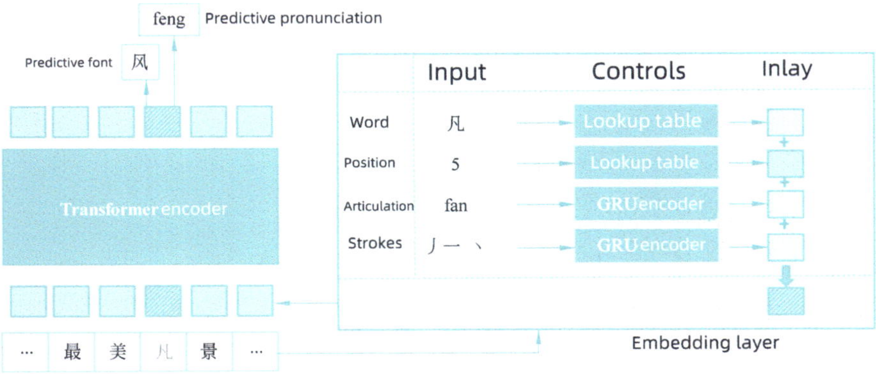

Fig. 17.6 PLOME model structure

to the confusion set instead of using the fixed token "[MASK]" in BERT. In addition to character prediction, PLOME also introduces phonetic prediction to learn phonetic-level spelling error knowledge.

Moreover, knowledge of phonetic and visual similarity is also important for this task. PLOME uses a GRU network based on character phonetics and strokes to model this knowledge. Corresponding experiments were conducted on widely used benchmarks. This method achieved excellent performance compared to the most advanced methods.

17.3 NLPIR KDN: Knowledge-Driven Multi-type Text Proofreading and Fusion Algorithm

This section presents a knowledge-driven multi-type text proofreading fusion algorithm, knowledge-driven network (KDN), aiming to integrate everyday tasks such as spelling, grammar, and language disorder proofreading into a unified theoretical model. Since spelling errors mainly occur in phonetically similar words, a phonetic code is added to the transformer-based encoder for a more flexible phonetic similarity calculation. Since the grammatical errors in the text are mainly redundancy and omission errors, a conditional random field (CRF) layer is added at the end of the transformer block to model label dependencies and complete non-autoregressive sequence prediction. External syntactic knowledge is used to identify semantic error areas and delete semantic repetitions for language disorder errors. Finally, the information in these partially correct sentences is selectively mixed to predict the correct output.

The structure of the KDN model is shown in Fig. 17.7. KDN uses a phonetic code mechanism to integrate phonetic and glyph similarity information for spelling error correction. It uses model or manually constructed language knowledge for

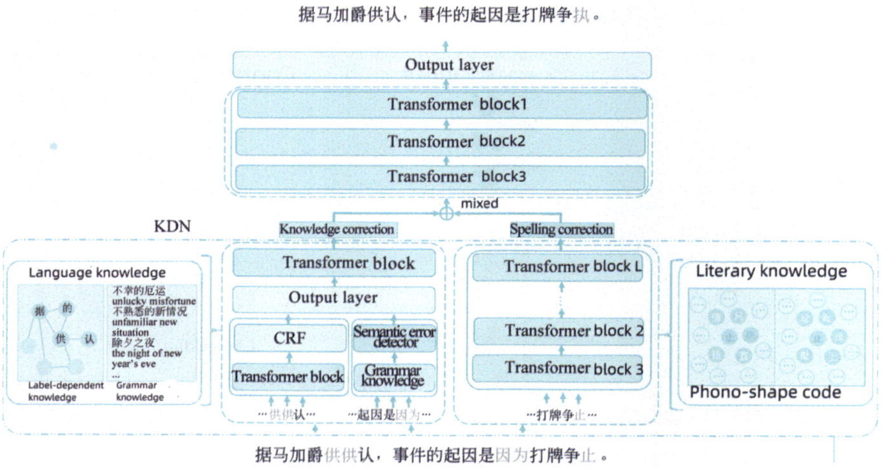

Fig. 17.7 KDN model structure

knowledge (including grammar and language disorder errors) correction, and finally selectively mixes some correct information to achieve text correction.

17.3.1 Grammar Proofreading

As shown in Fig. 17.7, the grammar checking unit locates error positions through token dependency knowledge and then makes grammar modifications based on Bert MLM. It mainly consists of a Bert encoding layer and a CRF layer. By modeling token dependencies, error prediction is implemented through non-autoregressive sequence prediction. Here, four types of token categories are defined: $\text{label}_{\text{keep}}$, $\text{label}_{\text{mistaken}}$, $\text{label}_{\text{missing}}$, $\text{label}_{\text{redundant}}$. Of these, $\text{label}_{\text{keep}}$ indicates that the character is correct and does not need to be modified; $\text{label}_{\text{mistaken}}$ indicates that the character $\text{label}_{\text{missing}}$ has a spelling error and needs to be replaced; $\text{label}_{\text{missing}}$ indicates that other characters need to be inserted after this character; and $\text{label}_{\text{redundant}}$ indicates that this character is redundant and can be directly deleted. The probability of each character in these four categories can be obtained through the CRF layer:

$$P_{i,\text{label}}\left(l_i = l|X\right) = \text{softmax}(w^T h^D\left(X\right) \tag{17.1}$$

where $P_{i,\text{label}}(l_i = l|X)$ represents the conditional probability that character x_i belongs to category l, w^T denotes the parameters learned by the model, h^D represents the hidden state vector of the last layer of the Bert Encoder module, and l belongs to the category label set [$\text{label}_{\text{keep}}$, $\text{label}_{\text{mistaken}}$, $\text{label}_{\text{missing}}$, $\text{label}_{\text{redundant}}$].

After obtaining the sequence prediction label L, rewrite the input X according to the following rules:

$$x_i' = \begin{cases} X_i, \text{ if } l_i = \text{label}_{\text{keep}} \\ [MASK], \text{if } l_i = \text{label}_{\text{mistaken}} \\ n, \text{if } l_i = \text{label}_{\text{redundant}} \\ X_i [MASK], \text{if } l_i = \text{label}_{\text{missing}} \end{cases} \tag{17.2}$$

According to the above formula, a new sequence $X' = (x_1', x_2', x_3', \ldots x_n')$, will be obtained, the length of which may be different from the original X sequence.

Finally, BERT is used to predict the [MASK] position. When predicting $\text{label}_{\text{mistaken}}$-type errors, use the phonetic code to calculate the phonetic similarity between the candidate and original characters. This step can effectively reduce the situation of misc-correction.

17.3.2 Grammatical Error Proofreading

The language errors check unit refers to the practice of Zhang and others [25]. It is mainly composed of a syntax knowledge base and a syntax error-checking module. The syntax knowledge base mainly contains errors such as semantic repetition and sentence mixing, as shown in Table 17.1. The templates of the syntax knowledge base mainly have three types: For the "day by day... becoming more perfect" type template, the modification method is to delete the word on the left, that is, to modify to "... becoming more perfect." For the "real knowledge... opinion" type template, the modification method is to delete the word at the right, that is, to modify to "real knowledge...." For the "lowest... above" type template, words at both the left and right can be deleted, that is, to modify to "lowest..." or "... above."

With the syntax knowledge base, you can quickly locate the language error fragments in the text. According to language habits, three modification methods are defined, namely $\text{action}_{\text{delete left}}$, $\text{action}_{\text{delete right}}$, $\text{action}_{\text{delete all right}}$. Then, it automatically decides which modification method to adopt based on the perplexity of the language model. ppl_{left} represents the perplexity of the sentence after deleting the word on the

Table 17.1 Examples of syntax knowledge base format

Template	Modification method
Day by day... improve day by day 一天天...日臻完善	Delete words at the left 一天天
With great insight... the advice of 真知灼见...的意见	Delete words at the ri ght 的意见
The lowest... above 最低...以上	Delete words at both left and right 最低 or 以上

left, and ppl_{right} represents the perplexity of the sentence after deleting the word on the right. Therefore:

$$\text{action} = \begin{cases} \text{action}_{\text{deleteleft}}, \text{ifppl}_{\text{left}} < \text{ppl}_{\text{right}} \\ \text{action}_{\text{deleteright}}, \text{ifppl}_{\text{right}} > \text{ppl}_{\text{right}} \\ \text{action}_{\text{deleteallright}}, \text{ifppl}_{\text{right}} = \text{ppl}_{\text{right}} \end{cases} \tag{17.3}$$

According to the above formula, we can determine the modification method for language errors. Together with grammar proofreading, we can obtain the intermediate output results.

17.3.3 Similarity Calculation Based on Phonogram Code

Since most phonetic or glyph spelling errors involve the misuse of similar characters, the phonogram code mechanism is used to handle the similarity between any characters, thereby improving the spelling correction ability. This method converts a Chinese character into an alphanumeric sequence, which, to some extent, retains the phonetic and glyph features of Chinese. The first four characters of the phonogram code are phonetic codes, and the last seven characters are glyph codes. The similarity of the phonetic code and the glyph code can be calculated separately using the formulae:

$$P = 0.4(\nabla p1) + 0.4(\nabla p2) + 0.1(\nabla p3) + 0.1(\nabla p4) \tag{17.4}$$

$$\begin{aligned} S = 0.25(\nabla s1) &+ \left(\frac{0.5(\nabla s2 + \nabla s3 + \nabla s4 + \nabla s5 + \nabla s6)}{5} \right) \\ &+ 0.25\left(1 - \frac{|s7 - s7'|}{\max(s7, s7')}\right) \end{aligned} \tag{17.5}$$

where P represents the similarity of the phonetic code and S represents the similarity of the glyph code. It represents the character comparison operation. If the two characters are the same, return 1, otherwise, return 0. For more details about the phonogram code, please refer to the relevant literature.

17.3.4 Proofreading Fusion Algorithm

After completing the spelling and knowledge correction, two partially corrected sentences have been obtained (more partially corrected sentences can also be obtained). In order to predict the final correct output, a proofreading fusion module is used to mix these partially corrected sentences. The outputs of multiple encoders are combined through a gate:

$$c_{\text{combined}} = \lambda c_{\text{knowledge}} + (1-\lambda) c_{\text{spelling}} \tag{17.6}$$

The gating variable is obtained by the following formula:

$$\lambda = \sigma\left(W\left[c_{\text{knowledge}}; c_{\text{spelling}}\right] + b\right) \tag{17.7}$$

where σ is the activation function, and W and b are learnable parameters. Then the combined representation c_{combined} is treated as a single input to the Transformer encoder.

17.4 Design and Application of NLPIR Automatic Text Proofreading System

In order to demonstrate and verify the function of the text proofreading algorithm, this section introduces a text proofreading test Web service developed based on the client/server architecture and a text proofreading plugin that can proofread the input text and give suggestions for modification after proofreading.

17.4.1 Automatic Proofreading Module

The automatic proofreading module is the core part of the NLPIR text automatic proofreading system. Its main function is to proofread the text input by the user. This section provides an overview of the knowledge-based proofreading algorithm and the address-level proofreading algorithm, and finally uses the proofreading fusion algorithm described in Sect. 17.3.4 to fuse the proofreading results.

The process of the knowledge-based proofreading algorithm is as follows:

1. The organized idioms, proverbs, and xiehouyu are stored using data structures such as trie trees.
2. The trie tree is used to find out the possible erroneous knowledge expressions according to the knowledge prefix.
3. The Trie tree is reused to find the candidate set of knowledge based on the knowledge prefix.
4. Methods such as edit distance are used to find the best candidate knowledge.

The proofreading algorithm based on address hierarchy is as follows:

1. Store the national and regional addresses obtained by crawling in the form of a multibranch tree.
2. Use address recognition tools to extract address information from the text.
3. Use the address multibranch tree to complete the recognized address.
4. Compare the address information in the text with the completed address information, the one with the highest match is the proofreading result.

17.4.2 Front-end and Back-end Design and Implementation

This proofreading system adopts a front-end and back-end separation design; the front-end interaction module is responsible for the front-end page display. The front-end page is based on the HTML design and uses some Bootstrap styles. The interaction between the front-end page and the back-end is implemented through JavaScript scripts. The data interaction between the front-end and back-end uses the JSON format. The HTTP request initiated by the front-end page will be distributed to different functions in the back-end according to the URL and method of the request. The system data flow is shown in Fig. 17.8:

The back-end interface of the automatic proofreading module is shown in Table 17.2.

The automatic proofreading module proofreads the text entered by the user and gives the error position and proofreading suggestions. This system obtains the content entered by the user into the text box through JavaScript scripts, submits it to the back-end through Ajax technology, and then draws the text result returned by the back-end into the text box through JavaScript scripts.

Automatic proofreading is divided into the following steps:

1. Obtain the text information entered by the user into the text box.
2. Divide by line breaks and pass them one by one to the back-end through Ajax technology.
3. The back-end divides the text by punctuation and processes the text using multi-threading technology.
4. Merge the final proofreading information according to the return value of multi-threading and return it to the front-end page.

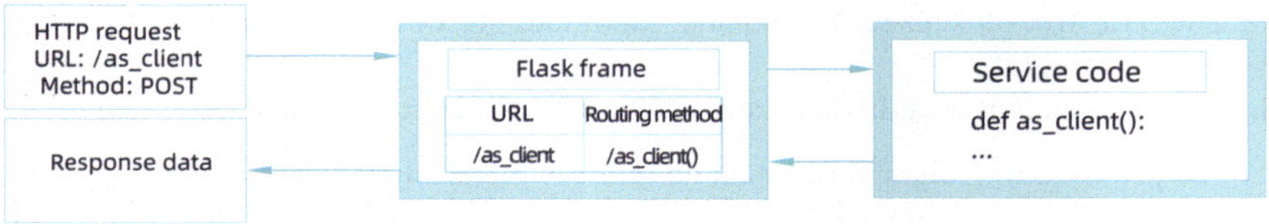

Fig. 17.8 System data flow

Table 17.2 Automatic proofreading module back-end interface

URL	Method	Back-end Function	Function
/as_client	POST	as_client()	Accept the text to be proofread and proofread text
/as_client	GET	As_client()	Render online proofreading pages

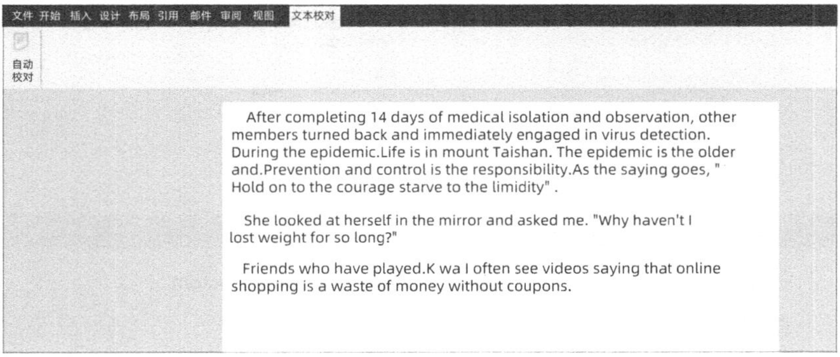

Fig. 17.9 Office plugin interface

17.4.3 Online Proofreading Plugin Office

The online proofreading plugin office is developed using Visual Studio Tools for Office (VSTO) technology, based on .NET Framework 4.0, and the development software is Visual Studio 2019. The interface of this plugin is shown in Fig. 17.9.

The algorithm flow of automatic proofreading is as follows:

1. Read the article content in sections, filter out pictures, tables, etc.
2. Call the call_spell_check_API (URL, Text) function, which sends the text to the backend server at the specified URL. This function then returns the proofread result.
3. Sort all errors by position.
4. Based on the error correction results, calculate the positions for deleting and inserting text.
5. Start the revision mode, and perform insert or delete operations on all errors.

The algorithm flow for error entry into the database is as follows:

1. Read the user ID, group, message type, importance, and note fields.
2. Read the text selected by the user, judge and handle errors, fill in the error situation, including the original data of the error expression, the original data of the correct expression, the error expression, the correct expression, the error type, and encapsulate the error situation into JSON data.
3. Return the JSON data to the back-end server via the POST method.

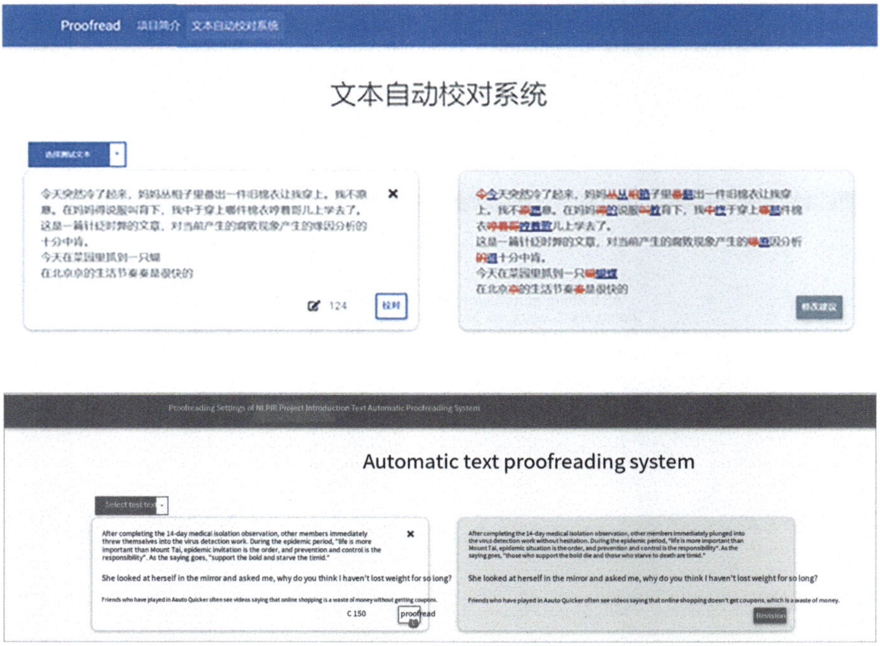

Fig. 17.10 Online proofreading example

17.4.4 Online Proofreading Function Example

As shown in Fig. 17.10, the user inputs the text to be corrected into the correction text box, then clicks the "Proofread" button, and the text to be corrected will be sent to the back-end in segments. The back-end splits the text into sentences, uses multithreading technology to complete the proofreading, then merges the results and returns them to the front-end page. Suspected error sentences will be crossed out in red, and candidate-recommended words marked in blue will be given afterward. It can be seen that this system can handle various types of errors, such as wrong characters, extra characters, and missing characters.

References

1. Li Liang. Unqualified proofreading books become "poisonous" industry insiders: the punishment is too light, it does not hurt [N]. China Youth Daily, 2014, (A21).
2. WeChat User & Ecological Research Report [EB/OL]. 2017.
3. Cui Guoping. The root cause and governance of the proliferation of typos in WeChat public accounts[J]. Today's Media (Academic Edition), 2018, 26(2): 34–36.

4. DAMERAU FJ. 1964. A Technique for Computer Detection and Correction of Spelling Errors[J]. Commun. ACM 7, 1964(3): 171–176.
5. CHANG C H. A New Approach for Automatic Chinese Spelling Correction[C]. Proceedings of Natural Language Processing Pacific Rim Symposium. 1995: 278–283.
6. Qian Yili. Research on automatic proofreading methods for Chinese text segmentation and part-of-speech tagging[D]. Taiyuan: Shanxi University, 2003.
7. Zhang Yangsen, Cao Yuanda, Yu Shiwen. A rule-based and statistical combined model and algorithm for automatic error detection in Chinese text[J]. Journal of Chinese Information Processing, 2006, 20(4): 3–9, 57.
8. CHOLLAMPATTS, NG H T. A Multilayer Convolutional Encoder-Decoder Neural Network for Grammatical Error Correction[C]. Thirty-Second AAAI Conference on Artificial Intelligence. 2018.
9. ELMAN JL. Finding structure in time[J]. Cognitive science, 1990, 14(2): 179–211.
10. HOCHREITER S, SCHMIDHUBER J. Long Short-Term Memory[J]. Neural computation, 1997, 9(8): 1735–1780.
11. PETERS M, NEUMANN M, IYYER M, et al. Deep Contextualized Word Representations[J]. 2018.
12. DEVLIN J, CHANG M W, LEE K, et al. 2019. BERT: Pre-trained of Deep Bidirectional Transformers for Language Understanding. Proceedings of the 2019 Conference of the North American Chapter of the Association for Computational Linguistics: Human Language Technologies, Volume 1(Long and Short Papers), 2019: 4171–4186.
13. Zhuoli Yan. Research on the method of automatic proofreading of Chinese text at the word level [D]. Zhengzhou: Zhengzhou University, 2018.
14. BENGIO Y, DUCHARME R, VINCENT P, et al. A Neural Probabilistic Language Model[J]. Journal of Machine Learning Research, 2003, 3(2): 1137–1155.
15. WANG Q, LIU M, ZHANG W, et al. Automatic Proofreading in Chinese: Detect and Correct Spelling Errors in Character-Level with Deep Neural Networks[J]. School of Information Science and Technology, Southwest Jiaotong University, 999 Xian Road, Chengdu, China.
16. Li Zhoujun, Fan Yu, Wu Xianjie. A Review of Pre-trained Techniques for Natural Language Processing[J]. Computer Science, 2020, 47(3): 162–173.
17. MIKOLOV T, SUTSKEVER I, CHEN K, et al. Distributed Representations of Words and Phrases and Their Compositionality[C]. Advances in Neural Information Processing Systems. 2013: 3111–3119.
18. PETERS M, NEUMANN M, IYYER M, et al. Deep Contextualized Word Representations[J]. 2018.
19. HONG Y, YU X, HE N, et al. FASPell: A Fast, Adaptable, Simple, Powerful Chinese Spell Checker Based on DAE-Decoder Paradigm[C]. Proceedings of the 5th Workshop on Noisy User-generated Text (W-NUT 2019). 2019: 160–169.
20. ZHANG S, HUANG H, LIU J, et al. Spelling Error Correction with Soft-Masked BERT[C]. Proceedings of the 58th Annual Meeting of the Association for Computational Linguistics. 2020: 882–890.
21. CHENG X, XU W, CHEN K, et al. SpellGCN: Incorporating Phonological and Visual Similarities into Language Models for Chinese Spelling Check[C]. Proceedings of the 58th Annual Meeting of the Association for Computational Linguistics. 2020: 871–881.
22. WANG B, CHEW, WU D, et al. Dynamic Connected Networks for Chinese Spelling Check[C]. Findings of the Association for Computational Linguistics: ACL-IJCNLP, 2021: 2437–2446.
23. XU H D, LIZ, ZHOU Q, et al. Read, Listen, and See: Leveraging Multimodal Information Helps Chinese Spell Checking[C]. ACL/IJCNLP(Findings), 2021.
24. LIU S, YANG T, YUE T, et al. PLOME: Pre-trained with Misspelled Knowledge for Chinese Spelling Correction[C]. Proceedings of the 59th Annual Meeting of the Association for

Computational Linguistics and the 11th International Joint Conference on Natural Language Processing(Volume 1: Long Papers), 2021: 2991–3000.
25. ZHANG Y, BAOZ, ZHANG B, et al. Technical report of suda-alibaba team on CTC-2021[R]. https://github.com/HillZhang1999/CTC-Report.

Chapter 18
Automatic Summarization

This chapter presents the fundamentals of automatic summarization technology, covering common algorithms such as keyword-based, topic-oriented, and Chinese document summarization. It highlights efficiency improvements, extraction techniques for high- and low-frequency words, and methods to reduce sentence redundancy. The chapter concludes with a practical application example, demonstrating the use of the Natural Language Processing and Information Retrieval - Institute of Computing Technology, Chinese Lexical Analysis System (NLPIR-ICTCLAS) for automatic summarization in real-world scenarios.

18.1 Overview of Automatic Summarization

Currently, the main method employed to obtain information from big data is the use of database queries and text searches. Databases are mainly used to process structured information, whereas information retrieval is mainly used to process unstructured information. The returned result set is not only generally very large, but it is also very difficult for users to find useful information from the retrieval results. Therefore, it is particularly important to automatically extract summaries from the retrieval results [1]. This section implements a keyword-based automatic summarization system and a topic-oriented automatic summarization system, and works on improving the processing efficiency of the summary, extracting summaries based on keywords, and removing redundancy between sentences in the summary. First, in terms of improving efficiency, the system introduces the inverted index structure of search engines to count the features of words and sentences. It also uses a double-array Trie tree to store the word segmentation dictionary and the user's topic word table, enhancing the efficiency of word lookup. These improvements are applied to enhance the quality of both high-frequency and low-frequency words. Additionally, the system combines keyword extraction with summary extraction, incorporating

H. Zhang, J. Shang, *Natural Language Processing and Applications*,
https://doi.org/10.1007/978-981-97-9739-4_18

word adjacency categories and word position locality into the keyword extraction process. Second, to reduce sentence redundancy, the concept of sentence inclusion is introduced. This approach effectively removes sentences with inclusion relationships, reducing redundancy in the summary and improving its overall quality. Based on the above research, a standard vertical search system is implemented, and topic-oriented automatic summarization technology is applied to this system.

The research on information retrieval originated from the document query and abstracting and indexing work in libraries. With the rapid development and popularization of computer network technology, the content of information retrieval research has expanded from traditional text retrieval to pictures, audio, video, and other multimedia information retrieval. The retrieval object has expanded from relatively closed, stable, and centrally managed information content by independent databases to open, dynamic, fast-updating, widely distributed, and loosely managed network content. Since the 1960s, researchers have proposed various retrieval models, such as the Boolean retrieval model, vector space retrieval model, and probabilistic retrieval model, etc., and have produced famous prototype systems such as SMART and OKAPI. These retrieval models and prototype systems have promoted the research and development of information retrieval.

Automatic summarization is a technology that uses computers to implement text preprocessing, semantic analysis, and automatic summary extraction. In 1958, the article published by Luhn [2] unveiled the research on automatic summarization using computers. Due to the limitations of computer hardware conditions and application requirements, the initial automatic summarization technology was not paid attention to. In 1995, Jones and others [3] published "Automatic Summarizing" in the internationally renowned academic journal *Information Processing and Management*, marking the arrival of the era of automatic summarization, and the research on automatic summarization technology had entered a new development period. Some scholars divide the abstract extraction process into three stages: the first stage is to analyze and process the source text, divide it into basic fragments for obtaining abstracts (such as sentences, paragraphs, etc.) and other fragments that affect these basic fragments, and count the corresponding information; the second stage calculates the basic fragments obtained in the first stage; the third stage selects abstract fragments based on the related information obtained earlier and combines them into an abstract according to certain rules. Some scholars divide automatic summarization into different types, and according to the different routes adopted, automatic summarization methods can be divided into two major categories: statistics-based methods and machine learning-based methods, which are briefly described below.

Statistics-based methods are divided into three types. The first method is based on term statistics, which directly uses term features for sentence or concept measurement and recognition, uses the weight of features contained in the sentence to measure the importance of the sentence, ranks the sentences according to importance, then selects the abstract sentences according to the ranking, and finally outputs the abstract in order of appearance of the abstract sentences in the document. The basic starting point of this method is that important sentences are composed of important words, and important sentences are more suitable to be selected as abstract sentences. This method rarely considers the relationship between sentences, focusing primarily on using word frequency information to determine the importance of content, as represented by the centroid method and the maximum marginal similarity method. Zhou Jinhua and others [4] proposed to use the word co-occurrence graph method for multi-document automatic summarization. The second method involves further discovery and mining of features, utilizing more extensive statistical features to identify important sentences or concepts in the document and extract key content units to form summaries. On one hand, it addresses the limitations of the vector space model, which relies solely on terms, by introducing N-tuples and other techniques to determine the statistical characteristics of sentences. On the other hand, it employs statistical correlation methods to identify topic marker features and other relevant information while incorporating additional linguistic features. This approach better measures the importance of sentences. The third method is the multi-document automatic summarization technique based on document structure relationships.

This method fully utilizes the structural information of the text to extract important concepts (sentences, subtopics, class centers) from the document, which is the current focus and hotspot of research. Erkan and others [5] proposed the LexRank method, using the idea of centrality in social networks to measure important concepts in the document set, introducing the idea of the PageRank algorithm into automatic summarization. In the defined graph, each point represents a sentence. If the similarity between two sentences is greater than a certain threshold, there is an edge connecting the two sentences. The importance of each sentence is determined by the importance of all sentences adjacent to it, that is, if a sentence is adjacent to many important sentences, it is also very important. In the multi-document automatic summarization method based on machine learning, methods such as latent semantic analysis (LSA) and support vector machines can be used, and multi-document summaries can be generated through supervised and unsupervised methods. For given text features, various current machine learning methods can be used to find a method suitable for abstract extraction based on the selection and combination of features. According to the number of input documents, the summary can be divided into single-document summary and multi-document summary.

According to the relationship, the summary can be divided into extractive summary (extract) and understanding summary (abstract). The former extracts fragments from the original text to form a summary, and the latter is formed after reorganizing the main content of the original text. According to the application of the summary, it can be divided into generic summary (generic) and query-oriented summarization. The former objectively reflects the main content of the document, and the latter focuses on the content of interest to the user. The above types of divisions are not mutually exclusive. For example, extractive summaries can be implemented in single documents as well as in multiple documents. Before the mid-1990s, the main research content in the field of automatic summarization was single-document summarization technology.

The main methods used this period were text extraction and information extraction. With the rapid development of Internet technology, the demand for cross-text information fusion is becoming increasingly popular. The research on multi-document summarization in the general field officially began in the mid-1990s. So far, significant achievements have been made in the research on English multi-document automatic summarization, among which the more excellent results include the lexical chain method, maximal marginal relevance (MMR) method, and centroid method. Domestic research on multi-document summarization started late, especially the research on Chinese automatic summarization, which is limited by large-scale test datasets and evaluation tools, and is currently still in its initial stage, waiting for further in-depth exploration. Some researchers have summarized automatic summarization methods into four types: automatic excerpting, understanding-based automatic summarization, information extraction, and structure-based automatic summarization. Among them, automatic excerpting is the basis of other methods and is the simplest and easiest to implement. This section does not adopt this classification but simply divides it into extraction-based automatic summarization and understanding-based automatic summarization.

18.1.1 Extraction-Based Automatic Summarization

Current research on automatic summarization methods focuses on extraction-based automatic summarization. Extraction-based automatic summarization views the text as a linear sequence of sentences (or paragraphs, or other text fragments, collectively referred to as sentences in this section) and views sentences as a linear sequence of words. Extraction-based automatic summarization usually proceeds in three steps:

1. Divide the original text according to certain rules to obtain the sentence sequence and word sequence expression of the original text.
2. Calculate the weights of words and sentences, arrange all sentences in the original text in descending order of weight, and the sentences with the highest weights are determined as summary sentences.
3. Output all summary sentences in the order they appear in the original text. The key issue of extraction summarization is the selection and ordering of sentences. There are many methods for sentence selection, and the principle of selection is consistent: try to use the most important sentences to reflect the topic of the document set, that is, the topic relevance is as high as possible, and the content redundancy between the extracted summary sentences is as small as possible. Therefore, sentence selection is divided into two problems: sentence weight calculation and sentence similarity calculation. The sentences of single-document summarization are generally arranged in the order in which they appear in the original text, while in multi-document summarization, most of them are arranged in the order of document writing (publication) [6].

18.1.1.1 Sentence Weight Calculation

Whether a sentence is selected as a summary sentence is often based on the weight of the sentence. Factors for calculating sentence weight include the weight of the words contained in the sentence, the weight of the paragraph where the sentence is located, the position of the sentence in the paragraph, and the similarity of the sentence to other sentences in the document. Therefore, calculating the weight of the sentence, the weight of the word, and the weight of the paragraph is a cross-process that affects each other [7]. From the perspective of the original text, the calculation of sentence weight mainly depends on the following features: word frequency, the distribution of terms in the document, the title, position, syntactic structure, clue words, indicative phrases, etc. Among them, term frequency, the distribution of terms in the document, and the title primarily and indirectly affect the sentence weight by influencing the term weight, while position, syntactic structure, clue words, and indicative phrases directly affect the sentence weight.

Below is a brief explanation of these seven factors affecting sentence weight:

1. Term Frequency. Luhn proposed the idea of using term frequency statistics for summarization when pioneering the field of automatic summarization, but relying solely on term frequency to indicate term weight is not enough. In 1995, Lisa F. Rau and others at the GE Research and Development Center in the United States completed the Automatic News Extraction System (ANES), which uses relative term frequency as the weight of words.
2. The Distribution of Terms in the Document. Whether a word is evenly distributed in most paragraphs of the document or appears concentrated in a few paragraphs has a significant impact on whether the word can reveal the theme.
3. Title. The keywords in the title play an important role in revealing the theme of the article, so the keywords appearing in the title have a higher weight.
4. Position. Surveys show that the probability of the topic of a paragraph being in the first sentence is 85%, and the probability of being in the last sentence is 7%, so it is necessary to increase the weight of sentences in special positions.
5. Syntactic Structure. There is some connection between sentence type and sentence importance. For example, most sentences in the summary are declarative sentences, while interrogative sentences, exclamatory sentences, etc., are not suitable for inclusion in the summary.
6. Clue Words. In H. E. Edmundson's summarization system, there is a pre-compiled clue word dictionary. The clue words in this dictionary are divided into three types: positive-value praise words (bonus word), negative-value derogatory words (stigma word), and zero-value invalid words (nullword). The weight of a sentence is equal to the sum of the weights of each clue word in the sentence.
7. Indicative Phrases. Paice from Lancaster University in the United Kingdom proposed a method to select summary sentences based on various indicative phrases. Compared with clue words, indicative phrases are much more reliable.

18.1.1.2 Sentence Similarity Calculation

The commonly used sentence similarity calculation models mainly include vector space model, query likelihood model, and translation model. The vector space model is widely used because of its simple calculation. The vector space model defines sentence similarity as the cosine value of the angle between two vectors. The specific calculation formula is as follows:

$$sim(d,q) = \frac{d * q}{|d| * |q|} = \frac{\sum_{i=1}^{t} w_{i,d} * w_{i,q}}{\sqrt{\sum_{i=1}^{t} w_{i,d}^2} * \sqrt{\sum_{i=1}^{t} w_{i,q}^2}} \quad (18.1)$$

where $|d|$ and $|q|$ represent the modulus of the document vector and the topic vector; $w_{i,d}$ and $w_{i,q}$ represent the weight of the ith word in the document vector and the topic vector, respectively. Aliguliyev [8] applied the vector space model to sentence similarity calculation and proposed a method based on the Google search engine, introducing the Normalized Google Distance (NGD) algorithm for calculating sentence similarity using returned search results.

18.1.1.3 Sentence Ordering

In order to ensure the consistency and coherence of the summary sentences, it is necessary to arrange the order of the summary sentences. Currently, the sorting methods are of two types: one is time sorting, which generally selects a certain time as a reference point, and then calculates the absolute time of other relative times, such as using the publication date as a reference point, and calculating the absolute time for dates within this week, past or future dates, and expressions such as "today, yesterday, last night." The other is expansion sorting, which aims to improve the coherence of the summary by putting topics with certain content relevance together.

The extractive automatic summarization method only performs limited depth analysis on useful text fragments, and its efficiency and flexibility are high, so it is suitable for large-scale texts such as network information, and cases that only focus on the document topic and do not require high coherence of the summary. At present, the extractive automatic method has been widely used and researched, and many effective methods have been proposed, such as summary generation in the search engine field. But overall, the quality of the summaries obtained by the extractive automatic method is still not satisfactory. The quality of the summaries obtained by efficient methods is poor, and the methods with good summary quality are difficult to apply in various situations due to their huge computational load.

18.1.2 Understanding-Based Automatic Summarization

The main difference between understanding-based automatic summarization and extractive automatic summarization is whether the summary content comes from the original text and the depth of semantic analysis of the document. The summary content obtained by the understanding-based automatic summarization method does not completely come from the original text. It uses linguistic knowledge to obtain language structure, uses domain knowledge to make judgments and inferences, obtains the meaning representation of the summary, and finally generates the summary from the meaning representation. Understanding-based automatic summarization combines existing knowledge bases to conduct in-depth analysis and mining of the grammar, semantics, etc., of the document for "understanding" the original text, while extractive automatic summarization only conducts shallow analysis of the vocabulary of the document.

The shortcoming of understanding-based automatic summarization is that the domain is strictly limited. The reasons for its domain limitation are as follows:

1. The technology of grammar and semantic analysis for large-scale real corpora cannot be done manually; therefore, if you want to obtain high-quality analysis results, you must limit the corpora to be processed within a certain range.
2. The basis of this method is knowledge representation such as frames. Frames need to be predetermined according to domain knowledge; therefore, if you want to extend the understanding-based automatic summarization system applicable from a certain domain to another domain, a new framework needs to be reestablished, and the heavy burden of filling and organizing domain knowledge makes this method difficult to port.

18.2 Automatic Summarization Based on Keyword Extraction

The overall process of automatic summarization based on keyword extraction is shown in Fig. 18.1. The principles and details of each algorithm in the process are introduced below.

18.2.1 Text Preprocessing

The content of web page text is generally obtained through a certain method of extracting the main text from the web page, so the obtained text inevitably contains many irregular, meaningless characters or strings (such as , etc.), and even some unprocessed HTML tags. Some web pages themselves have very complex

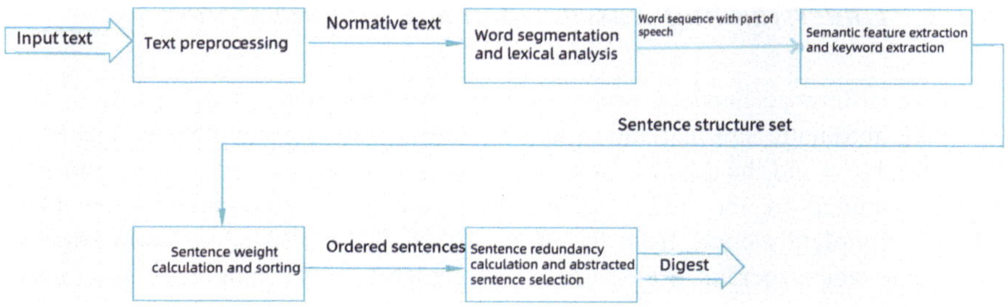

Fig. 18.1 Overall process of automatic summarization based on keyword extraction

Fig. 18.2 Web page content example

content formats. For example, the >> in Fig. 18.2 not only has no meaning for summary generation, but may also cause a large number of meaningless characters to appear when segmenting words. Since the dictionary does not contain these words, this can increase the processing time of subsequent stages such as summary word frequency and part-of-speech statistics. Therefore, in the preprocessing stage, these meaningless strings need to be removed. The numbers appearing in the text have full-width and half-width differences, and direct segmentation will cause the same.

Numbers due to full-width and half-width differences produce different words, which will also affect the extraction of summaries. In addition, other characters also have full-width and half-width differences, English letters have case differences, and so on. In the preprocessing stage, these characters need to be converted into a unified format, and then the next step is taken. This stage divides these special characters into two categories: one is meaningless strings, and the other is special format strings. For meaningless strings, they are directly deleted from the original text, and for special format strings, they are converted into the standard format used in the summary system.

18.2.2 Stop Word List

Words, as the basic units expressing the meaning of sentences and documents, can express different sentences and documents. However, the expressive ability of different words varies greatly. In the process of generating summaries and extracting keywords, filtering stop words can improve the system's computational efficiency and the quality of the summary. The construction method of the stop word list is as follows.

Manually Compile a Stop Word List Similar to a Blacklist. If a word in the document appears in the stop word list, it will be deleted. A word becomes a stop word mainly based on the following two rules. One is that the word appears frequently in the text, is widely distributed, and has a very "general" meaning. For example, words like "I" and "just" will appear in almost every document. Such words are not significant in distinguishing the importance of different sentences, and may even interfere with the true meaning of the sentence, so these common words should be ignored. Of course, if too many stop words are specified, it may also cause the sentence's meaning to be inaccurately expressed, so, the addition of such words needs to be done very carefully. Second, the frequency of this word is very high, but the actual meaning is not significant. These stop words mainly include modal particles, prepositions, adverbs, conjunctions, etc., which usually have no clear meaning, such as the common Chinese words "的," "在," "和," "接着," etc.

When using the stop word list, pay attention to the following two points:

1. Retain nouns, verbs, adjectives, and adverbs.
2. Retain the stop words that appear in the document title or topic description, and remove all other stop words.

Filtering and storage of the stop word list will be introduced in Sect. 18.2.3.

18.2.3 Double Array Trie Tree

Trie tree, also known as retrieval tree, word search tree or dictionary tree, is a tree-like structure used to store a large number of strings. Trie tree is a common implementation of the dictionary in the Chinese word segmentation algorithm. Its essence is a definite finite state machine (definite finite machine, DFA), where each node represents a state of the automaton. These states in the dictionary include "word prefix," "word formed," and so on. Its advantages are: it saves storage space by using the common prefix of strings, minimizes unnecessary string comparisons, and has higher query efficiency than hash tables. Trie tree has three basic characteristics:

1. The root node does not contain characters, and each node other than the root node contains only one character.
2. Connect the characters passed on the path from the root node to a certain node to form the string corresponding to that node.
3. The characters contained in all child nodes of each node are different.

Figure 18.3 is a Trie tree constructed for the five words—say, she, shr, he, and her.

There are two types of searches in the Trie tree: one is to find whether a word exists in the Trie tree, which is used for stop word filtering, and the other is to find the substring that can match the pattern string in the Trie tree in a string, mainly used for word segmentation. The general process of finding a word in the Trie tree is as follows: starting from the root node, go down layer by layer along the nodes corresponding to the characters in the word, until the last character of the word or

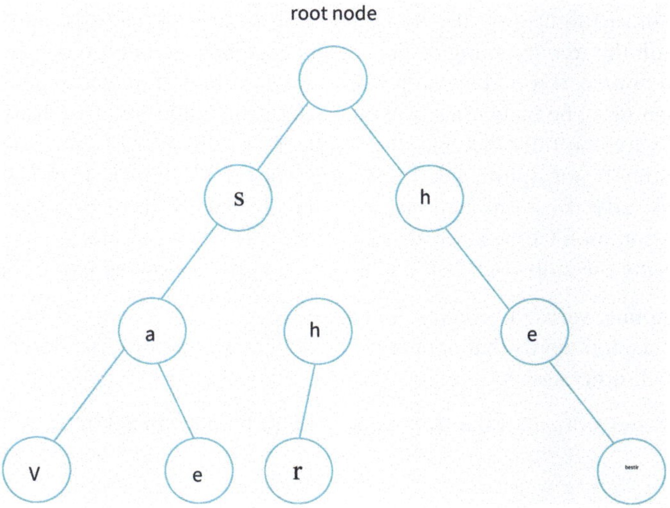

Fig. 18.3 Trie tree constructed for five words

the leaf node is reached. If the leaf node is not reached and this node is the end node of the word, the search is successful. If the leaf node is reached and the character in it matches the last character of the word, the search is successful. If a certain character in the word cannot find the corresponding node in the tree, or the Trie tree's leaf node has been reached, but the word has not matched the last character, the search fails. The difference between the two matching situations lies only in the handling of reaching the Trie tree leaf node: if the leaf node is reached and the word has not matched the end character, record this position in the string, then move the pointer back, and start searching from the root node of the Trie tree.

A Trie tree is a simple and efficient data structure that is easy to understand, but its internal storage consumption is quite large. To reduce the space waste of the Trie tree structure while maintaining search efficiency, some researchers have proposed a method to represent the Trie tree with three linear arrays. They further improved this method by using two arrays to represent the Trie tree, resulting in the double-array Trie tree (Double-Array Trie), as shown in Fig. 18.4.

The double-array Trie is a simple and effective implementation of the Trie tree, which consists of two integer arrays, one is base [], and the other is check []. The elements in these two arrays correspond to each other one-to-one. Suppose the array index is i. If both base[i] and check[i] are 0, it means the position is empty. If base[i] is a negative value, it indicates that the state is a word. check[i] represents this.

Relatively speaking, the Trie tree is a simple and efficient data structure, easy to understand, but it consumes a lot of memory. In order to reduce the space waste of the Trie tree structure and ensure the efficiency of the Trie tree query, researchers have proposed three linear array methods to represent the Trie tree, and further improved it on this basis, using two arrays to represent the Trie tree, which is the double-array Trie tree, as shown in Fig. 18.4.

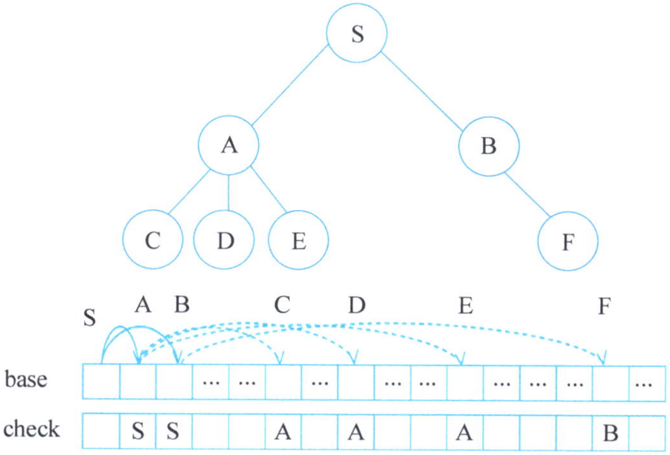

Fig. 18.4 Double-array Trie tree

If the update cycle is very long and the data scale is very large, then the double-array Trie tree is undoubtedly the best choice. The following example illustrates the process of constructing a word segmentation algorithm dictionary with a double-array Trie tree.

In addition to using the double-array Trie tree structure to store the dictionary when segmenting words, the management of the stop word list will greatly improve the efficiency of stop word filtering after using the double-array Trie tree. If you use the STL container to store the stop words of length m, then it takes m string comparison to judge whether a word is a stop word. In this way, for a document containing n words, its time complexity is O(mn). If the double-array Trie tree is used for storage, since it is a complete match for the entire word, and the word has been segmented, there is no need to backtrack when performing pattern matching, that is, the number of string comparisons required to judge whether a word is a stop word is only the same as the length of the word itself. Therefore, for a document containing n words, its time complexity is O(an), where a is the average word length, generally 2–4.

18.2.4 Keyword Extraction

When extracting keywords, in addition to using the commonly used TF-IDF features, the accessor variety (AV), adjacency diversity value of the word, the locality of the word's position, and the position of the sentence where the word is located can also be reflected in the weight of the word as very important factors.

18.2.4.1 TF-IDF Feature of the Word

In Chinese, weight is a relative concept. The weight of a certain indicator is the relative importance of this indicator in the overall evaluation, and the weight of a word is the relative importance of a word in the corpus. The more classic algorithm for calculating word weight is the TF-IDF algorithm. In information retrieval, TF-IDF is a numerical statistic, aimed at reflecting the relative importance of a word to a document set or a document in a corpus.

TF represents the proportion of a word in all words in a document. The larger the TF value, the higher the word frequency. DF is the ratio of the total number of documents to the number of documents containing a certain word. The larger the IDF value, the fewer documents contain that word, indicating that the word has a stronger document classification ability.

The TF-IDF feature is represented by the product of TF value and IDF value. Its meaning is: if a word only appears in one or a few documents in a document set or corpus, and its frequency in the document is very high, it indicates that the word is important for the document. It has strong representativeness, so it can be used as the keyword of the document, and it can also play an important role as a feature word in document classification.

The common calculation formula of TF-IDF feature is as follows:

$$w_{(t,d)} = \log_2 \left(TF\right)_{(t,d)} + 1.0) * \log_2 x \left(\frac{N}{n} + 1.0\right) \tag{18.2}$$

where, $w_{(t,d)}$ is the weight of word t in document *d*, $TF_{(t,d)}$ is the frequency of word t in document *d*, *N* is the total number of documents, and n_t is the number of documents in the document set containing word *t*.

In automatic abstract extraction, sometimes due to the small number of documents, or even when using a single document abstract method, the number of documents is 1, so the IDF differentiation is not large. Neto and others [9] proposed to use inverse sentence frequency (ISF) instead of IDF for sentence-level word weight measurement, which is essentially still a word weight calculation method of TF-IDF. Because sentences are very short and contain few valid words, some short sentences cannot even express complete meanings independently; Hence, the effect of using ISF is not ideal, so word frequency is generally used to replace TF-IDF features.

18.2.4.2 AV Value of Words

In real society, there are such people: they often go in and out of different occasions, and have some kind of connection with the people in each occasion. In this social circle, such people are usually considered very active and suitable as representatives of this social circle. The AV value of a word is a weight calculation method applied to words by analogy with the above life prototype. Feng and others [10] used the concept of adjacency diversity to describe the flexibility of word usage. If an article is compared to a social circle in a social network, then the words that make up the basic unit of the article can be compared to different people in the social network. According to the previous description, the more diverse the context in which a word appears, the more important the word is in the entire article.

Here, the n words before and after a word are used to represent the context in which the word appears (the context of the word). The left AV value of a word refers to the number of types of words or words that appear on the left of the word, the right AV value of the word refers to the number of types of words or words that appear on the right of the word, and the AV value of the word is defined as the number of types of words or words that appear on its left and right. The larger the AV value of a word, the more flexible its use, the greater the probability of its use in different contexts, and the greater the probability of it becoming a keyword. Take, for example, the following text:

If the context window size is set to 2, then the left adjacency set of the string "National Day" excluding single-word words is {The set of adjacent right words for

"National Day" includes {convene, participate, ensure, all, attend, make it, comply}, and the set of adjacent right words includes {traffic, safety, activities, units, during}. Therefore, the left AV value of "National Day" in this corpus is 7, the right AV value is 5, and the AV value is 12. Similarly, the left AV value of "traffic" is 5, the right AV value is 5, and the AV value is 10.

From the above examples, it can be seen that the use of AV values is quite effective for discovering high-frequency keywords.

18.2.4.3 Locality of Word Position

Word frequency and AV values can effectively extract high-frequency, active keywords, but it is difficult to effectively extract low-frequency keywords. Zhang Qingguo and others [11] extracted keywords based on features such as word diameter and word distribution deviation. The word diameter refers to the distance between the first and last appearance of a word in the text. The word distribution deviation considers the statistical distribution of words in the article.

The variance of the word's position in the document can be used to represent the locality of the word. Suppose the candidate word T appears n times in the corpus, and the positions are P_1, P_2, …, P_n, then the position variance $D(T)$ of T is

$$D(T) = \frac{\sum_{i=1}^{n}(P_i - P)^2}{n-1} \tag{18.3}$$

where P represents the mean value of the candidate word T's position. The position variance of a word represents the sparsity of the word's distribution. The smaller the position variance of a word, the higher the concentration of each position. Therefore, the smaller $D(T)$, the better the locality, on the contrary, the worse the locality. Suppose locality estimation (LE) represents the measure of the word's locality, then the calculation formula of locality is as follows:

$$\mathrm{LE}(T) = \frac{1}{D(T)} \tag{18.4}$$

18.2.4.4 Position of the Word in the Sentence

A survey result shows that the probability that the topic of a paragraph is the first sentence is 85%, and the probability that it is the last sentence is 7%. Therefore, a piecewise function can be used to quantify the impact of sentence position (SP) on word weight.

The calculation formula of sentence position (SP) is as follows:

$$\mathrm{SP}=\begin{cases}-\alpha\left(x-\dfrac{1}{2}\right), & x<\dfrac{1}{2}\\ \beta\left(x-\dfrac{1}{2}\right), & x>\dfrac{1}{2}\end{cases} \tag{18.5}$$

where x represents the position of the sentence in the paragraph; l represents the number of sentences in the paragraph; α represents the smoothing factor of the sentence position in the upper half of the paragraph; and β represents the smoothing factor of the sentence position in the lower half of the paragraph.

18.2.4.5 Calculation of Word Weight

The weight of a word is obtained by summing the weighted measures of three features: the AV value of the word, the local position of the word, and the position of the sentence where the word is located. The weighting coefficients of each feature measure are obtained through manual training methods using a training dataset.

18.2.5 Sentence Segmentation

In the lexical analysis stage, the punctuation in the article is marked as the "/w" category and segmented. When scanning the segmented and marked text string, if a segment with the "/w" category is encountered, it is judged whether it meets the sentence segmentation rules. If it does, the sentence is segmented at this point. If a punctuation mark meets the following conditions, sentence segmentation is performed.

1. The punctuation is ".", "?", "!", "...".
2. The punctuation is ",", ";", "(",")", "—", and the distance from the beginning of the sentence to the punctuation is greater than the sentence length threshold. For paragraphs without punctuation at the end, a period is automatically added, and the sentence is segmented from this position.

In addition, due to the complexity of the content of web documents, a sentence length threshold needs to be set. For long sentences without punctuation, if it exceeds this threshold, further processing is performed on this sentence, specifically as follows:

1. If the previous content is a string composed of repeated single characters (meaning that they are all single characters after segmentation), discard it.
2. If the previous content is all non-repeated single characters (that is, each character is segmented into a single character, not a word) composed of strings, discard it.

3. Look for sentence-ending modal particles, such as "了," "吗," "等." If present, segment the sentence from this position, if not, discard this sentence and continue scanning from this position.

18.2.6 Sentence Similarity Calculation

The vector space model views sentences as vectors. Each dimension of the vector represents a segment of the sentence. The granularity of this segment can be a single character, word or phrase, n-tuple, etc., collectively referred to as terms (Term), represented by t. Each element of the vector represents the weight of the term, denoted by w. Thus, sentence S can be represented as a vector (t_1, w_1; t_2, w_2;...; t_N, w_N), where term t is called a feature, and its weight w is called feature weight, simply denoted as $S(w_1, w_2,\ldots, w_N)$. To represent a sentence as a vector space model, it is necessary to determine the features and feature weights. Here, words are used as features. There are many factors affecting the weight, and TF-IDF is chosen as the main calculation basis. The simplest method is to calculate the inner product of the sentence vectors.

$$\mathrm{sim}(Q,D)=\sum_{t\in D\cap Q} w_{d,t} * w_{q,t} \tag{18.6}$$

The sentence redundancy between candidate summary sentences is represented by cosine similarity. The sentence redundancy of a candidate summary sentence s with the sentence set S of the summary is composed of two parts: The first is the maximum redundancy between the candidate summary sentence and the sentences in the summary sentence set:

$$R_{\max}=\max_{s_j\in S}\mathrm{sim}\left(s_i,s_j\right) \tag{18.7}$$

The second is the average redundancy of the candidate summary sentence with all sentences in the summary sentence set:

$$\overline{R}=\frac{\sum_{j=1}^{n}\mathrm{sim}\left(s_{1i},s_j\right)}{n} \tag{18.8}$$

where n is the total number of sentences in S. Therefore, the weight of the candidate summary sentence is recalculated as

$$w'=\lambda w-(1-x)\left(\alpha R_{1\max}+(1-\alpha)R(-)\right) \tag{18.9}$$

where W is the sentence weight calculated only based on the weight of the words in the sentence, which does not change when calculating sentence redundancy multiple times; λ and α are estimation coefficients, generally obtained through training on a certain training set.

According to the calculated weight of the candidate summary sentence, all candidate summary sentences are re-sorted, the candidate summary sentence with the highest weight is found, and it is judged whether it meets the summary length requirement. If it does, the sentence is added to the summary sentence set, otherwise, the sentence is marked as a noncandidate summary sentence, and the next candidate summary sentence is taken for judgment.

18.3 Topic-Oriented Automatic Summarization

In the field of information retrieval, document content relevance as the main measurement method in document retrieval has been deeply studied over the past few decades. The granularity of a document can be any level of information content (such as document sets, documents, paragraphs, sentences, etc.). In the task of automatic summarization, the "document" here can be directly represented by sentences, so the relevance becomes the similarity between the sentence and the query.

When a search engine calculates document summaries, in addition to reflecting the main content of the document, it also needs to be relevant to the query, so that it can provide the information that users really care about. A query submitted by a user can be regarded as a topic, and the summary extracted according to the query is called a topic-oriented summary. The structure of the topic-oriented document summary system is shown in Fig. 18.5.

The understanding of the query directly determines the final performance, and a wrong understanding cannot produce correct retrieval results. In the description of the query topic, the words after stop word processing can be divided into query words and auxiliary words. Query words are the words that truly express the intention of the user's query. Although auxiliary words can independently express actual meanings, they only play a role in making the tone coherent and appropriate in the topic, and they have no actual meaning to the content to be queried. Their function is similar to the interjections in English such as "by the way," and removing these words does not affect the grasp of the semantics truly expressed by the query. For example, the query statement is "Please help me look up the book 'Survey of Chinese Farmers,'" the real query request is "Survey of Chinese Farmers," and words like "please," "help," and "look up" are auxiliary words.

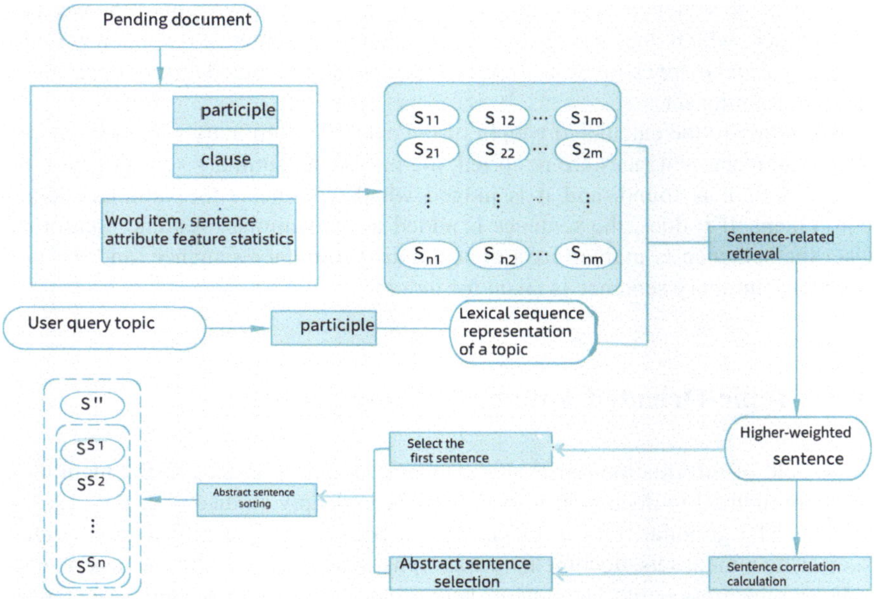

Fig. 18.5 Structure of a topic-oriented document summary system

18.3.1 Improved Maximum Marginal Relevance Method

Carbonell and others [12] proposed the concept of marginality for the selection of abstract sentences, and based on the concept of marginality, they proposed a maximum marginal relevance (MMR) abstract method. Its essence is: when selecting abstract sentences, make the sentence to be selected into the abstract sentence set, both highly relevant to the topic and make the redundancy between this sentence and the selected abstract sentence set as small as possible, thereby ensuring the relevance between the sentence and the query, while reducing the redundancy of the abstract, increasing the coverage of the content, and thus obtaining a higher quality of the abstract. The calculation formula of maximum marginal relevance is as follows:

$$\text{MMR} = \arg \max_{s_i \epsilon (D-S)} \left\{ \lambda \text{sim}_1 \left(s_i, q \right) - \left(1 - \lambda \right) \max_{s_j \epsilon S} \text{sim}_2 \left(s_i, s_j \right) \right\} \tag{18.10}$$

where q represents the query topic; D represents the document composed of a set of sentences; S represents the selected abstract sentence set, which is a subset of D; $D{-}S$ represents the difference between sets D and S; s represents the candidate abstract sentence; sim_1 represents the relevance of the candidate abstract sentence to the query; sim_2 represents the redundancy between the candidate abstract sentence and the selected abstract sentence set. The parameter λ is used to adjust the weights of sim_1 and sim_2.

The maximum marginal relevance method can obtain a summary that is relatively relevant to the query, but it is not sufficient to reflect the main content of the document. Therefore, it is necessary to consider the weight of the sentence itself; then formula (18.10) becomes

$$W_{q,s_i} = \alpha W_{s_i} (1-\alpha)\left(\lambda \text{sim}_1 (s_i, q) - (1-\lambda) \max_{s_j \epsilon S} \text{sim}_2 (s_i, s_j) \right) \tag{18.11}$$

The redundancy between s_i and s_j can be represented as the weighted sum of the maximum redundancy and the average redundancy, which better reflects the true redundancy between the candidate summary sentence and the summary sentence set.

18.3.2 Domain Topic Word List

For certain specific application fields, such as the extraction of summaries in vertical search engines for the automotive industry, the addition of industry knowledge can improve the quality of the summary. Generally, the weight of industry terms is increased by loading the topic word list, thereby improving the professionalism of the summary. The topic word list includes words and weights. Table 18.1 is an example of a judicial topic word list.

The topic word list is used in the final step of word weight calculation. After calculating the weight W_t based on the statistical features of the word, the domain topic word weight W_u is weighted to obtain the final weight of the word.

$$W_t^{'} = \lambda W_t + (1-\lambda) W_u \tag{18.12}$$

If the topic word list is introduced after segmentation, unexpected situations may occur: there are fragmented words in the segmentation system, and even if these word fragments form words that appear in the topic word list, they are meaningless. However, if the topic word list is loaded into the segmentation dictionary during the segmentation stage, this situation will not occur.

The word dictionary will not have this problem.

Table 18.1 Example of judicial topic word representation

Word	Weight
Politics and law	0.8
Legal system	0.6
Law	0.65
Court	0.5

18.3.3 Sentence Inclusion Relationship

It is also common for one sentence to be included in several other sentences. For example:

"中国未来20年房价上涨的压力仍然是很大的." 姜伟新说.
中央政府一直房价过快上涨的决心更大.
住建部部长姜伟新明确表示, 中国未来20年房价上涨压力仍然很大, 但中央政府抑制房价过快上涨
的决心更大.

"China will still face great pressure to raise housing prices in the next 20 years," said Jiang Weixin. The central government is more determined to curb the rapid rise in housing prices. Jiang Weixin, former Minister of Housing and Urban-Rural Development, made it clear that China will still face great pressure to raise housing prices in the next 20 years, but the central government is more determined to curb the rapid rise in housing prices.

The above example is obtained by sentence segmentation into three sentences, where the third sentence includes the first and second sentences. The first and second sentences are referred to as the subclauses of the third sentence, and the third sentence is referred to as the long sentence. When the length of the subclause is very small, the redundancy between the subclause and the long sentence, calculated using cosine similarity, will be relatively small and cannot represent their inclusion relationship. Therefore, another method is needed to represent this relationship. The concept of word overlap can be defined, where two sentences are seen as word sets *A* and *B*, with *A* before *B*. The simplest information redundancy can be quantified as $A \cap B$. For example, the formula for calculating the word overlap of sentences *A* and *B* relative to *A* is as follows:

$$\text{Overlap}\,B_A = \frac{A \cap B}{B} \tag{18.13}$$

Therefore, the inclusion between sentences *A* and *B* can be defined as

$$\text{Contain}(A, B) = \max\left(\text{Overlap}B_A, \text{Overlap}A_B\right) \tag{18.14}$$

Consider the three sentences in the above example as word sets *A*, *B*, and *C*, the process of calculating the inclusion between the first and third sentences is as follows:

1. Calculate the word overlap $\text{Overlap}C_A$ between sentences *A* and *C* relative to *A*.
2. Calculate the word overlap $\text{Overlap}A_C$ between sentences *A* and *C* relative to *C*.
3. Calculate the inclusion between sentences *A* and *C*:

$$\text{Contain}(A, C) = \max\left(\text{Overlap}C_A, \text{Overlap}A_C\right) \tag{18.15}$$

4. If Contain (A ， C) is greater than or equal to the predefined inclusion threshold Contain (Thresh), then C includes A. OverlapC_A is greater than OverlapA_C or A includes C OverlapC_A is greater than OverlapC_A, otherwise, there is no inclusion relationship between A and C.

In the 245,890 news corpora collected by the author's research group, a total of 19,048 pairs of sentences with inclusion relationships were extracted. From these, 100 pairs of sentences were randomly selected for manual evaluation, of which 28 pairs had no inclusion relationship. The accuracy of the method in this small test set was 72%.

In the above-mentioned process of calculating word overlap, each word in the sentence is assigned the same weight of 1. At this time, if the subordinate clause has a few more or fewer general meaning words than the long sentence, and these words are not stop words, it will generate a lot of noise, making it difficult to calculate their inclusion relationship. Therefore, if different weights are assigned to each word when calculating word overlap, the noise caused by word differences can be effectively eliminated. The definition of weighted word overlap is as follows:

$$\text{Overlap}B_A = \frac{\sum_{t \in A \cap B} w_t}{\sum_{t \in B} w_b} \tag{18.16}$$

18.4 Research on Chinese Document Automatic Summarization Technology Based on Topic Model and Information Entropy

In order to apply topic information to the extraction of summary sentences, this section proposes a method based on information entropy to measure the importance of sentences [13]. This method establishes a probability model for sentences as random variables, calculates the occurrence probability of sentences based on this model, calculates the information entropy of sentences, and finally measures the importance of sentences based on information entropy. At the same time, the influence of the number of words in a sentence on the weight of the sentence is also considered, and a threshold for the minimum number of words in a sentence is set to filter out sentences below this threshold.

Depending on whether the automatic summary originates from the original text, automatic summaries can be divided into extractive summaries and abstractive summaries. Extractive summaries directly extract representative sentences from the original text as the document summary. Usually, extractive summaries view the document as a collection of sentences, and select sentences from this collection as the document summary through algorithms. The result of extractive summaries mainly depends on the choice of the algorithm, and a good algorithm can usually accurately find the main sentences of the article, generating document summaries.

In addition, extractive summaries often have the characteristic of not being limited by domain. Summarizing summaries first deeply analyze the original text, extract information based on the domain knowledge base, then use natural language processing technology for analysis, and finally use linguistic knowledge and natural language processing technology to generate document summaries.

The research on automatic document summarization began more than 50 years ago, when Luhn calculated the weight of words by counting word frequency, calculated the weight of sentences through the weight of words, and selected specific sentences as the summary of the document according to the weight. Foreign research on document summary technology has a long history and has made significant progress. Some scholars have tried to introduce the chapter structure features of the document and use similarity analysis and other means to select the document summary. Salton and Gerard analyzed the document in units of paragraphs through the structure of the article, and measured the importance of paragraphs through the similarity between paragraphs. But this method relies on the chapter structure of the article, and its applicability to documents with simple chapter structures is poor. Sasha and others adopted a method called SC at DUC2004; the core idea of this algorithm is: First, measure the importance of different categories through the results of sentence clustering. The category containing more sentences is considered more important. Then extract representative sentences from the category as the summary of the document. In the clustering process, the SVM model is used to represent sentences, and the cosine value between vectors is used to measure the similarity of the sentences. This method focuses on the lexical analysis of the document, and using the SVM model to represent sentences may lead to a dimensionality disaster, with the training cost being too high.

Domestic research on automatic summarization technology started late, i.e., in the 1980s. In 1988, Shanghai Jiaotong University developed an experimental system for automatic compilation of Chinese literature abstracts. This system has been able to extract summaries of scientific and technological literature and has achieved certain results. By the twenty-first century, Chinese automatic summarization technology has made significant progress. Wang Jicheng and others [14] proposed a method based on the chapter structure of Chinese Web documents for automatic summarization, that is, sequentially performing chapter structure analysis, word weight calculation, keyword extraction and sentence weight statistics, and finally generating a summary. Zhang Qi et al. [15] proposed a document summary extraction method based on sentence similarity. When measuring sentence similarity, the method considers unigram, bigram, and trigram information, and combines these similarities through a regression approach. This method uses statistical machine learning techniques and takes into account the position information of words, which is helpful for keyword mining. However, it does not effectively recognize the importance of certain valuable words that appear less frequently, such as names and place names, leading to a lower probability of extracting topic sentences related to these words.

In order to overcome some of the shortcomings of some methods in extracting summaries, such as the need for many rules and poor universality, the author's research group proposed an unsupervised document summary algorithm based on the topic model. The LDA model obtains the topic distribution of each document in the document set, along with the word distribution corresponding to each topic. At the same time, the most relevant topics are selected based on the weight of the topic distribution to mine the shallow semantics of the text.

18.4.1 Topic Model

The topic model is a statistical model used in machine learning and natural language processing to discover abstract topics in a series of documents. Intuitively, if an article has a central idea, then some specific words will appear more frequently. For example, if an article is about dogs, then words like "dog" and "bone" will appear more frequently; if an article is about cats, then words like "cat" and "fish" will appear more frequently. However, some words, such as "this" and "and," should appear roughly equally in both articles. But the reality is an article usually contains multiple topics, and each topic occupies a different proportion. Therefore, if an article is 10% about cats and 90% about dogs, then the number of times keywords related to dogs appear will be about nine times the number of times keywords related to cats appear. A topic model tries to represent this feature of documents in a numerical framework, automatically analyzing each document, counting the words in the document, and determining which topics are present in the current document and what proportion each topic occupies based on the statistical information.

Latent Dirichlet Allocation (LDA) is currently a popular topic model and a typical bag-of-words model, which assumes that an article is a collection of words, with no order or sequence between the words. A document can contain multiple topics, and each word in the document is generated by one of these topics, i.e., it is assumed that the multiple topics in a document follow a multinomial distribution, and all the words within a topic also follow a topic distribution. In addition, the Bayesian estimation method is used, assuming that the prior distribution of the document topic follows a Dirichlet distribution, and the prior distribution of the words in the topic also follows a Dirichlet distribution. The structure of the LDA model is shown in Fig. 18.6.

The document generation process of the topic model is as follows.

Sample from the Dirichlet distribution α to generate the topic distribution $\boldsymbol{\theta}_i$ of document *i*.

1. Sample from the multinomial distribution $\boldsymbol{\theta}_i$ of the topic to generate the topic $z_{i,j}$。 of the *j*th of document *i*.
2. Sample from the Dirichlet distribution β to generate the word distribution $z_{i,j}$ of the topic $\phi_{z_{i,j}}$.
3. Sample from the multinomial distribution $\boldsymbol{\theta}_i$ to finally generate the word $\boldsymbol{w}_{i,j}$.

Therefore, the joint distribution of all visible and hidden variables in the entire model is

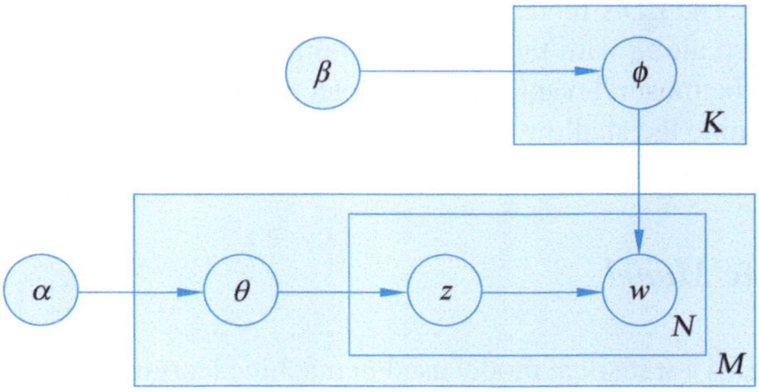

Fig. 18.6 LDA model structure

$$P(w_i, z_i, \theta_i, \phi|\alpha, \beta) = \prod_{j=1}^{N} P(\theta_i|\alpha) P(z_{i,j}|\theta_i) P(\phi|\beta) P\left(w_{i,j}|\theta_{z_{i,j}}\right) \quad (18.17)$$

The specific process of Gibbs sampling is as follows.

First, traverse all the words in all documents and randomly assign a topic to each, i.e., $z_{m,n} = t \sim Mult\left(\frac{1}{K}\right)$, where m represents the mth article, n represents the nth word in the document, t represents the topic, k represents the total number of topics, and the corresponding $n_m^{(k)} + 1$, $n_m + 1$, $n_k^{(k)} + 1$, $n_k + 1$ represent the number of times the kth topic appears in the mth document, the sum of the number of topics in the mth document, the number of times the kth topic corresponds to t, and the total number of words corresponding to the kth topic.

Then repeat the following operations.

Traverse all the words in all the documents. If a word in the current document mm corresponds to topic k, then $n_m^{(k)} - 1$, $n_m - 1$, $n_k^{(k)} - 1$, $n_k - 1$, i.e., first take out the current word, and then according to the Topic in LDA. The probability distribution of the sample is used to sample a new topic, and the corresponding counts are $n_m^{(k)}$, n_m, $n_k^{(k)}$, n_{k^o}, and n_k on this new topic k are incremented by one. The calculation formula for the probability distribution of the topic sample is as follows:

$$P\left(z_i = k|z_{-i}, w\right) \infty \left(n_{k,-i}^{(t)} + \beta_t\right)\left(n_{m,-i}^{(t)} + \alpha_k\right) / \left(\sum_{t=1}^{V} n_{k,-i}^{(t)} + \beta_t\right) \quad (18.18)$$

Finally, after the iteration ends, estimate the parameters of the model according to the distribution of the obtained topics. The estimation formula for the parameters is as follows.

$$\phi_{k,t} = \left(n_k^{(t)} + \beta_t\right) / \left(\sum_{t=1}^{V} n_k^{(t)} + \beta_t\right) \tag{18.19}$$

$$\phi_{m,t} = \left(n_m^{(k)} + \alpha_k\right) / \left(\sum_{k=1}^{V} n_m^{(k)} + \alpha_t\right) \tag{18.20}$$

18.4.2 Information Entropy

To measure the importance of a sentence, the unit of information entropy is introduced. In information theory, entropy is used to measure the expected value of a random variable. The entropy value H of a random variable X with a value domain of $\{x_1, x_2, \ldots, \boldsymbol{x}_n\}$ is defined as

$$(X) = E\left(I(X)\right) \tag{18.21}$$

where $I(X)$ is the self-information of the random variable X. Meanwhile, according to the definition of expectation and the formula of self-information, another expression of entropy value H is obtained as

$$H(X) = \sum_{i=1}^{n} P(x_i) I(x_i) = -\sum_{i=1}^{n} P(x_i) \log_b P(x_i). \tag{18.22}$$

In this section, information entropy is used to measure the average expected value of a sentence appearing in a certain combination of words, which is a random variable. This random variable is modeled using an indicator random variable approach, where its value range is limited to two possible values: {appear, not appear}. The information entropy of the sentence is then calculated based on the probability distribution of this random variable across the specified value range.

18.4.3 Calculation Method of Sentence Information Entropy

Text uses information entropy to measure the weight of sentences. This section makes an independence assumption for the words in the document, that is, it is assumed that the appearance of each word is unrelated to the appearance of other words. Therefore, the probability of a sentence appearing in a document is

$$P\left(\text{sentence}|\text{topic}_{(j)}\right) = \prod_{i=1}^{m} P\left(\text{token}_{(i)}|\text{token}_{(j)}\right) \tag{18.23}$$

where $\text{token}_{(i)}$ is the ith word of the sentence, m is the number of words in the current sentence, $topic_{(j)}$ is the topic with the highest probability in the current document topic distribution, $P(\text{topic}_{(i)}|\text{topic}_{(j)})$ is the probability value of a specific word appearing under the current topic obtained by training the LDA model, that is, $\varphi_k = \text{topic}_{(j)},\ t = \text{topic}_{(i)}$.

Considering the appearance of a sentence in a certain combination of words as a random variable, the value range of this random variable is {appear, not appear}, and the calculation formula of the information entropy of this random variable is

$$E(\text{sentence}) = P\left(\text{sentence}|\text{topic}_{(j)}\right) * \log\left(\frac{1}{P\left(\text{sentence}|\text{topic}_{(j)}\right)}\right) + \overline{P}\left(\text{sentence}|\text{topic}_{(j)}\right) * \log\left(\frac{1}{\overline{P}\left(\text{sentence}|\text{topic}_{(j)}\right)}\right) \tag{18.24}$$

where $E(\text{sentence})$ is the information entropy of a sentence, $P(\text{sentence}|\text{topic}_{(j)})$ is the probability value of the sentence appearing under the current topic, and $\overline{P}\left(\text{sentence}|\text{topic}_{(j)}\right)$ is the probability value of the sentence not appearing under the current topic.) is the probability value of the sentence not appearing under the current topic, given the current combination of words the sentence does not appear under the current topic with the current combination of words.

18.4.4 Introduction of Algorithm

18.4.4.1 Algorithm Proposal

Traditional document summarization systems usually calculate the weight of words and sentence similarity, ignoring the topic information of the document. Summarization algorithms that consider the topic information of the document usually judge the topic of a document based on the topic sentence, and the judgment of the topic depends on the determination of the topic sentence. A document collection can be seen as composed of documents generated by different topics, where the topic is similar to the category of the document. For a document in the document collection, its central idea usually comes from one or a few topics, so in the process of extracting document summaries, it should mainly focus on this one or these few topics. At the same time, for different topics, the weight of words under each topic is different. For example, words related to sports should have a high weight under a sports topic. In addition, some sports-specific terms, such as names of people, names of events, etc., should also be recognized by this sports topic and given a higher generation probability than other topic words. The LDA model can perfectly solve these problems. It can accurately obtain the topic distribution of the document and display the priority of the topic in the form of probability, and at the same time, it can correctly obtain the word distribution under each topic, so that some uncommon words belonging to this topic do not lose their weight and recognition due to their rare occurrence. When identifying the priority of words under a certain topic, it is also represented in the form of word generation probability, where words with higher probabilities have higher weights. In this way, the weights of words with clear categories but rare occurrences can also be well estimated.

Based on this, this section uses the LDA model to perform a shallow semantic analysis on the document collection and documents, and obtains the topic distribution of each document in the document collection and the regional distribution of the corresponding topics. By filtering the topic to obtain the weight of the words in the document, at the same time, it is necessary to consider the situation where the weight given to the ultra-short sentence when measuring the sentence weight with information entropy is too high and does not match the actual situation. Here, a threshold is set for the number of words in the main sentence of the document, and only sentences with a number of words not lower than this threshold are selected as the main sentences of the document.

18.4.4.2 Algorithm Process

The summary algorithm process based on the topic model and information entropy is as follows:

1. Preprocess the documents in the document collection, such as Chinese word segmentation, stop word removal, etc., and convert the document into a word space vector.
2. Perform Gibbs sampling on the above space vector to obtain the topic distribution of the document.
3. For each document, select the topic with the highest probability in the topic distribution, and obtain the corresponding word probability distribution based on the selected topic.
4. Calculate the information entropy of each sentence in the document and obtain the weight of the sentence.
5. According to the weight of the sentence and the word limit, obtain the summary of each article.

The flowchart of the summary algorithm based on the topic model and information entropy is shown in Fig. 18.7.

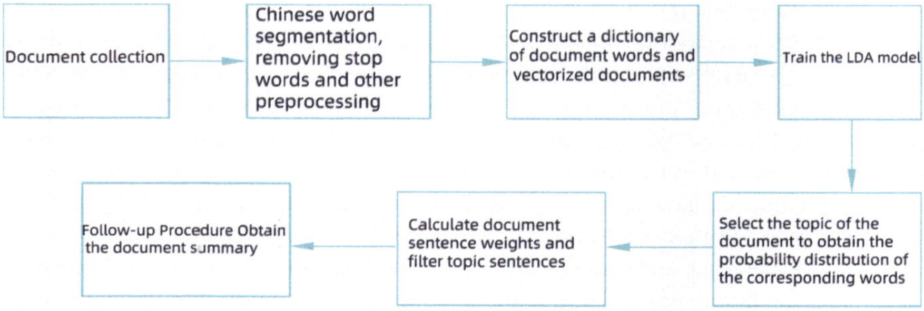

Fig. 18.7 Flowchart of the summary algorithm based on the topic model and information entropy

Table 18.2 Automatic summary test results

Classification result	Number of documents	Proportion %
Accurately reflecting the topic	184	61.33
Basically reflecting the topic	90	30.00
Not reflecting the topic well	26	8.67

18.4.5 Example of Automatic Summarization Application

This section uses NLPIR-ICTCLAS for Chinese document segmentation, focusing on the trend of natural language processing towards real-world materials and instantiation. The test is conducted with 300 real documents, which are various news articles randomly crawled from the Sina News channel using a web crawler. The number of topics k in the LDA model here is set to 200. The hyperparameters are set empirically, α is 0.25, and β is 0.01. The number of iterations is set to 200. The threshold for the number of words in a sentence is set to 10, and sentences below this threshold are not selected as candidate sentences for the final document summary. The upper limit of the number of words in the summary is set to 200. After calculating the weight of the sentence, one or more sentences are selected in order of weight from high to low as the document summary. The number of sentences selected depends on the number of words in the sentences already selected, so that the total number of words in the final document summary does not exceed the set upper limit. Finally, each document's summary is compared with the content and title of the document, and the relevance of the summary to the document content is judged.

Manual scoring is used to evaluate the summary results, divided into three levels, which are accurately reflecting the topic, basically reflecting the topic, and not reflecting the topic well. In order to reduce the impact of human factors on the evaluation results, the final result is the average after removing the highest and lowest scores. The specific test results are shown in Table 18.2.

The NLPIR Big Data Semantic Intelligence Analysis Platform provides an automatic summary function, which can automatically extract the summary of single or multiple texts, making it convenient for users to quickly browse the text content.

Fig. 18.8 Automatic summary example

Figure 18.8 is an example of this function. This example selects a sports news report from 2022 as the corpus and automatically summarizes this corpus.

As can be seen from Fig. 18.8, this function automatically extracts the key information of this sports news report and reorganizes it into a summary.

This section introduces the improvement of summary processing efficiency, the extraction of summaries based on keywords, and the removal of redundancy between sentences in the summary. In terms of improving processing efficiency, the inverted list structure of the search engine is introduced to count the features of words and sentences, and the double-array Trie tree is used to store the segmentation dictionary and user topic word table, in order to improve the efficiency of word lookup. When combining keyword extraction with summary extraction, the diversity of adjacent words and the locality of word positions are introduced in keyword extraction to improve the extraction quality of high-frequency and low-frequency words, respectively. In terms of removing sentence redundancy, the concept of inclusion between sentences is introduced, which can effectively deduplicate multiple sentences in the article that have an inclusion relationship, reducing the information redundancy of the summary and improving the quality of the summary.

Combining summary extraction, the diversity of adjacent words and the locality of word positions are introduced in keyword extraction to improve the extraction quality of high-frequency and low-frequency words. In terms of removing sentence redundancy, the concept of inclusion between sentences is introduced, which can effectively deduplicate multiple sentences in the article that have an inclusion relationship, reducing the information redundancy of the summary and improving the quality of the summary.

References

1. Zhang Huaping, Gao Kai, Huang Heyan, etc. Big Data Search and Mining[M]. Beijing: Science Press, 2014.
2. LUHN H P. The Automatic Creation of Literature Abstracts[J]. IBM Journal of Research and Development, 1958, 2(2): 159–165.
3. JONES KS, et al. Introduction: Automatic Summarizing[J]. Information Processing and Management, 1995, 31 (5): 625–630.
4. Zhou Jinhua, Liu Guiquan. Research on Multi-document Summary Based on Decaying Word Co-occurrence Graph[J]. Small and Micro Computer Systems, 2009, 30(1): 173–177.
5. ERKAN G, DRAGOMIR R. LexRank: Graph-based Lexical Centrality as Salience in Text Summarization[J]. Journal of Artificial Intelligence Research, 2004, 22: 457–479.
6. ZHANGJ, CHENGXQ, WU G W, et al. AdaSum: An Adaptive Model for Summarization[C]. Proceedings of the ACM 17th Conference on Information and Knowledge Management (CIKM2008), 2008
7. Liu Ting, Wang Kaizhu. Four main methods of automatic text summarization[J]. Journal of Information Science, 1999, 18(1): 10–19.
8. ALIGULIYEV R M. A New Sentence Similarity Measure and Sentence Based Extractive Technique for Automatic Text Summarization[J]. Expert Systems with Applications, 2009, 36: 7764–7772.
9. NETOJL, SANTOSA, et al. Generating Text Summaries Through the Relative Importance of Topics[C]. Proceedings of IBERAMIA-SBIA, Brazil, 2000:300–309.
10. FENG H D, CHEN K, DENGXT, et al. Accessor Variety Criteria for Chinese Word Extraction[J].Computer Linguistics, 2004, 30(1):75–93
11. Zhang Qingguo, Xue Dejun, Zhang Zhenhai, et al. Automatic keyword extraction from massive data sets based on feature combinations[J]. Journal of Information Science, 2006, 25(5):7–11
12. CARBONELL J, GOLDSTEIN J. The Use of MMR, Diversity-based Reranking for Reordering Documents and Producing Summaries[C]. Proceedings of the 21st ACM-SIGIR International Conference on Research and Development in Information Retrieval, Australia, 1998: 335–336.
13. Li Ran, Zhang Huaping, Shang Jianyun, et al. Research on Chinese Document Automatic Summarization Technology Based on Topic Model and Information Entropy[J]. Computer Science, 2014, 41(11): 298–301
14. Wang Jicheng, Wu Gangshan. Research on Automatic Summarization of Web Document Guided by Discourse [J]. Computer Research and Development, 2003. 40(3): 398–405.
15. Zhang Qi, Huang Chajing, Wu Lide. A New Method for Calculating Similarity Between Sentences and Application on Automatic Text Summarization[J]. Journal of Chinese Information Processing, 2005, 19(2): 93–99.

Part V
Application

Chapter 19
Natural Language Processing Application Projects

This chapter summarizes the excellent student assignments from the Beijing Institute of Technology's quality course "Big Data Analysis and Applications," and introduces natural language processing application projects for the government and businesses. Big data mining of features and behaviors of Weibo bloggers, semantic disambiguation systems for subtitles, big data analysis for postgraduate entrance examinations, and customer service call text summarization and extraction. The key technologies involve data mining and analysis, knowledge graphs, semantic disambiguation, text summarization, sentiment analysis, sensitive information detection, privacy protection, and other natural language processing technologies.

19.1 Big Data Mining of Weibo Bloggers' Characteristics and Behaviors

19.1.1 Introduction

With the rapid development of social networks on the Internet and mobile platforms, a large amount of personal information of network users is available online. Originally fragmented information is integrated in the big data environment [1]. Statistical analysis and data mining methods for social network data have become one of the important tools for commercial applications and scientific research. At the same time, big data mining capabilities also threaten users' privacy.

Currently, based on privacy content, the privacy and protection issues of social networks can be divided into three categories: one user's basic attributes, identity, and social relationship information, including real name, gender, age, affiliated institutions, friend relationships, and social influence, etc. This information can be used to locate social network users in real life. Second user's behavioral attributes,

H. Zhang, J. Shang, *Natural Language Processing and Applications*,
https://doi.org/10.1007/978-981-97-9739-4_19

including posting, forwarding, commenting, following time, and frequency, etc., which reflect the user's daily routine and behavioral trajectory in real life, further constituting the user's behavioral characteristics. Third user's mental characteristics, which can be calculated through latent semantic analysis of user speeches, including personality traits, value orientation, self-cognition status, and social needs, etc. This type of information carries a strong personal tone and reflects the user's internal psychological state.

In terms of the epistemology and methodology of big data, Zhang Huaping and others proposed the argument of "knowing, seeing the micro, understanding the meaning." Among them, knowing refers to the use of big data to understand the objective world as a whole, quickly obtain macro characteristics and structures, and is a fast and effective method to understand the objective world as a whole. Seeing the micro refers to the targeted study of representative micro data under the guidance of macro characteristics and structures. Here, it is not necessary to calculate every micro data. Understanding the meaning refers to the meaning of big data language content, which is the understanding and cognition of semantics, belonging to the category of natural language understanding. This section will discuss the social network big data mining work for the above three types of user privacy from the three dimensions of "knowing, seeing the micro, understanding the meaning," and look at the privacy protection of social networks from the perspective of data mining.

First, for the user's basic attributes, macro feature analysis is oriented toward user groups. Here, the concept of the Weibo ecosystem is introduced, which includes Weibo users, user posts, and other user activities. By combining the data of 17 million Sina Weibo users with real identities, an in-depth analysis of the Weibo ecosystem is conducted, including basic statistical feature analysis, digital feature analysis, and text feature analysis, to fully grasp various macro features of Sina Weibo users. Based on this, a user influence model is constructed, and the user's intentions are deeply studied.

Second, in terms of user behavior attributes, we start from the micro level and extract specific behavior patterns from the behavior of social network users (original microblogs, forwarded microblogs, following microblog users, posting comments, etc.). Studies have shown that the group behavior of microblog users shows a two-step distribution rule. However, due to the irregularity and randomness of user behavior records, coupled with the constraints of the user's own habits, life, learning, or work and other objective factors, research on individual behavior is currently mainly limited to the study of writing style and text features, the study of one of the objective factors, and simple statistical research, etc. Based on the above problems, this study proposes a behavior matrix model to describe the behavior of microblog users and designs a behavior matrix analysis method. This research deepens the understanding of user behavior and has important significance for research and application in areas such as friend recommendation, identity inference, group analysis, and precision marketing.

Finally, in terms of the user's mental characteristic attributes, a method is proposed to automatically assess the values of social network users using semantic analysis. Values, as an important aspect of personality that reflects social needs and desires, are widely used in various fields such as e-commerce, social networks, organizational behavior analysis, and public opinion monitoring and prediction. Traditional value assessments use scale-based questionnaires, which have high time and economic costs. This study uses the linguistic connection between values and word usage to automatically evaluate the values of microblog users based on their public statements on social networks, thereby grasping the user's behavioral preferences and social needs.

The large amount of publicly available personal data in social networks provides convenient conditions for the above three analyses. Here, taking Sina Weibo as an example, through data crawling, model analysis, and case study methods, it shows how to obtain users' basic attributes, behavioral attributes, and mental characteristic attributes and other personal privacy information through big data mining methods in the social network environment.

19.1.2 Macro Feature Big Data Mining

This section mainly analyzes the results of Sina Weibo data mining from a macro perspective, focusing on basic statistical information analysis of Weibo data, numerical feature analysis, user tendency analysis, etc. From the perspective of privacy protection, macro features reflect the overall characteristics of a country's social network. From the perspective of national security, the statistical data of a super-large-scale population has macro strategic security value.

The dataset used in this study was collected from Sina Weibo, and after a large amount of screening and processing, the cleaned dataset contains information of 17 million users (excluding a large number of machine-generated zombie users and dormant users). The dataset contains multiple fields, such as Weibo ID, gender, nickname, birthday, region, self-introduction, number of Weibo posts, number of fans, number of followings, blog address, educational experience, and certification level, etc.

1. Basic Statistical Feature Analysis
2. Numerical Feature Analysis and Influence Model
3. Research on User Behavior Feature Model

In the basic statistical feature analysis, we focused on studying the geographical distribution, gender distribution, educational background, and age distribution of three indicators, from which we obtained the answers to the following questions:

1. Which regions have the highest user density?
2. What is the relationship between male and female users?
3. What is the distribution of users' educational background and age?

1. Geographical Distribution

Among the 17 million users, about 16.5 million users have filled in geographical location information. Analysis reveals that there are approximately 10.8 Sina Weibo users per thousand people nationwide. The region with the highest user density is Beijing, with 79 users per 1000 people, and the region with the lowest user density is Gansu Province, with only 3.9 users per 1000 people. At the same time, among the 34 provincial administrative regions nationwide, 10 have user densities above the average.

2. Gender Distribution

The results of the gender distribution analysis show that female users account for 55% of Sina Weibo, and male users account for 45%, which is not as close to 1:1 as people think. The main reason for this phenomenon may be the difference in occupations between male and female users, which allows women to spend more time on Weibo.

3. Educational Background and Age Distribution

The statistical results of educational background and age distribution are shown in Fig. 19.1. In the dataset, about 662,000 users filled in their educational background, accounting for only 3.8% of all users. Among these users, nearly 83.2% have a bachelor's or graduate degree. From the perspective of age distribution, users aged 21~40 account for about 75% of all users. These data fully demonstrate that young people are more receptive to new things.

2. Numerical Feature Analysis and Influence Model

This study uses the method of function regression to analyze the number of Weibo posts, the number of fans, and the number of followers, the three numerical features.

Analysis was conducted, and the fitting function was obtained.

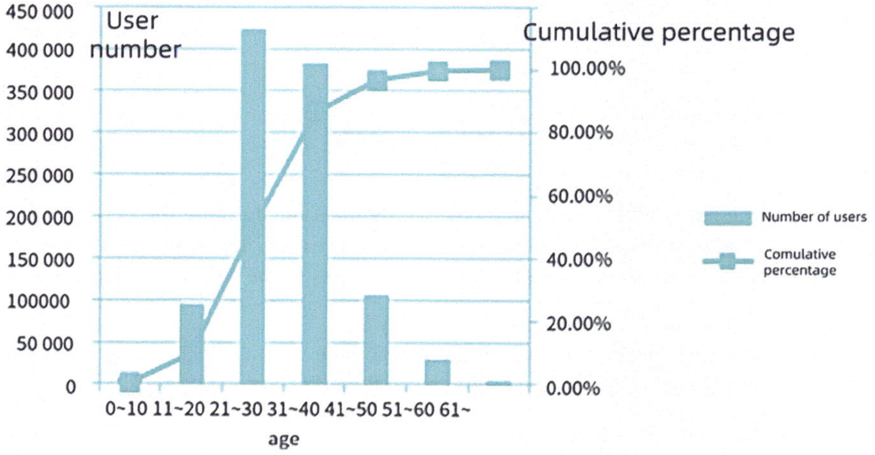

Fig. 19.1 Statistics of Educational Background and Age Distribution

1. Analysis of the Number of Weibo Posts—User Count

By counting the number of users corresponding to the number of Weibo posts in the dataset, a data graph is drawn on the double logarithmic coordinates, as shown in Fig. 19.2.

Previous studies have suggested that the relationship between the number of Weibo posts and the number of users follows a power law distribution, and therefore the data distribution in a double logarithmic coordinate should be close to linear. However, as can be seen from Fig. 19.2, which is derived from existing data, the results do not completely fit a linear model. This indicates that in the Sina Weibo dataset, the relationship between the number of Weibo posts and the number of users does not completely follow a power law distribution. However, from a local perspective of Fig. 19.2, the number of Weibo posts and the number of users do fit a linear model in certain areas.

By performing regression fitting in segments, it is shown that the distribution of the number of Weibo posts and the number of users follows a segmented step power law distribution. If the segmentation point is taken as (6.2,6.543911846), then the linear function of the first segment is as shown in Eq. (19.1), and the corresponding power law distribution function is as shown in Eq. (19.2).

$$y = -0.5226x + 9.8307 \tag{19.1}$$

$$y = 18\,595.97x^{-0.5226} \tag{19.2}$$

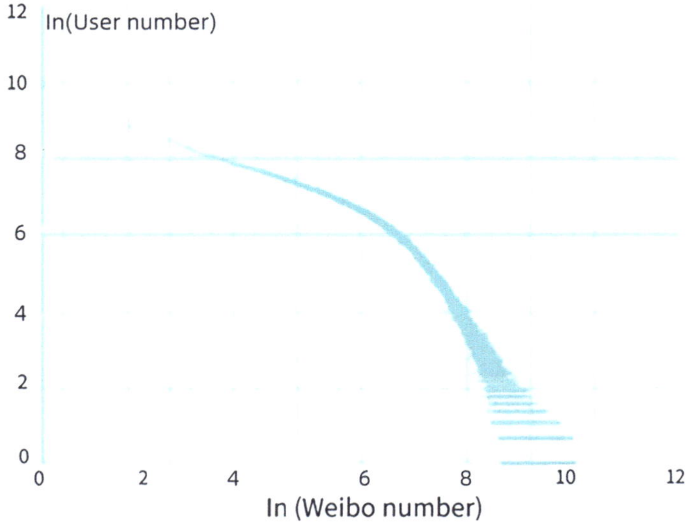

Fig. 19.2 Logarithmic relationship between the number of Weibo posts and the number of users

The linear function of the second segment is as shown in Eq. (19.3), and the corresponding power law distribution function is as shown in Eq. (19.4).

$$y = -1.977x + 19.04 \tag{19.3}$$

$$y = 185766301.8x^{-1.9771} \tag{19.4}$$

2. Analysis of the Number of Followers and the Number of Users

Using the abovementioned method, the double logarithmic relationship between the number of followers and the number of users is analyzed, and it is found that the number of followers and the number of users also follow a segmented power law distribution. If the segmentation point is taken as (6.2,8.147288), these two segments respectively follow the two power law distributions: the linear function of the first segment is as shown in Eq. (19.5), and the corresponding power law distribution function is as shown in Eq. (19.6). The linear function of the second segment is as shown in Eq. (19.7), and the corresponding power law distribution function is as shown in Eq. (19.8).

$$y = -0.5192x + 13.286 \tag{19.5}$$

$$y = 588893.1x^{-0.5192} \tag{19.6}$$

$$y = -1.5214x + 15.775 \tag{19.7}$$

$$y = 7095703x^{-1.5214} \tag{19.8}$$

3. Analysis of the Number of Followings and the Number of Users

Using the same method of analysis, it is found that the relationship between the number of followings and the number of users follows a power law distribution. Its linear function and power law distribution function are as shown in Eqs. (19.9) and (19.10).

$$y = -1.9541x + 18 \tag{19.9}$$

$$y = 168088301x^{-1.9541} \tag{19.10}$$

4. User Influence Model

The analysis results shown in Table 19.1 are obtained from the dataset. It is found that the average values of all parameters of verified users are higher than those of common users. Based on the existing data, a relevant user influence calculation model is derived:

Table 19.1 User influence analysis

Index	Value	Verified User	Common User	Male User	Female User	Verified Male User	Verified Female User
Average score	512.23	6965.11	337. 23	600.75	440.42	7887.36	5833.1
Average microblog	774.92	1435. 41	704.89	685. 26	854.06	1202.99	1524.11
Average subscription	176.66	34. 19	171.08	181.32	172.41	362.67	313.81
Influence	0. 433 038	4.613 95	0.235 711	0.612 074	0.313 807	6.25499	3.62132

$$\text{Influence}(\alpha) = \frac{(\text{followers} - \alpha \cdot \text{following})}{\text{posts}} \tag{19.11}$$

Among them, followers are the number of fans of the user who posts the microblog, following is the number of users the microblog poster is following, α is the probability of reciprocation, that is, when a user follows another user, the probability that the followed user will follow back.

5. User Preference Analysis

User preference analysis plays an important role in macro perspective analysis. Words reflecting hot topic tendencies are mined from the user's self-introduction information, and the top 10 words are taken, in order of frequency from high to low, such as "life," "self," "love," "like," "follow," etc. These words often occupy an important position in the analysis of user behavior intentions.

3. Research on User Behavior Feature Model

This study aims to establish a personality and behavior model based on an individual's microblog content and behavior matrix. Users on the network platform usually have many behaviors, such as logging in, clicking, etc. Specifically, on the microblog platform, the main behaviors of users can be manifested as logging in, uploading photos, posting microblogs, commenting, etc. The microblog user behavior here refers to the behavior of a user during a specific time period on the microblog platform for social or other autonomous activities, including posting original microblogs, forwarding microblogs, following microblog users, being followed by microblog users, logging in to accounts, logging out of accounts, commenting, and liking. The microblog user behavior defined here is the user's micro behavior. The pattern analysis of individual user behavior features precisely reveals a large amount of personal private information, such as personal work and activity patterns (frequency of nightlife, frequency of business trips, etc.). When combined with regional distribution, it can also clearly track each person's whereabouts, and when combined with the methods described in this section, it can even reveal whether the user belongs to the working class or the wealthy and leisure class and discover the individual's work status and economic situation from another perspective.

In the user behavior feature model, user behavior is represented by the variable Bi, where i represents a specific behavior. Corresponding to various behaviors, Bi can represent the number of original microblogs posted, the number of microblogs forwarded, the number of other users followed, the number of other users who follow this user, the number of account logins, the number of account logouts, the number of comments posted, and the number of likes, etc.

In order to describe and analyze user behavior patterns, this study has established an original behavior matrix model, and on this basis, individual behavior matrix models, group behavior matrix models, and weekly behavior matrix models have been constructed. These models aim to effectively record the fragmented and irregular behaviors of microblog users over a certain period of time. By adopting certain rules and algorithms for measurement and statistics, the behavior patterns of microblog users during this period are described and depicted. Through further analysis and deduction of these matrices, the user's potential attributes and deep-level information can be targeted and mined.

1. Original Behavior Matrix and Individual Behavior Matrix

In order to clearly describe the behavior of Weibo users and analyze their basic rules, this study has established an original behavior matrix model. The basic idea of this model is to depict the behavior rules of a user by describing the user's behavior activities over a certain period of time.

The original behavior matrix mainly describes the behavior quantity of user k within m observation days and n time steps. The original behavior matrix is used to store user behavior data, which is the basis for other behavior matrices and the original form of other behavior matrices.

The individual behavior matrix is mainly used for comparative analysis of behavior rules between different individuals. In order to further analyze user behavior in detail, this study establishes an individual behavior matrix based on the original behavior matrix. The individual behavior matrix mainly describes the behavior quantity of user k in n time steps, and the behavior quantity of each time step is the summary of m observation days.

2. Behavior Matrix Analysis Method

In statistical research, it is found that the discrete observation samples of individual behavior show random characteristics, but there are still identifiable rules. The behavior matrix analysis method proposed here is just helpful for analyzing these rules. The core idea of the behavior matrix analysis method is using the user behavior matrix to generate a behavior vector space, thereby using the core algorithm of principal component analysis to mine the most representative feature behaviors of users.

1. Behavior Vector Space Model

The basic idea of the behavior vector space model is to use vectors in high-dimensional space to represent the behavior rules and habits of different users. From a larger perspective, user behavior habits can be represented by distribution vectors divided by a certain time. The vector that represents the characteristics of user behavior is defined as the behavior vector, and the vector space where this vector is located is called the behavior vector space; thus, we can obtain the behavior matrix by calculating the similarity coefficient between different behavior vectors, and the similarity of behavior rules between different users or user groups can be compared.

2. Feature Behavior Analysis Method

Principal component analysis is mainly used to mine the feature behaviors in the behavior vector, and the feature behavior analysis method mainly draws on this idea. The basic principle of principal component analysis is as follows.

Suppose Σ is the covariance matrix of the n-dimensional random vector X (see formula (19.12)).

$$\mathbf{X} = \left[\mathbf{X}_1, \mathbf{X}_2, \cdots, \mathbf{X}_p\right] \tag{19.12}$$

The eigenvalue-eigenvector pairs are (λ_1, e_1), (λ_2, e_2), $\cdots$, (λ_p, e_p), , where $\lambda_1 \geq \lambda_2 \geq \cdots \geq \lambda_p \geq 0$. The ith principal component can be represented by Eq. (19.13). The covariance matrix $\sum$ of the standardized random vector is the Pearson correlation coefficient matrix before its standardization.

$$Y_i = \mathrm{X}e_i = e_{i1}\mathrm{X}_1 + e_{i2}\mathrm{X}_2 + \cdots + e_{ip}\mathrm{X}_p$$

$$i = 1, 2, \cdots, p. \tag{19.13}$$

After standardization, the covariance matrix Σ of the random vector is the Pearson correlation coefficient matrix before its standardization.

The feature behavior analysis method is used to calculate the eigenvectors (i.e., feature behaviors) of the correlation coefficient matrix of the behavior matrix. It is believed that the vectors that obtain larger weights are often repeated behavior rules, such as more active specific behaviors at specific times, that is, the principal components are a set of vectors that determine the behavior space. The correlation coefficient matrix obtained through calculation can be used to describe.

It describes the correlation between the variables in the behavior matrix. The algorithm for calculating each principal component is as follows:

Input: Behavior matrix.
Output: Eigenvectors and their eigenvalues.
Step 1: Calculate the correlation coefficient matrix R of the behavior matrix.
Step 2: Calculate the eigenvectors and their eigenvalues of matrix R.
Step 3: Sort the eigenvalues in descending order.

Step 4: Calculate the proportion and cumulative proportion of each eigenvalue.
Step 5: Output the eigenvectors, eigenvalues, and the proportion and cumulative proportion of the eigenvalues of the behavior matrix.

3. Normalization

This study also conducted comparative research on the patterns of different user behaviors, which involves dealing with the difference in the amount of user behavior data within different observation dates. The observation day with a large amount of data has a greater impact on the results of the behavior matrix than the observation day with a small amount of data. The use of multiple response normalization can eliminate this impact. Assign the same weight (all 1) to each observation day, and normalize the number of behaviors on the day from 0 to 1, thereby achieving the comparison of each principal component under dimensionless conditions.

19.1.3 Experiment and Analysis

The dataset includes all the Weibo posts from 4027 users with Weibo posting records from March 5, 2012 to May 20, 2012, totaling 410,618 Weibo posts. The behavior matrix analysis method is used to compare user relevance and analyze their potential daily behaviors. Using 3019 Weibo posts from seven verified Weibo users (from March 14, 2011 to October 16, 2011) as the dataset, and using 60 min as a time step (a day is divided into 24 time steps), the feature behavior analysis method is used to perform behavior matrix analysis, and its correlation coefficient matrix is obtained as Table 19.2.

From this matrix, it can be seen that the correlation coefficient of user U8 is much lower than that of the other eight users, indicating that this user's behavior activity pattern is unrelated or negatively related to other users. It can be inferred that this user's daily routine may be different from that of most people. In this matrix, the highest correlation coefficient is between U4 and U1, indicating that

Table 19.2 Seven verified Weibo users' potential daily behaviors

	U1	U2	U3	U4	U5	U6	U7	U8	U9
U1	1								
U2	0.4473	1							
U3	0.7469	0.7608	1						
U4	0.847	0.6130	0.8184	1					
U5	0.6988	0.6441	0.810	0.7045	1				
U6	0.6035	0.3439	0.6025	0.8133	0.4829	1			
U7	0.7731	0.4658	0.7587	0.7651	0.8436	0.7008	1		
U8	0.0740	−0.0296	0.1910	−0.0610	−0.1400	−0.2574	−0.2136	1	
U9	0.7529	0.2211	0.6480	0.7165	0.6794	0.6610	0.7491	0.0110	1

these two users have similar behavior activity patterns. It can be inferred that they have similar work habits, lifestyle habits, and daily routines.

19.1.4 Automatic Evaluation Method of Weibo Blogger's Values

This study proposes an automatic evaluation method based on social networks, automatic estimation of Schwartz value (AESV), which can automatically evaluate people's values based on their speeches or forwarding activities on social networks, and can adapt to different social backgrounds, including changes in language and time.

Values are an individual's or organization's understanding of the importance of different things in life, and they play a significant role in guiding an individual's or organization's behavior and attitudes. Schwartz's model of values has a high degree of universal applicability. A large amount of research has confirmed the strong relationship between values and behavioral choices; therefore, it can be used as an important basis for predicting individual or organizational behavioral choices, and can be applied to various fields of business or government. Evaluating an individual's values will reveal deeper aspects of an individual's privacy, such as whether the mind is healthy, whether there are biases in values, etc.

1. AESV Model

This study proposes a method for evaluating the values of social network users based on their public statements. Schwartz defined ten types of values from the basic needs of human life, namely, hedonism, benevolence, universalism, power, achievement, tradition, conformity, security, self-direction, and stimulation, each of which includes several specific values. Existing research has shown that different value priorities can lead to significant differences in the themes and wording of individual statements. Based on this, this study proposes the AESV model. This model includes the following three steps.

1. Generate the Feature Index of the Value Vector Space

Use keyword extraction technology based on classified documents to obtain specific vocabulary related to specific value concepts under the current social background. Use Baidu News search engine to build a dynamic corpus, with search keywords as category labels, to ensure that the corpus can be automatically classified and synchronized with the current social background. To ensure a high correlation between the search results and the value concepts, first construct a three-layer tree structure of search words, in which the first layer contains ten words representing value concepts, the second layer contains 56 words representing specific values, and the third layer contains 198 synonyms extended from the second layer. Each time you search, use three search words at the same time, use a value concept, a specific value, and a synonym according to the tree structure. Because the ranking of Baidu News is consistent with people's attention and discussion of the news, this study extracted the titles and abstracts of the top 160 news within a specified time window.

This study chooses the square root of the statistic as the basis for feature extraction, which shows the association between vocabulary and value types. A higher value indicates a strong correlation between the vocabulary and the value type, which can be used as the feature vocabulary of this value type. Extracting feature vocabulary specifically requires the following three steps: first, use the NLPIR-ICTCLAS word segmentation system for new word discovery and word segmentation; second, calculate the square root of the statistic of each word in each category, and finally, select the top 200 words in each value type, remove duplicate words and summarize, to generate the feature index of the vector space.

2. Calculate the Dynamic Value Vectors

Calculate the weight of each feature word in the ten value types in the feature index to generate dynamic value vectors. The weight of each feature word is the document frequency of the word in the value type, the product of three statistical measures: the inverse document frequency logarithm of the word and the information gain of the word. Based on this, we can obtain vector representations of ten value types, which are the specific mappings of value concepts to social network language in the current social context, and are used for subsequent personal value assessment.

3. Assessing Priorities of Personal Value

The assessment of priorities of personal value mainly consists of the following three steps.

1. Obtain personal corpus. Crawl the 200 latest personal statements from multiple social networking platforms (such as Sina Weibo, Tencent Weibo, forums).
2. Calculate the personal value vector.
3. Calculate the similarity between the personal value vector and the ten dynamic value vectors. By calculating the cosine similarity between the personal value vector and the ten dynamic value vectors, ten similarity values can be obtained for each person. These ten similarity values are sorted from high to low, and the personal value assessment results are obtained. To confirm the substitutability of this method for traditional scales, this study conducted reliability and validity tests. The test results show that the AESV model has high stability and accuracy.

2. Experiment

To confirm the simplicity and feasibility of the AESV model, this study used the AESV model to analyze 92 Sina Weibo users, including 30 randomly selected verified users and 62 ordinary users, and obtained the value assessment results of each user. The values of verified users and ordinary users were compared and analyzed, as shown in Table 19.3. It can be seen from the table that there are significant

Table 19.3 Comparison of values between V users and ordinary users

Value	V Subscriber	Regular Users	*P*-value
Enjoyment	−0.3142	0.6779	0.000 02
Benevolent love	1.1008	1.6939	0.001 82
Tradition	−0.41559	−0.1659	0.089 17
Stimulate	−0.5968	−0.2315	0.113 63
Safely	−0.2835	−0.4744	0.196 44
Achievement	0.2127	0.1363	0.633 11
Universal	0.4339	−0.2108	0.001 20
Might	0.5944	−0.1748	0.000 07

differences in various values between verified users and ordinary users. Using the method proposed in this study, according to other application and research needs, a large-scale user value analysis can be conducted on social networks to meet different business and policy needs.

19.2 Semantic Disambiguation System for Subtitles

19.2.1 Introduction

In order to achieve the goal of enhancing China's cultural influence by promoting Chinese TV dramas, it is necessary to improve the quality of machine translation of Chinese TV drama scripts. One of the key issues involved is understanding the Chinese spoken language semantics carried by the TV drama scripts and eliminating the ambiguity in them. Based on the above background, this study focuses on the key technology of Chinese semantic disambiguation applicable to the above specific field and constructs a semantic disambiguation system that has practical value and can reduce manual costs [2].

The corpus obtained in this study includes the original sentences of the script and the disambiguation sentences that have been manually modified for disambiguation. By constructing a knowledge graph, it provides ambiguity-related knowledge for the semantic disambiguation algorithm. The disambiguation knowledge provides the replacement relationship between the ambiguous words and the disambiguation words, and the attributes provide the context information when the replacement occurs. After completing the preprocessing work including cleaning, word segmentation, and removing stop words, rule-based knowledge extraction and knowledge fusion based on semantic similarity were carried out.

This study proposes a semantic disambiguation algorithm based on knowledge graphs and semantic features. The text to be disambiguated is input into this algorithm at the sentence level. After completing the preprocessing work, ambiguous words found in the sentence are annotated. The BERT pretrained language model is used to extract the semantic feature vectors of the context of the ambiguous words.

The cosine similarity and normalization function of the new context and historical context are used to calculate the semantic feature scores of the candidate disambiguation words. Based on the scores, output rules for modification suggestions are established, and disambiguation modification suggestions and their recommended intensity are output for each input sentence. Finally, the effectiveness of this algorithm is verified through experiments. This algorithm supports semantic disambiguation of a large number of ambiguous words and has adaptability to specific domains.

This study combines the construction of a semantic disambiguation knowledge graph and a semantic disambiguation algorithm to complete the overall design of the semantic disambiguation system and develop a semantic disambiguation system with a visual interface. This system can automatically acquire disambiguation knowledge and provide corresponding modification suggestions for the text to be disambiguated, reducing the dependence of disambiguation work on manual labor and effectively improving the efficiency of semantic disambiguation work.

19.2.2 Construction of Semantic Disambiguation Knowledge Graph

This study belongs to the category of supervised semantic disambiguation. However, unlike the traditional supervised semantic disambiguation that uses a corpus with semantic annotations, the corpus obtained in this study consists of original sentences and sentences modified by manual disambiguation. Therefore, this study aims to utilize the relevant concepts and technologies of knowledge graphs to mine knowledge from the corpus that can be used for semantic disambiguation. This will enable the construction of a semantic disambiguation knowledge graph and the realization of a semantic disambiguation algorithm based on knowledge graphs and semantic features. The construction process of the semantic disambiguation knowledge graph is illustrated in Fig. 19.3. The corpus used in this study is divided into a training set and a test set. The training set contains 1,191,191 original sentences and corresponding disambiguation sentences, and the test set contains 21,863 original sentences and corresponding disambiguation sentences.

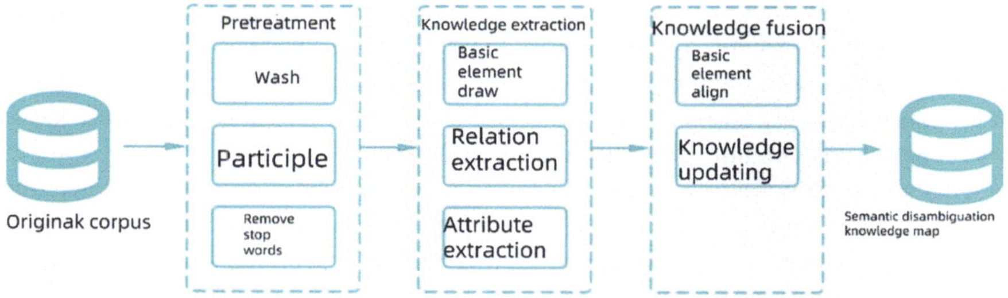

Fig. 19.3 Construction process of the semantic disambiguation knowledge graph

1. Definition of the Semantic Disambiguation Knowledge Graph

The knowledge source of the semantic disambiguation knowledge graph is a corpus composed of original sentences and sentences modified by manual disambiguation. First, the corpus needs to be analyzed. The manual semantic disambiguation method for the original text is: replace words or phrases that may cause ambiguity in the context of the original text with words or phrases that will not cause ambiguity in that context, thereby forming the modified disambiguation text. If it is believed that there are no words or phrases in the original text that could potentially cause ambiguity, then no modifications are made to the original text.

The specific representation of the disambiguation knowledge graph is explained as follows:

The representation of a piece of disambiguation knowledge is a triplet V_a is the ambiguous word, represented as a list of one or more words (the length of the list is greater than or equal to 1). R is the relation "can be replaced by," meaning V_a can be replaced by V_b. V_b is the disambiguation word for V_a, also represented as a list of one or more words (the length of the list is greater than or equal to 1). Additionally, V_b has two attributes; the frequency of T in the corpus. and the set of contexts in which T appears in the corpus.

2. Corpus Preprocessing

In order to build a high-quality disambiguation knowledge graph, preprocessing is a necessary step before knowledge extraction. The preprocessing performed for each sentence in the corpus mainly includes cleaning, word segmentation, and stop word removal.

3. Rule-Based Disambiguation Knowledge Extraction

Rule-based disambiguation knowledge extraction includes three contents, namely basic element extraction, relation extraction, and attribute extraction.

1. Basic Element Extraction

The aim of basic element extraction is to discover ambiguous words and disambiguation words. The idea of discovering ambiguous words and disambiguation words is to compare the segmented sentences before and after disambiguation, ignoring the words that appear in both sentences (i.e., no change), treating the words that only appear in the pre-disambiguation sentence as ambiguous words, and the words that only appear in the post-disambiguation sentence as disambiguation words.

2. Relation Extraction

The aim of relation extraction is to discover the relationship between ambiguous words and disambiguation words, usually achieved by replacing an ambiguous word with a disambiguation word. Considering the semantic disambiguation modification method mainly based on replacement in the corpus, this study believes that the ambiguous words and disambiguation words that can establish a relationship should be in the same position in their respective sentences.

3. Attribute Extraction

An ambiguous word being replaced by a disambiguation word is called a replacement, and this replacement as a whole corresponds to two attributes, namely the number of times the replacement occurs and the set of contexts in which the replacement occurs. The extraction of these two attributes needs to be carried out in sync with relation extraction. The definition of context involved in this study is as follows: a disambiguation word that establishes a relationship with an ambiguous word corresponds to multiple contexts. A context includes two situations: one is the text of the sentence excluding the ambiguous word, called a single-sentence context, the other includes the text of the previous sentence, the current sentence, and the next sentence excluding the ambiguous word, called a three-sentence context. The single-sentence context and the three-sentence context will be merged according to their similarity after semantic similarity calculation.

4. Knowledge Fusion Based on Semantic Similarity

Rule-based knowledge extraction enables the acquisition of basic elements, relationships, and attributes from unstructured corpora, which will serve as elements for constructing a knowledge graph. However, these results may contain redundant or duplicate information, necessitating cleanup and integration. This requires knowledge fusion to enhance the quality of knowledge. The knowledge fusion operations involved in this study include basic element alignment and knowledge updating. Through basic element alignment, various repetitions that may occur due to ambiguous words are eliminated. Through knowledge updating, information that has not been replaced by ambiguous words in the corpus is integrated into the knowledge graph.

5. Knowledge Representation Based on Semantic Feature Vectors

After completing preprocessing, knowledge extraction, and knowledge fusion, a complete semantic disambiguation knowledge graph is obtained. However, the representation of text cannot directly calculate the semantic similarity between basic elements. To improve the efficiency of the semantic disambiguation algorithm, it is necessary to represent the ambiguous words, disambiguation words, context, and other basic elements or attributes in the semantic disambiguation knowledge graph in advance.

The vectors representing basic elements or attribute values are stored in their respective matrices. To quickly correspond and read the vectors in the matrix with the basic elements or attribute values, it is necessary to index the basic elements or attribute values first. When reading the vector, the vector in the corresponding row of the matrix can be obtained according to the index number.

19.2.3 Semantic Disambiguation Algorithm Based on Knowledge Graph and Semantic Features

1. Semantic Feature Vector Extraction Based on BERT

Because the semantics of the text cannot be directly calculated for similarity, it is necessary to convert the semantics of the context text into a directly calculable vector, which involves the extraction of text semantic feature vectors. Here, the BERT pretrained language model in the field of deep learning is chosen for semantic feature vector extraction. The input of the BERT model is a token sequence obtained from sentences. Unlike the English BERT model, which uses words as tokens, the Chinese BERT model inputs a character-level token sequence. By replacing one or several characters in a sentence with [MASK] and inputting it into the BERT pretrained language model, the vector corresponding to [MASK] primarily contains the contextual semantics of the [MASK] position, integrating potential information at the current location. Therefore, by replacing the ambiguous word with the same number of [MASK]s, the vectors corresponding to [MASK] can be obtained. These vectors can be considered as the context semantic feature vectors of the ambiguous word. This study categorizes the context of ambiguous words into two cases: single-sentence context and three-sentence context, which only differ in the length of the contextual text. The specific process of obtaining the semantic feature vector of the context is as follows: obtaining the sentence where the ambiguous word is located, forming a token sequence with characters as units, replacing the token at the position of the ambiguous word with [MASK] to form a single-sentence context token sequence, inputting this sequence into the BERT pretrained language model, obtaining the vector representation of each token in the sequence, calculating the average value of the vectors corresponding to all [MASK] tokens, and using this as the semantic feature vector representation of the single-sentence context at the position of the ambiguous word.

2. Calculation of Semantic Similarity between Multi-Feature Vectors

Based on obtaining the semantic feature vector of a certain context, the cosine similarity algorithm is used to calculate the semantic similarity between two contexts or between one context and multiple contexts. The cosine similarity uses the cosine value of the angle between vectors as a measurement index; the smaller the angle, the higher the similarity. Because the semantic feature vector of a context will not be negative, the range of cosine similarity of the two vectors discussed here is [0,1].

Suppose there are two vectors, A and B; the similarity between these two vectors can be calculated using formula (19.14):

$$\text{Similarity} = \cos(\theta) = \frac{A \cdot B}{\|A\|\|B\|} = \frac{\sum_{i=1}^{n} A_i \times B_i}{\sqrt{\sum_{i=1}^{n}(A_i)^2} \times \sqrt{\sum_{i=1}^{n}(B_i)^2}}. \tag{19.14}$$

In order to calculate the similarity between a target vector and multiple candidate vectors, multiple candidate vectors can be concatenated into a matrix to improve calculation efficiency. Suppose the target vector is f, the ith candidate vector is c_i, and both the target vector and candidate vector are d-dimensional, then e vectors c can be concatenated into matrix F using Eq. (19.15).

$$F = \left[c_1, c_2, \cdots, c_i, \cdots, c_e\right]^T_{e \times d} \tag{19.15}$$

Next, the similarity between vector f and each vector in matrix F can be calculated according to Eq. (19.16), resulting in a similarity vector g, where the value of the ith dimension in g is the similarity between f and the ith candidate vector.

$$\boldsymbol{g} = \frac{\boldsymbol{f} \cdot \boldsymbol{F}^T}{\|\boldsymbol{f}\| \|\boldsymbol{F}\|} \tag{19.16}$$

Similarly, in order to calculate the cosine similarity between vectors belonging to two vector sets, we can first combine the two vector sets into two matrices according to the method shown in Eq. (19.16) to improve computational efficiency. Suppose the two matrices obtained by combination are C and D, then we can calculate the similarity between any vector in matrix C and any vector in matrix D according to Eq. (19.17) and obtain the similarity matrix G.

$$\mathrm{G} = \frac{\boldsymbol{C} \cdot \boldsymbol{D}^T}{\|\boldsymbol{C}\| \|\boldsymbol{D}\|} \tag{19.17}$$

3. Semantic Disambiguation Algorithm Based on Knowledge Graph and Semantic Features

Corpus-based semantic disambiguation methods include: graph-based or context-based, unsupervised learning methods of word clustering, supervised learning methods of training classifiers based on machine learning or deep learning, and semi-supervised learning methods using small-scale annotated corpora. These methods have achieved good results in general semantic disambiguation tasks, but there may be some problems in the specific field of semantic disambiguation of Chinese film and television subtitles: the number of ambiguous words that can be disambiguated is limited, and it is impossible to deal with a large number of ambiguous words. Training a model for each ambiguous word not only requires each ambiguous word to have enough knowledge but also may lead to excessive algorithm time complexity. The context of ambiguous words may need to be extracted from a large range of texts, which is not suitable for spoken dialogue and semantic rapid change scenarios.

The semantic disambiguation algorithm based on knowledge graph and semantic features proposed in this study (hereinafter referred to as this algorithm) can solve the above problems to a certain extent, taking into account both operational efficiency and accuracy, and improving the applicability to specific fields.

Since the semantic disambiguation knowledge required by this algorithm is stored in the semantic disambiguation knowledge graph, it is necessary to load the knowledge graph.

Load the text to be disambiguated into the semantic disambiguation algorithm in the form of a sentence list. After completing the preprocessing work, Eq. (19.18) is obtained.

$$S_i = (1-\alpha) \times \max(g_i) + \alpha \times \max(g_i') \tag{19.18}$$

$$P_i = \frac{e^{R_i}}{\sum_k e^{R_K}} \quad R_i = \frac{e^{\frac{n_i}{N_i}}}{\sum_k e^{\frac{n_k}{N_k}}} \times S_i. \tag{19.19}$$

Here, i represents the ith candidate disambiguation word $V_{d,i}$ that the ambiguous word V_a found in the text to be disambiguated can be replaced with. g_i is a vector composed of the semantic similarity between the three-sentence context of ambiguous word V_a in the text to be disambiguated and each historical three-sentence context corresponding to the ith candidate disambiguation word $V_{d,i}$, calculated using the introduced method. max (g_i) represents the maximum value in g_i. g' is the similarity vector between single-sentence contexts. S_i is the mixed maximum similarity of the ith disambiguation word $V_{d,i}$ of V_a. α is the proportion of single-sentence similarity in the mixed similarity, and α is set to 0.7. n_i represents the number of similarities in the historical three-sentence context of $V_{d,i}$ that exceed the threshold (set to 0.9 in this paper) with the new three-sentence context. N_i represents the total number of three-sentence contexts in the history of $V_{d,i}$. R_i is the weighted result of the mixed similarity of the context of V_a and $V_{d,i}$, where the weight is the SoftMax value of the ratio of the number of contexts exceeding the threshold to the total number of contexts for each candidate disambiguation word of V_a. P_i is the result of normalizing R_i through SoftMax, which can be regarded as the probability of V_a being replaced by the ith candidate disambiguation word $V_{d,i}$, and serves as the semantic feature score of the ith candidate disambiguation word $V_{d,i}$ of V_a.

3. Output Disambiguation Modification Suggestions

The disambiguation modification suggestions for a sentence to be disambiguated include two parts: one is the disambiguation sentence corresponding to the original sentence, and the other is the preferred disambiguation word and alternative disambiguation word corresponding to each ambiguous word found in the original sentence. For an original sentence, if no ambiguous words that already exist in the semantic disambiguation knowledge graph are found in it, the original sentence will be output directly as the disambiguation sentence. If several ambiguous words are found in it, for one of the ambiguous words, the candidate disambiguation word with the maximum semantic feature score corresponding to it will be taken as the preferred disambiguation word for this ambiguous word, and other candidate disambiguation words will be taken as alternative disambiguation words.

19.2.4 Experimental Results and Analysis

In order to evaluate the effectiveness of the actual Chinese film and television dialogue semantic disambiguation work, and verify the superiority of this algorithm in a specific field, this study set up a comparative experiment, comparing the accuracy indicators of several different methods on the dialogue dataset.

In order to make the comparative experiment more reasonable, this study selects other methods for comparison that can disambiguate thousands of ambiguous words with a reasonable time complexity. First, the definition and rationality of the baseline are introduced. Since there are a certain number of manually disambiguated sentences in the script dataset used in this study, these sentences are considered unambiguous during disambiguation, and no modifications are made to them. Therefore, the baseline is defined by directly outputting each sentence in the test set without any modifications. Next, the five methods chosen for comparison in this study are presented:

1. Using one-hot encoding to encode the text, a semantic disambiguation algorithm based on the constructed knowledge graph and context similarity, and ranking the candidate disambiguation words directly according to similarity
2. Using Word2Vec to encode the text, a semantic disambiguation algorithm based on the knowledge graph and context similarity
3. Using Word2Vec to encode the text, a semantic disambiguation algorithm based on the knowledge graph and semantic features, and ranking the candidate disambiguation words according to the calculated semantic feature scores
4. Using the BERT pretrained language model to encode the text, a semantic disambiguation algorithm based on the knowledge graph and context similarity
5. Using BERT to encode the text, a semantic disambiguation algorithm based on the knowledge graph and semantic features, which is the focus of this study

Using the above methods, a semantic disambiguation knowledge graph is constructed based on the training set, obtaining a total of 4230 ambiguous words, 9689 disambiguation words, and 999,125 context scenarios. On average, one ambiguous word establishes a relationship with 2.29 disambiguation words, and one ambiguous word corresponds to 236.20 context scenarios. Based on this knowledge graph, this algorithm supports semantic disambiguation of a large number of ambiguous words, making it more suitable for the practical work of semantic disambiguation in Chinese film and television drama scripts than previous algorithms that only support semantic disambiguation of the order of 10. At the same time, this algorithm has a certain scalability and can be applied to semantic disambiguation of other Chinese texts or other types of short texts.

Table 19.4 shows that this algorithm has a 7.61 percentage point improvement in sentence-level accuracy compared to the baseline method, proving that this algorithm has excellent results in the specific field of semantic disambiguation of Chinese film and television drama scripts, and can provide help for practical work, reducing the dependence of semantic disambiguation work in specific fields on manual work to a certain extent, which is a major advantage and goal of this algorithm.

Using this algorithm, the 21,863 original sentences in the test set are input in the order of the plot, and semantic disambiguation modification suggestions are obtained for each sentence, as shown in Table 19.5.

As shown in Table 19.5, this algorithm finds ambiguous words in the original example sentences that exist in the knowledge graph. Reasonable modification suggestions were given for each ambiguous word, so that the newly generated sentence can unambiguously and accurately express the semantics of the original sentence, proving that the construction method of the knowledge graph and the semantic disambiguation algorithm proposed in this study are both reasonably designed and can achieve the desired effect.

19.2.5 Semantic Disambiguation System

The semantic disambiguation system primarily consists of three modules: the construction of a disambiguation knowledge graph, semantic disambiguation, and disambiguation result evaluation. Among these, the visualization user interface involved in the semantic disambiguation module is implemented using web technology, as shown in Fig. 19.4.

The operation steps of the semantic disambiguation module are as follows.

Step 1: Data loading. This step is executed in the backend when the Web service starts, automatically reading the semantic disambiguation knowledge graph and BERT pretrained language model.

Table 19.4 Comparative experiment results

Method	Word Granularity Accuracy	Word Granularity Recall	Sentence Granularity Accuracy
Baseline	–	–	86.08
One-Hot + Similarity	56.89	67.53	86.97
Word2Vec + Similarity	63.21	89.86	88.43
Word2Vec + Semantic feature scoring	71.23	93.47	90.56
BERT + Similarity	69.34	92.45	89.78
BERT + Semantic feature scoring	76.02	98.17	93.69

Table 19.5 Test set disambiguation result examples

Original Sentence	Status	Be Changed To	Suggestion
Mom's gonna have grandchildren	Suggestion modification	Mom wants grandkids	[Suggestion modification] have -> want [Nomodificationrequired] grandchildren->grandkid
What do I tell my parents	Can be modified	How do I explain this to my parents	[Can be modified] tell->explain
Here's his background does have a problem	Can be modified	Here's his background a suspect	[Can be modified] problem->suspect
I wonder if you could find someone a little more serious	Suggestion modification	I wonder if you could hire someone a little more serious	[Suggestion modification] find->hire

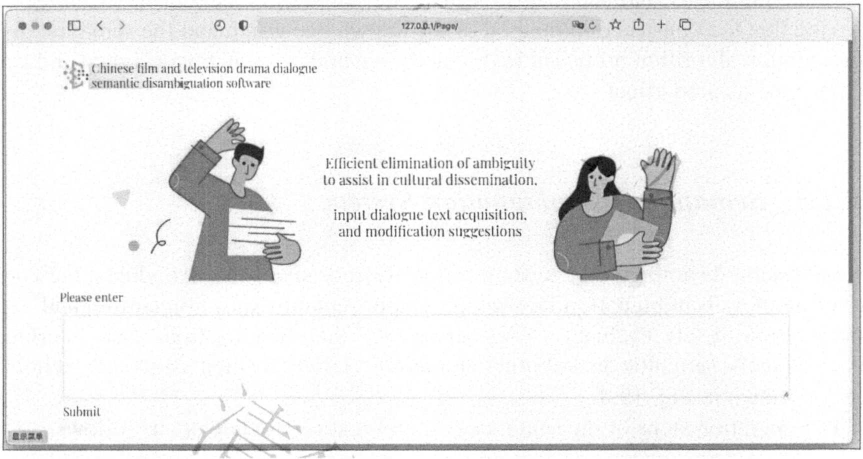

Fig. 19.4 Semantic disambiguation module Web page

Step 2: Obtain text. Enter the text that needs to be semantically disambiguated in the form of one sentence per line into the "Please enter" text box on the Web page and click the "Submit" button. The Web page will pass the text to the server side for subsequent steps.

Step 3: Semantic disambiguation. Run in the backend, for each sentence in the read text, on the basis of preprocessing, find out which ambiguous words are included in this sentence. If there are no ambiguous words in this sentence, give the result of "no ambiguity." Otherwise, calculate the scores of the candidate disambiguation words, and then give modification suggestions. The results will be passed to the front-end Web page in JSON format.

Step 4: Disambiguation result display. Display the disambiguation results in the form of a table on the Web page. Among them, the "Suggestion" column by default displays each ambiguous word and its corresponding preferred disambiguation word. When moving the mouse pointer to the "" icon at the end of each line, all alternative disambiguation words are displayed.

19.3 Big Data Analysis of Postgraduate Entrance Examination

19.3.1 Introduction

In 2021, the number of postgraduate entrance examination registrants reached 3.77 million, setting a new historical record. Nowadays, postgraduate entrance examination has become the choice of more and more people. However, the current postgraduate entrance examination book and training market is mixed, which can make newcomers who have just embarked on the journey of postgraduate entrance examination feel confused. This project aims to comprehensively analyze postgraduate entrance examination information from multiple aspects such as postgraduate entrance examination teachers, postgraduate entrance examination material, postgraduate entrance examination institutions, and examinees, objectively depict the current situation of postgraduate entrance examination from a large amount of data, analyze and compare the advantages and disadvantages of postgraduate entrance examination teachers, postgraduate entrance examination materials, postgraduate entrance examination institutions, conduct sentiment evaluation of these postgraduate entrance examination teachers, postgraduate entrance examination materials, postgraduate entrance examination institutions from the perspective of examinees. Newcomers can choose suitable teachers, materials, and institutions based on these objective analysis results. At the same time, this project also analyzed the emotions of the examinees, summarizing the emotional changes of the examinees in different years and changes in the emotions of candidates in 2021. These conclusions can help candidates to grasp the rhythm of postgraduate entrance examination emotional fluctuations well and help candidates can make a calm analysis and timely adjustment.

19.3.2 Module Design

The system developed by this project consists of the following four modules.

1. Postgraduate Entrance Examination Expert Analysis Module

The postgraduate entrance examination expert analysis module processes content related to Zhang Yu, Li Yongle, Li Lin, Tang Jiafeng, Xiao Xiurong, and Xu Tao, which is published by Weibo users, analyzes the results of data processing, and thus gives reasonable and effective suggestions to postgraduate students.

The data source of this module is the content published by Weibo users. First, data related to the abovementioned experts is obtained through web crawler technology, and data cleaning and de-duplication are carried out. Then, Jieba word segmentation software is used for word segmentation, and keywords are extracted based on TF-IDF. A sentiment word cloud is generated, and the SnowNLP library is used for text sentiment analysis. Finally, the main themes in the data are analyzed through the LDA model.

2. Postgraduate Entrance Examination Material Analysis Module

The postgraduate entrance examination material analysis module counts the sales of popular postgraduate entrance examination materials and selects several from the categories of English, Mathematics, and Politics for sentiment analysis. First, crawl the related comments about postgraduate entrance examination materials on the Zhihu website, and after data cleaning (including removing noise, stop words, invalid characters, and data), use word segmentation tools for word segmentation processing. After counting the word frequency, extract the keywords related to the postgraduate entrance examination materials, calculate the sentiment score of the keywords, and finally use visualization tools to display the results.

3. Postgraduate Entrance Examination Institution Analysis Module

The postgraduate entrance examination institution analysis module includes three sub-modules: data acquisition, hotspot institution statistics, and institution comment analysis.

The data acquisition sub-module uses the selenium framework to crawl a large amount of user comments on different postgraduate entrance examination institutions from portals such as Zhihu, Tieba, and Sina, conducts data analysis, and returns statistical data to system maintenance personnel. System maintenance personnel process data in JSON format and analyze data using methods such as word cloud maps. Users can interact with the content on the website through the browser. This sub-module mainly uses the following programming languages:

1. Python. For the development of the backend selenium and data analysis modules.
2. HTML5. To get the XPath path structure and implement the crawler function.

The hotspot institution statistics sub-module is used to analyze a large amount of postgraduate entrance examination institution data obtained by the data acquisition sub-module. This sub-module comprehensively uses technologies such as TF-IDF, word frequency statistics, Jieba word segmentation, keywords, and word cloud extraction to analyze the data crawled by the data acquisition sub-module, and get the heat of all postgraduate entrance examination institutions in the data. The few postgraduate entrance examination institutions with the highest heat are the hotspot institutions.

The institution comment analysis sub-module is used to calculate the positive rate in the comments under the topics of each postgraduate entrance examination institution. This sub-module comprehensively uses sentiment analysis technology to analyze the comment data of each postgraduate entrance examination institution separately, calculates the sentiment tendency, and gets the positive rate and negative rate in the comments of each postgraduate entrance examination institution.

2. Postgraduate Entrance Examination Material Analysis Module

The sales of popular postgraduate entrance examination books are statistically analyzed, and basic sentiment analysis is conducted by selecting from English, Mathematics, and Politics. The specific process of sentiment analysis is as follows. First, crawl the related comments about books on the Zhihu website, clean the data including removing noise, stop words, invalid characters and data, and then use the word segmentation tool for word segmentation processing. After counting the word frequency, extract the keywords related to the book, then analyze the sentiment score of the keywords, and finally use visualization tools to display the results.

4. Candidate Analysis Module

For the crawled Weibo comment data, first use Python to standardize the database data, then classify the data, get several types of files such as different months in the same year and different years in the same month, perform word segmentation separately, use the word segmentation results for sentiment analysis, and use animation and Pandas for dynamic visualization of data. For the chat data obtained from QQ, measure the chat data volume on a monthly basis, analyze the changes, use word segmentation and stop word removal to process the basic data, provide source data for the LDA topic model and word cloud, and then perform SA analysis on the data for 12 months.

19.3.3 Results and Analysis

1. Analysis of Postgraduate Entrance Examination Experts

Overall, the evaluation of the analysis results of the postgraduate entrance examination experts by the candidates is mostly positive. The words that appear in the keyword cloud are mostly positive words such as thank, like, happy, cute, and gratified. The sentiment analysis results also show that positive emotions account for the vast majority. The specific analysis of each postgraduate entrance examination expert is not detailed here.

2. Analysis of Postgraduate Entrance Examination Materials

Taking English materials as an example, as can be seen from Table 19.6, Hone Bao Book and Lian Lian You Ci are more prominent in words, while Huang Pi Book is more prominent in reading and writing, which is also in line with the public's subjective cognition.

Taking mathematics materials as an example, as can be seen from Table 19.7, the review book focuses more on the basics. In the dimension of exercises, Li Yongle's review book scored the highest. At the same time, when mentioning postgraduate

Table 19.6 English material sentiment analysis

	Key Words				
Profile	Word	Read	Vocabulary	Writing	Choice Questions
Hong Bao Book	1	0.6	0.62	0.18	
Nian Lian You Ci	1	0.5	0.39	0.24	0.12
Huang Pi Book	0.85	0.96	0.24	0.53	0.33

Table 19.7 Mathematics material sentiment analysis

	Key Words								
Book	Basics	Math	Zhang Yu	Video	Materials	Strengthen	Handouts	Tang Jiafeng	Exercises
Li Yongle Whole Review Books	1	0.85	0.31	0.41	0.47	0.29	0.31	0.23	0.31
Tang Jiafeng Whole Review Books	1	0.95	0.98	0.57	0.16	0.41	0.45	0.38	0.22
Zhang Yu 18	0.91	0.95	1	0.59	0.34	0.44	0.18	0.44	0.15

entrance examination materials, people often compare it with similar things, such as when students comment on Tang Jiafeng, they often mention Zhang Yu and Li Yongle.

3. Analysis of Postgraduate Entrance Examination Institutions

This project uses the word cloud third-party library and generates intuitive word clouds based on the statistical data obtained from multidimensional analysis. According to the weight of the postgraduate entrance examination institutions in the analysis results, they are displayed in the word cloud with corresponding sizes. The top five hottest postgraduate entrance examination institutions are Wendu, Wangdao, Kaicheng, Kaochong, and Qihang.

Based on the analysis results, the postgraduate entrance examination institutions are displayed in the word cloud with their sizes corresponding to their weights.

4. Candidate Result Analysis

The candidate analysis used 150,000 pieces of data, including postgraduate entrance examination expert Weibo comment data and postgraduate entrance examination QQ group chat data. There are two purposes for this: one is to analyze the different stages of the candidates' postgraduate entrance examination process. The analysis focuses on changes in emotions and feelings during different stages of the candidates' postgraduate entrance examination process. Based on the analysis results, corresponding suggestions are provided to help candidates alleviate the pressure caused by changes in the external environment. Additionally, the study aims to reflect differences in the state of candidates across different years through a dynamic emotional graph. For example, the mood of candidates in even years is generally lower than that in odd years, which indirectly indicates that the difficulty of math problems in even years is relatively higher.

From the emotional changes in QQ group chat records, it can be seen that most candidates start preparing for the exam from May to June. They enter the climax of postgraduate review from September to October. The number of chats decreases from November to December, indicating that all candidates are focused on sprinting.

19.4 Customer Service Call Text Summary Extraction

19.4.1 Introduction

This project is a competition entry for "Customer Service Call Text Summary Extraction," sponsored by the China Computer Society and China Unicom.

The customer service center needs to handle a large number of customer calls every day. Customer calls need to be converted from voice to text, and the text needs to be summarized to extract the core demands of customers. However, completing these tasks manually will increase the workload of customer service and reduce work efficiency. Therefore, it is worth considering using artificial intelligence algorithms to automatically generate text summaries.

The task of this project is to extract summaries from customer service call data. Because customer service calls belong to specific domain call data, there are certain differences from general text summary extraction, and the main difficulties are as follows:

1. The voice call is transcribed into text content through a third-party service, and there are certain transcription errors.
2. The text length is not fixed, and there may be cases where the text is too long.
3. Because it involves specific domain customer service call text, there may be more professional vocabulary.

19.4.2 Data Description

The data comes from the customer service center call text database. First, the call is recorded, and then a third-party service is used to convert the voice into text. The text is mainly used for data analysis, customer service personnel indicator evaluation, and keyword analysis under normal circumstances.

The dataset uses CSV format files, including index, call text, and summary. The text format is UTF-8, with a total of 25,000 training data.

19.4.3 Evaluation Criteria

This project uses the ROUGE evaluation metric.

The ROUGE metric compares the automatically generated summary with the reference summary. Among them, ROUGE-1 measures the unigram matching situation, ROUGE-2 measures the bigram matching situation, and ROUGE-L records the longest common subsequence. All three only use the F1 score. The weighted calculation expression of the ROUGE metric is as follows:

$$0.2 * f - \text{score}(\text{R1}) + 0.4 * f - \text{score}(\text{R2}) + 0.4 * f - \text{score}(\text{RL}) \quad (19.20)$$

19.4.4 Experimental Method

1. Model Selection

In the research of Chinese natural language generation technology, pretrained language models are still blank. Although MT5 supports Chinese, it does not design data processing methods and build pretrained corpora according to the characteristics of Chinese, so it cannot improve the effect of Chinese natural language generation to a satisfactory level. In order to promote the development of Chinese natural language generation technology, the technical team of Zhuiyi Technology Company, combined with the characteristics and needs of Chinese natural language generation technology research, built an open-source T5 PEGASUS model customized for Chinese (https://github.com/ZhuiyiTechnology/t5-pegasus). This project uses that model.

2. Data Preprocessing

In the data preprocessing stage, text correction technology is used to correct common text error types (such as typos) in the training set.

3. Adjusting Appropriate Hyperparameters

First, t5-base (12 layers) and t5-small (eight layers) are evaluated. Due to the sparsity of the data, t5-base does not perform as well as t5-small, so this project chose t5-small. Next, the appropriate max_len and output_len are selected. By calculating the average length, maximum length, and minimum length of the text, the values of max_len and output_len are continuously adjusted and experimented with.

After evaluating different values, it was finally determined that the effect is best when top_p = 0.9 and beam_search = 3.

Table 19.8 Comparison of the highest ROUGE values of several models

Model Used	Highest Rouge Value
Pointer-Generator-Network	1376.92
Unilm	1656.84
Bart	1762.77
T5-base_27Epochs(Ours)	1815.03
T5-small_35Epochs(Ours)	1891.18

4. Experimental Results and Display

In the experiment, the highest ROUGE values of several models were compared, and the experimental results are shown in Table 19.8. The experiment proves that the method used in this project is better than the effects of several baseline models.

Here is an example.

The call is as follows:

[Agent] Sir, hello. [Customer] Ah, hello, I called to schedule the broadband the day before yesterday, he didn't answer my call, then I didn't get through, can you help me urge it? I want to cover this as soon as possible. Peace on. [Agent] Please wait a moment, is the contact phone number you left when applying still your current mobile number? Hmm, wait a moment. [Customer] Can you turn on your phone? He didn't answer the phone though. [Agent] Hmm, what you applied for, did you apply through our customer service? If so, we should be able to contact you again, you can wait for us to contact you again. We only see one electronic medical order here, was it applied for on January 26th? Yesterday? Yes, yes, that's right, well, let me expedite it for you, then you wait for a reply, okay, you're welcome, ah, thank you for calling, goodbye.

References

1. Zhang Huaping. Sun Mengshu, Zhang Ruiqi, etc. Big data mining of Weibo bloggers' characteristics and behaviors [J]
2. Liu Ziyu. Research on Key Technologies of Chinese Disambiguation Based on Knowledge Graph and Semantic Features [D]. Beijing: Beijing Institute of Technology, 2021.

Chapter 20
Natural Language Processing Application Cases

This chapter continues to summarize the excellent student assignments in the National Quality Course "Big Data Analysis and Applications" at Beijing Institute of Technology. Compared with the application projects introduced in Chap. 19, the application cases introduced in this chapter are specific, primarily involving the authorship analysis of the first 80 chapters and the last 40 chapters of "A Dream in Red Mansions," the network public opinion analysis of Ding Zhen's popularity, involving key natural language processing technologies including web crawlers, named entity recognition, machine reading comprehension, sentiment analysis, text summarization, and machine translation, etc.

20.1 Analysis of the Author Identity of the First 80 Chapters and the Last 40 Chapters of "A Dream in Red Mansions"

20.1.1 Introduction

Were the first 80 chapters and the last 40 chapters of "A Dream in Red Mansions" written by the same author? People generally believe that there are two authors of "A Dream in Red Mansions." Cao Xueqin wrote the first 80 chapters, and Gao E continued the last 40 chapters. However, there has been a lot of controversy in the Redology community about the author of "A Dream in Red Mansions," and there are many different views. This project aims to analyze the authorship of the first and last parts of "A Dream in Red Mansions" [1].

The claim that Gao E continued the last 40 chapters of "A Dream in Red Mansions" first came from Zhang Wentao, a literary scholar of the Qing Dynasty. Zhang Wentao's self-commentary on the poem "Gift to Gao Lanshu (E) Tongnian"

H. Zhang, J. Shang, *Natural Language Processing and Applications*,
https://doi.org/10.1007/978-981-97-9739-4_20

states: "The legendary 'A Dream in Red Mansions' was all supplemented after the 80th round."

20.1.2 Input Data

The data source for this case is the complete text of "A Dream in Red Mansions," and the data format is UTF-8.

20.1.3 Analysis Tools and Methods

This case uses NLPIR-Parser as the analysis tool.

This case starts with function words in the text and analyzes the similarity of the text by calculating the KL distance between the data.

1. NLPIR-Parser

NLPIR-Parser is a technology that integrates natural language understanding, web search, and text mining. It is a text search and mining development platform for internet content processing needs (the online demonstration platform address of NLPIR is http://ictclas.nlpir.org/nlpir/), which provides a basic toolset for secondary development.

2. Selection of Function Words

Everyone has a specific text style of writing. Although the content of a work will differ from beginning to end, the habit of using words is not easy to change. Due to the different plots of the first 80 chapters and the last 40 chapters, the words involved are different, but a person's way and frequency of using function words may have a certain pattern.

This case uses the 47 function words listed in the book *A New Theory on the Completion of A Dream in Red Mansions*, published by Li Xianping in 1987, classified into the following five categories:

1. Thirteen classical Chinese function words: 之、其、或、亦、方、于、即、皆、因、仍、故、尚、乃.
2. Nine sentence-ending function words: 呀、吗、咧、罢咧、啊、罢、罢了、么、呢.
3. Thirteen commonly used colloquial function words: 了、的、着、一、不、把、让、向、往、是、在、别、好.
4. Ten function words indicating contrast, degree, comparison, etc.: 可、便、就、但、越、再、更、比、很、偏.
5. The suffix "儿" after nouns and the suffix "儿" after adverbs, adjectives, and verbs (considered as two different function words).

3. KL Distance

KL distance can measure the distance between two random distributions of data. When the data of two random distributions are the same, their relative entropy is 0. As the difference between the data of two random distributions increases, their relative entropy also increases. Therefore, KL can be used distance to compare the similarity of the text. Its formula is

$$(P\|Q) = \sum_{x \in X} P(x) \log \frac{P(x)}{Q(x)} \tag{20.1}$$

20.1.4 Results and Analysis

This case divides the 120 chapters of "A Dream in Red Mansions" into 3, 6, 12 equal parts in order, and these three division methods are respectively named "3 groups," "6 groups," and "12 groups." Each group of data is used as a corpus, and NLPIR is used to perform batch word segmentation analysis on each group of data, then the frequency of function words is counted, and finally, the KL distance between each group of data is calculated.

Next, take "3 groups" as an example for a detailed introduction, "6 groups" and "12 groups" are the same. The 120 chapters are divided into three equal parts in order, namely the 1st to 40th chapters, the 41st to 80th chapters, and the 81st to 120th chapters, as three groups of data. The frequency and probability of 47 function words in each group are counted. The frequency and probability of some function words in these three groups of data are shown in Table 20.1, where the probability is the ratio of the number of a certain function word in this group of data to the total number of function words in this group of data.

The results obtained from the KL distance calculation formula are shown in Table 20.2. The probability values of each function word in the row number of Table 20.2 are denoted as $P(x)$, and the probability values of each function word in the column number of Table 20.2 are denoted as $Q(x)$. Other group experiments are the same. For example, when calculating the KL value of the 1st to 40th chapters and the 41st to 80th chapters, x in the KL distance calculation formula represents a certain function word; $P(x)$ represents the probability of x in the 1st to 40th chapters; and $Q(x)$ represents the probability of x in the 41st to 80th chapters. It should be noted that $D(P\|Q)$ and $D(Q\|P)$ are different.

From Table 20.2, it can be observed that the KL value of the 1st to 40th chapters and the 81st to 120th chapters in the first row is about 10 times the KL value of the 1st to 40th chapters and the 41st to 80th chapters. As the difference between two groups of random distribution data increases, their relative entropy will also increase. Therefore, the similarity between the 1st to 40th chapters and the 81st to 120th chapters is lower than the similarity between the 1st to 40th chapters and the 41st to 80th chapters.

Table 20.1 Frequency and probability of some function words in three groups of data

Word	Chapters 1–40		Chapters 41–80	Chapters 81s–120t		
	Frequency	Probability	Frequency	Probability	Frequency	Probability
了	5981	0.199 712 836	7740	了	5981	0.199 712 836
的	3854	0.128 689 729	5156	的	3854	0.128 689 729
不	3063	0.102 277 281	3805	不	3063	0.102 277 281
Is	2293	0.076 566 048	2975	Is	2293	0.076 566 048
One	2202	0.073 527 448	2750	One	2202	0.073 527 448
With	1607	0.053 659 700	1855	With	1607	0.053 659 700
Convenient	1075	0.035 895 552	1272	Convenient	1075	0.035 895 552
In	1026	0.034 259 383	1089	In	1026	0.034 259 383
Just	935	0.031 220 783	1101	Just	935	0.031 220 783

Table 20.2 KL values of three groups of data

The number of chapters corresponding to $P(x)$	The number of chapters corresponding to $Q(x)$		
	Chapters 1–40	Chapters 41–80	Chapters 81–120
Chapters 1–40	0	0.008	0.082
Chapters 41–80	0.007	0	0.06
Chapters 81–120	0.051	0.049	0

As shown in Fig. 20.1, it can be observed that the similarity between the 1st to 40th chapters and the 41st to 80th chapters is relatively high, and there is a significant change in the similarity between the 1st to 40th chapters and the 41st to 80th chapters compared to the similarity between the 1st to 40th chapters and the 81st to 120th chapters.

The '6 groups' divide the 120 chapters into 6 equal parts in sequential order: the 1st to 20th chapters, the 21st to 40th chapters, the 41st to 60th chapters, the 61st to 80th chapters, the 81st to 100th chapters, and the 101st to 120th chapters.

The KL distance calculation results of these six groups of data are shown in Table 20.3, and the corresponding graph is shown in Fig. 20.2. In Fig. 20.2, from the right end of each line, from top to bottom, they are 1~20, 61~80, 41~60, 21~40, 81~100, 101~120.

When the difference between two groups of randomly distributed data increases, their KL values also increase. From the results, it can be found that the KL values of the first four groups of data are significantly larger when compared with the last

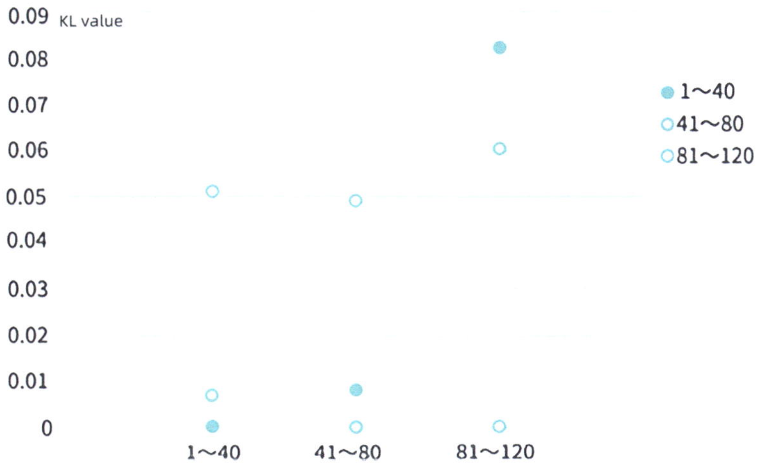

Fig. 20.1 KL values of three groups of data

Table 20.3 KL values of six groups of data

$P(x)$ number	$Q(x)$ number value					
	1~20	21~40	41~60	61~80	81~100	101~120
1 ~20	0	0.088	0.078	0.023	0.155	0.177
21~40	0.065	0	0.01	0.026	0.027	0.054
41~ 60	0.052	0.01	0	0.015	0.046	0.065
61~ 80	0.019	0.028	0.018	0	0.07	0.087
81~ 100	0.104	0.018	0.033	0.057	0	0.041
101~ 120	0.11	0.038	0.045	0.068	0.018	0

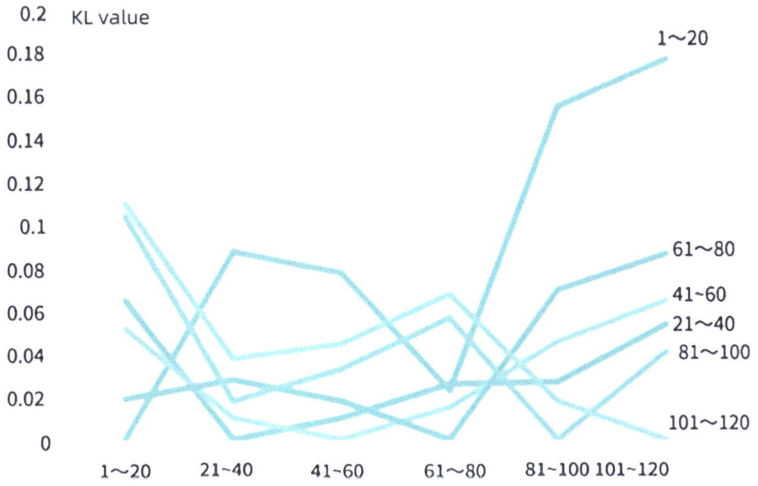

Fig. 20.2 KL values of six groups of data

two groups of data, and the KL values of the last two groups of data are significantly smaller when compared with the first four groups of data.

The '12 groups' divide the 120 times evenly into 12 parts: the 1st to 10th times, the 11th to 20th times, ..., and the 111th to 120th times. The KL distance calculation results for these 12 groups of data are shown in Table 20.4.

From Table 20.4, it can be observed that the KL values of any group of data in the first 80 times are higher than the KL values of other groups of data in the first 80 times when compared with each group of data in the last 40 times.

Figure 20.3 shows the KL values of the 12 groups of data and the corresponding four groups of data in the last 40 times. It can be seen that there is a varying degree of difference between the KL values of the first 80 times' eight groups of data and the last 40 times' four groups of data. The KL values between the last 40 times' four groups of data are smaller than the KL values between the first 80 times' eight groups of data and the last 40 times' four groups of data, indicating that the similarity of the last 40 times' four groups of data is higher.

A series of analyses reveals that there are indeed significant differences in the use of function words between the first 80 times and the last 40 times. Based on this, it can be boldly speculated that the last 40 times were written by another person.

20.2 Analysis of Internet Public Opinion on Ding Zhen's Popularity

20.2.1 Introduction

In November 2020, Ding Zhen became unexpectedly popular due to his innocent and simple smile, becoming a new top influencer on the internet. Female netizens cheered for him, even causing some male netizens to be jealous. He is also one of the few grassroots celebrities to have been reposted by the spokesperson of the Ministry of Foreign Affairs and blessed by a CCTV anchor, combining handsomeness and popularity in one.

For such a hot topic that has sparked widespread discussion and even controversy, this project collects a large amount of data and conducts text processing and analysis by crawling ordinary user comments and Weibo topic discussions under the official media bloggers of Sina Weibo, all questions and answers about Ding Zhen and all user comments under the topics on Zhihu, and Bilibili video bullet comments, thereby realizing network public opinion analysis based on big data.

20.2.2 System Structure and Method

This project mainly includes data crawling and processing modules and data analysis modules. The main technologies and tools used by each module are shown in Fig. 20.4.

Table 20.4 KL values of 12 groups of data

	$Q(x)$ number value											
$P(x)$ number	1～10	11～20	21～30	31～40	41～50	51～60	61～70	71～80	81～90	91～100	.101～110	111～120
1～10	0	0.024	0.109	0.102	0.128	0.048	0.069	0.03	0.192	0.186	0.193	0.188
11～20	0.027	0	0.07	0.644	0.82	0.028	0.041	0.018	0.149	0.151	0.154	0.15
21～30	0.08	0.057	0	0.013	0.016	0.024	0.019	0.056	0.033	0.042	0.05	0.062
31～40	0.094	0.066	0.018	0	0.01	0.023	0.025	0.059	0.034	0.049	0.055	0.07
41～50	0.1	0.066	0.019	0.007	0	0.025	0.024	0.057	0.031	0.042	0.048	0.056
51～60	0.057	0.037	0.032	0.025	0.033	0	0.017	0.03	0.086	0.083	0.086	0.072
61～70	0.053	0.035	0.017	0.018	0.022	0.009	0	0.035	0.055	0.06	0.066	0.064
71～80	0.027	0.016	0.06	0.052	0.063	0.023	0.038	0	0.12	0.105	0.117	0.105
81～90	0.133	0.105	0.025	0.02	0.022	0.057	0.046	0.099	0	0.023	0.035	0.041
91 ～ 100	0.115	0.094	0.029	0.02	0.021	0.04	0.044	0.097	0.022	0	0.011	0.15
101 ～ 110	0.123	0.096	0.036	0.026	0.025	0.042	0.056	0.079	0.035	0.011	0	0.012
111 ～ 120	0.135	0.109	0.051	0.042	0.039	0.055	0.063	0.089	0.044	0.016	0.013	0

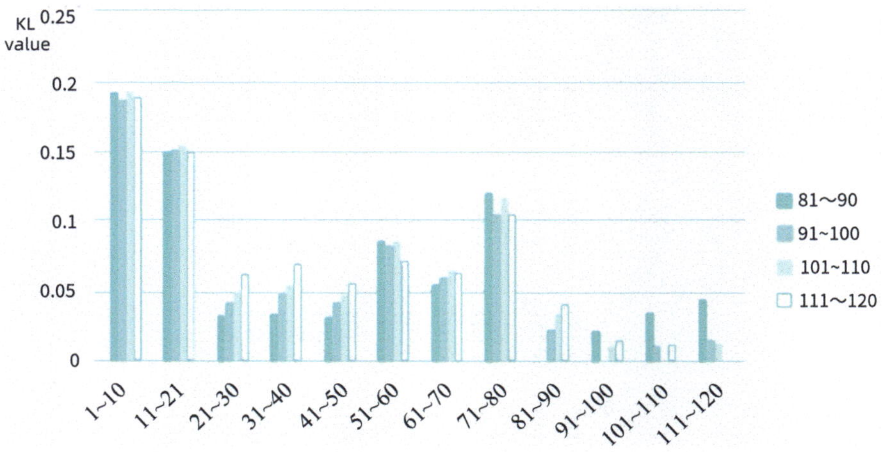

Fig. 20.3 KL values of each group of the 12 groups of data and the last four groups of data

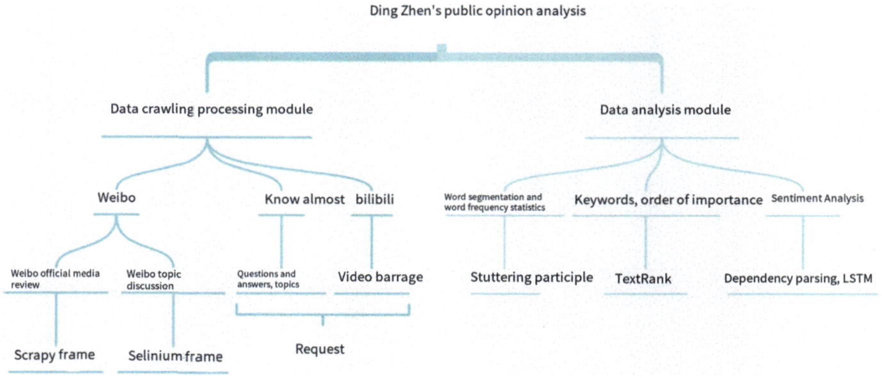

Fig. 20.4 System Structure and Method of This Project

1. Data Crawling

This project collects data from Weibo, Zhihu, and Bilibili. Since the user groups of these three platforms have different characteristics, the data from these three platforms can analyze the public's comments on the Ding Zhen event and the reasons behind it from different angles comprehensively. Data crawling on the Weibo platform is divided into two directions. The first direction is the comments under the official Weibo related to Ding Zhen. This part of the data can reflect the public's attitude under the mainstream media. This project crawled 60 Weibos related to Ding Zhen from Litang Ding Zhen, People's Daily, Sichuan TV News Scene, Sichuan Observation, and CCTV News, and obtained 30,000 comments. The second direction is the topic related to Ding Zhen. This project searched for 20 topics and collected a total of 2000 Weibo data points.

The topics crawled on Weibo are as follows: "#Ding Zhen#," "#Litang#," "#Some straight men's malice towards Ding Zhen#," "#Some men's attitude towards Ding Zhen#," "#Ding Zhen used to herd cattle and ride horses and now he has to work#," "#Ding Zhen's face is like the sloppy king#," "#All over the country are inviting Ding Zhen#," "#Ding Zhen's highland Tibetan makeup#," "#Ding Zhen appears in Litang tourism promotion#," "#Du Dong says Ding Zhen is like the eyes of Litang#," "#Ding Zhen responds to the 2020 internet explosion#," "#Ding Zhen says he wants to come to Beijing to watch the flag raising#," "#How hard Sichuan is for Ding Zhen#," "#Ding Zhen under the lens of passers-by#," "#How beautiful Ding Zhen's hometown is#," "#Invite Ding Zhen to my hometown#," "#Young Wolf Lord Ding Zhen#," "#Why Ding Zhen exploded#," "#How much Ding Zhen loves his little horse Pearl#," "#Ding Zhen didn't take care of the cattle and didn't achieve his goal#."

On Zhihu, the keywords "Ding Zhen" and "Ding Zhen poverty alleviation" were used to crawl related questions and answers. A total of 36,343 answers were crawled in 65 questions related to "Ding Zhen," and a total of 44,965 answers were crawled in 84 questions related to "Ding Zhen poverty alleviation."

On Bilibili, 986 videos related to Ding Zhen were crawled, with a total of 11,3921 bullet comments.

The data from the above three sources are stored as CSV files and JSON files, which serve as the basis for subsequent processing and analysis.

2. Generate Keywords and Rank Them by Importance

First, organize the data crawled from various platforms, extract all comment content, perform word segmentation and keyword extraction, for subsequent data analysis.

Use the TextRank algorithm to extract keywords and rank them, input the ranked keywords into the word cloud generation program. To better reflect the evaluation, this project also marked the part of speech and filtered out stop words, only retaining specified parts of speech, such as nouns, verbs, adjectives. Rank the keywords of specific parts of speech and draw word clouds, and analyze from the perspectives of external evaluation, associated reasons, negative attitudes, etc.

3. Overview of Public Opinion on Ding Zhen

People's evaluations of a person usually involve external and internal aspects, and the words of these evaluations are usually adjectives. To accurately extract evaluations, this project uses Weibo comment data, filters adjectives, considering that different gender groups may have different opinions. The crawled data is divided into male and female groups for analysis. Women's evaluations of Ding Zhen are almost all positive and have positive words, such as beautiful, kind, simple, etc., while men's evaluations also have positive words such as beautiful and clear, but frequently appearing are also backward, poor, hard, ignorant, etc. It can be seen that their evaluation of Ding Zhen takes into account his life background and local development status. This phenomenon will be further explored later.

4. Sentiment Classification of Weibo Topic Discussion

Through the long short-term memory artificial neural network to build a deep learning model for text sentiment classification, this project classified the sentiment orientation of 2084 Weibo topic discussions.

Analysis of Negative Comments on the Ding Zhen Phenomenon

In fact, for Ding Zhen's popularity, a considerable number of people hold negative attitudes, and there are often posts on Tieba such as "men generally despise and insult Ding Zhen." The typical two groups are "problem solvers" and "men."

In the keyword cloud of the Zhihu topic "Why do 'problem solvers' vent their grievances on Ding Zhen," we can see keywords such as cheap, poverty, bright, dark, etc., indicating that some people have doubts about Ding Zhen's sudden popularity. They believe that fame and fortune obtained easily without relying on hard work is not worth promoting.

In the keyword cloud of the Zhihu topic "Why do males generally despise and insult Ding Zhen," words such as outrageous, disgust, unacceptable, and disgusting appear. This also shows that Ding Zhen's sudden popularity is unacceptable to some males, and they believe that this phenomenon is abnormal.

5. Analysis of the Associated Reasons for the Ding Zhen Phenomenon

The reasons behind Ding Zhen's sudden popularity are far more than his appearance and the initial video shot by the photographer. To explore its deep reasons, remove adjectives from the data, extract nouns and verbs, and make the data more convincing. Since the comments on Weibo are relatively objective and the cultural level of Zhihu users is generally high, and their ability to think is strong, this project extracts keywords from the comments on Weibo and the answers on Zhihu.

In addition to the praise for Ding Zhen, the Ding Zhen phenomenon is closely tied to his hometown, Litang, as well as promotion, society, poverty alleviation, and tourism. This aligns with the tourism promotion efforts that Litang County has consistently pursued. Combined with the Litang grassland and other tourist attractions that Litang County has been promoting in recent years, Ding Zhen's sudden rise to fame has had a significant positive impact on the tourism promotion of Litang County.

This kind of promotion is also in line with the national poverty alleviation policy for poor areas, so it has received the support of the government and society. Through the above analysis, it can be seen that in the online world, the sudden popularity of a person will not have only a single reason. This phenomenon must be caused by multiple factors working together.

Through big data analysis, digging deep into its inherent factors and associations, we can see a more realistic situation.

Reference

1. Zhang Huaping, Shang Jianyun, Liu Zhaoyou. Big Data Intelligent Analysis[M]. Beijing: Tsinghua University Press, 2019.

The manufacturer's authorised representative in the EU is Springer Nature Customer Service Centre GmbH, Europaplatz 3, 69115 Heidelberg, Germany. If you have any concerns regarding our products, please contact ProductSafety@springernature.com

Printed and bound by CPI Group (UK) Ltd, Croydon, CR0 4YY
01/10/2026
02225941-0014